AF389466

Sample Preparation in Metabolomics

Sample Preparation in Metabolomics

Editors

Julia Kuligowski
Guillermo Quintás

MDPI • Basel • Beijing • Wuhan • Barcelona • Belgrade • Manchester • Tokyo • Cluj • Tianjin

Editors
Julia Kuligowski
Health Research Institute La Fe
Spain

Guillermo Quintás
Leitat Technological Center
Spain

Editorial Office
MDPI
St. Alban-Anlage 66
4052 Basel, Switzerland

This is a reprint of articles from the Special Issue published online in the open access journal *Metabolites* (ISSN 2218-1989) (available at: https://www.mdpi.com/journal/metabolites/special_issues/sample_preparations).

For citation purposes, cite each article independently as indicated on the article page online and as indicated below:

LastName, A.A.; LastName, B.B.; LastName, C.C. Article Title. *Journal Name* **Year**, *Volume Number*, Page Range.

ISBN 978-3-03943-813-6 (Hbk)
ISBN 978-3-03943-814-3 (PDF)

Contents

About the Editors

Julia Kuligowski, Ph.D., is a postdoctoral researcher at the Health Research Institute La Fe, in Valencia, Spain, where her main research focus is on clinical metabolomics and advanced data analysis tools. She studied biotechnological processes at the University of Applied Sciences Wiener Neustadt, Tulln (Austria) and was trained as an analytical chemist at the Vienna University of Technology (Austria) and the University of Valencia (Spain) where she earned her Ph.D. in 2011. As part of the Neonatal Research Unit, her research focused on the quantification of metabolites and biomarkers in biological samples obtained from clinical trials and animal models employing mass spectrometry-based techniques and vibrational spectroscopy. At present, she continues her research activities in the field of neonatology, focusing on the discovery of molecular biomarkers for an early, minimally invasive assessment of brain injury secondary to hypoxic-ischemic encephalopathy and the impact of nutrition on growth and development of preterm infants.

Guillermo Quintas, Ph.D., is a researcher at the LEITAT Technological Center (Barcelona, Spain). After completing his Ph.D. at the University of Valencia (Spain) in 2004, he obtained a post-doctoral fellowship at the Public Health Laboratory Valencia (Spain), the Vienna University of Technology (Austria), and the University of Valencia, where his research focused on hyphenated systems and infrared spectroscopy. In 2009, he made the transition to a position as senior researcher in Advancell (Barcelona, Spain) and at the Analytical Unit of the Health Research Institute La Fe (Valencia), where his activities were focused on providing tailored analytical solutions for clinical and pre-clinical research studies aiming at biomarker discovery and validation and drug development. In 2011, he moved to the Health and Biomedicine division of LEITAT. Currently, his main research is directed toward the joint analysis of information from multiple platforms and clinical data, developing novel chemometric approaches.

Preface to "Sample Preparation in Metabolomics"

Metabolomics aims at the comprehensive analysis of all low molecular weight metabolites present in a biological system under consideration. This is a demanding task due to the chemical, physicochemical, and biological diversity of metabolites; the broad range of rapidly changing concentrations; and the limited stability of some compound classes. Furthermore, the choice of the sample preparation method needs to be made in alignment with the study aim and the selected analytical approach. To generate high-quality data, sample preparation is a particularly important aspect, as, generally, a major source of error in analytical results is associated with pre-analytical steps. Despite its importance, sample preparation is often an overlooked aspect of studies involving metabolomics, so the spotlight of this book is on the latest developments in sample collection, preparation, and optimization for targeted and untargeted analysis using different analysis platforms in a diverse field of applications.

Expanding metabolome coverage to include complex lipids and polar metabolites is critical for the generation of well-founded hypotheses in biological assays. The performance of single-step sample preparation methods for the simultaneous extraction of the complex lipidome and polar metabolome from human plasma was evaluated [1]. Method performance was assessed using high-coverage hydrophilic interaction liquid chromatography-ESI coupled with tandem mass spectrometry (HILIC-ESI-MS/MS) targeting a panel of 1159 lipids and 374 polar metabolites. Among the tested methods, the isopropanol (IPA) and 1-butanol:methanol (BUME) mixtures were selected as the best compromises for the simultaneous extraction of complex lipids and polar metabolites, allowing for the detection of 584 lipid species and 116 polar metabolites in plasma.

The analysis of highly polar and small molecules in complex biolfuids is especially troublesome as widely employed reversed-phase liquid chromatography methods do not achieve adequate retention and co-elution with matrix compounds, eventually leading to strong ion suppression effects when using mass spectrometry detection. A smart set-up involves the electromembrane extraction of five polar metabolites from several plasma samples, in parallel, dramatically increasing the extraction throughput. Extracted compounds were analyzed using a multi-segment injection—capillary electrophoresis—mass spectrometry (MSI-CE-MS) method and fast-liquid chromatography—tandem mass spectrometry, achieving recoveries of up to 100% with variability as low as 2% [2].

Nuclear magnetic resonance (NMR) is one of the principal spectroscopic methods employed in metabolomics studies for the analysis of biofluids. Recently, high-resolution micro-magic angle spinning (HR-μMAS), has been introduced, paving the way for investigating microscopic specimens (<500 μg) with NMR spectroscopy. Sample preparation for HR-μMAS is challenging due to the μgscale specimen. This book addresses this topic with a special emphasis on three specimen types: biofluids, fluid matrices, and tissues [3]. In biomedical research, the use of cell lines is frequently employed to gain insights into mechanisms of diseases, as well as new treatment approaches. Dedicated protocols for cell harvesting, disruption, and metabolite quenching and extraction were recently developed and critically assessed [3–6]. Using Caulobacter crescentus as a model for Gram-negative bacteria, eight different sample preparation techniques were evaluated using a full-factorial experimental design in combination with the ANOVA multiblock orthogonal partial least squares (AMOPLS) algorithm for decomposing the contribution of each studied factor. All of the main effects of the studied factors were found to significantly contribute to the total observed variability, with cell retrieval having the biggest impact. The protocol showing the best performance in terms of recovery, versatility, and variability was centrifugation for cell retrieval in combination with MeOH:H2O (8:2) as quenching and extraction solvent, and freeze–thaw cycles as the cell disrupting mechanism [4].

Another study evaluated the effect of the choice of drying technique, i.e., centrifugal evaporation under vacuum vs. lyophilization, on the retrieved NMR spectroscopic profiles of hydrophilic extracts of three human pancreatic cancer cell lines [4]. Between 40 and 50 metabolites showed statistically significant differences in abundance in redissolved sample extract depending on the drying technique employed. Metabolite coverage was also differing, with a set of metabolites being exclusively detected in samples subjected to each drying technique. Hence, these observations showcase the impact of sample preparation on the final result. A three-dimensional multicellular tumor spheroid (3D MTS) model was the focus of another study [6]. Careful optimization of multiple steps of sample preparation enabled the probing of the metabolome of single MTSs in a highly repeatable manner at a considerable throughput and provided absolute concentrations with average biological repeatability of <20%. In a proof-of-principle study, distinct metabolic shifts upon MTS exposure to two metal-based anticancer drugs, which exhibit distinctly different modes of action, were retrieved. Therefore, biological variation among single spheroids can be assessed using the presented analytical strategy that is applicable for in-depth anticancer drug metabolite profiling. Finally, a sample preparation strategy for gas chromatography–mass spectrometry (GC-MS)-untargeted metabolomics of adherent cells grown under high (20%) fetal calf serum conditions was developed [7]. The reproducibility of using different proportions of methanol for the quenching of cells was compared for sample harvest, and the efficiency and reproducibility of intracellular metabolite extraction were tested by employing different volumes and ratios of chloroform. Additionally, the use of total protein amount versus cell mass for normalization purposes was assessed.

The last chapters of this book provide an overview of the state of the art in a selection of fields of applications. Cell-secreted extracellular vesicles (EVs) have rapidly gained prominence as sources of biomarkers, owing to their ubiquity across human biofluids and physiological stability. Many studies have been devoted to their protein, nucleic acid, lipid, and glycan content, but more recently the metabolomic profile of EV content has also gained attention. Beyond clinical applications, metabolomics has also elucidated possible mechanisms of action underlying EV function. There are challenges inherent to working with EVs, particularly concerning sample production and preparation. This chapter outlines recent advances in EV metabolomics whilst highlighting practical pitfalls in applying metabolomics to EV studies [8].

Human milk (HM) is the gold standard for infant nutrition, and, as such, it is an outstandingly complex biofluid with a dynamically changing composition. The use of novel, cutting-edge techniques involving different metabolomics platforms has permitted the expansion of knowledge on HM composition. This chapter presents the state of the art in untargeted metabolomic studies of HM, with an emphasis on sampling, extraction, and analysis approaches with a special focus on the achievable metabolome coverage. Finally, current knowledge gaps and potential future research directions are pointed out [9].

Plant-derived natural products are a valuable source of active compounds. Natural extracts are rich in different classes of metabolites, whereby the bioactivity of natural extracts can be a synergistic effect of several compounds. In recent years, metabolomics has emerged as an indispensable tool for the analysis of crude natural extracts, leading to a paradigm shift in natural products drug research. In this chapter, current advancements in plant sample preparation, analysis, and data processing are presented alongside several case studies of the successful applications of these processes in plant natural product drug discovery [10].

Finally, the last chapter is focused on the peculiarities of forest tree metabolomics. Tree tissues are intrinsically complex matrices, and the presence of several compounds, such as oleoresins and cellulose, might interfere during their analysis. Additionally, in this field of application, experimental design, tissue harvest conditions, and sample preparation are crucial factors to ensure consistency and reproducibility among datasets. This chapter discusses the main challenges when setting up a forest tree metabolomics experiment for mass spectrometry (MS)-based analysis covering technical aspects of all stages of the workflow. The importance of forest tree metadata standardization in metabolomics studies is also highlighted [11].

References

[1] Single-Step Extraction Coupled with Targeted HILIC-MS/MS Approach for Comprehensive Analysis of Human Plasma Lipidome and Polar Metabolome

[2] Electromembrane Extraction of Highly Polar Compounds: Analysis of Cardiovascular Biomarkers in Plasma

[3] General Guidelines for Sample Preparation Strategies in HR-μMAS NMR-based Metabolomics of Microscopic Specimens

[4] Choosing an Optimal Sample Preparation in Caulobacter crescentus for Untargeted Metabolomics Approaches

[5] Influence of Drying Method on NMR-Based Metabolic Profiling of Human Cell Lines

[6] Single Spheroid Metabolomics: Optimizing Sample Preparation of Three-Dimensional Multicellular Tumor Spheroids

[7] Modified Protocol of Harvesting, Extraction, and Normalization Approaches for Gas Chromatography–Mass Spectrometry-Based Metabolomics Analysis of Adherent Cells Grown Under High Fetal Calf Serum Conditions

[8] Metabolomics Applied to the Study of Extracellular Vesicles

[9] Current Practice in Untargeted Human Milk Metabolomics

[10]Metabolomics in the Context of Plant Natural Products Research: From Sample Preparation to Metabolite Analysis

[11]Experimental Design and Sample Preparation in Forest Tree Metabolomics

Julia Kuligowski, Guillermo Quintás

Editors

Article

Single-Step Extraction Coupled with Targeted HILIC-MS/MS Approach for Comprehensive Analysis of Human Plasma Lipidome and Polar MetabolomeElectromembrane Extraction of Highly Polar Compounds: Analysis of Cardiovascular Biomarkers in Plasma

Jessica Medina [1], Vera van der Velpen [1], Tony Teav [1], Yann Guitton [2], Hector Gallart-Ayala [1,*] and Julijana Ivanisevic [1,*]

[1] Metabolomics Platform, Faculty of Biology and Medicine, University of Lausanne,
    CH-1005 Lausanne, Switzerland; Jessica.medina@unil.ch (J.M.); vera.vandervelpen@unil.ch (V.v.d.V.);
    tony.teav@unil.ch (T.T.)
[2] Laboratoire d'Etude des Résidus et Contaminants dans les Aliments (LABERCA), Oniris, INRAE,
    F-44307 Nantes, France; yann.guitton@oniris-nantes.fr
* Correspondence: hector.gallartayala@unil.ch (H.G.-A.); julijana.ivanisevic@unil.ch (J.I.);
    Tel.: +41-21-692-5409 (H.G.-A.); +41-21-692-5098 (J.I.)

Received: 25 September 2020; Accepted: 27 November 2020; Published: 2 December 2020

Abstract: Expanding metabolome coverage to include complex lipids and polar metabolites is essential in the generation of well-founded hypotheses in biological assays. Traditionally, lipid extraction is performed by liquid-liquid extraction using either methyl-tert-butyl ether (MTBE) or chloroform, and polar metabolite extraction using methanol. Here, we evaluated the performance of single-step sample preparation methods for simultaneous extraction of the complex lipidome and polar metabolome from human plasma. The method performance was evaluated using high-coverage Hydrophilic Interaction Liquid Chromatography-ESI coupled to tandem mass spectrometry (HILIC-ESI-MS/MS) methodology targeting a panel of 1159 lipids and 374 polar metabolites. The criteria used for method evaluation comprised protein precipitation efficiency, and relative MS signal abundance and repeatability of detectable lipid and polar metabolites in human plasma. Among the tested methods, the isopropanol (IPA) and 1-butanol:methanol (BUME) mixtures were selected as the best compromises for the simultaneous extraction of complex lipids and polar metabolites, allowing for the detection of 584 lipid species and 116 polar metabolites. The extraction with IPA showed the greatest reproducibility with the highest number of lipid species detected with the coefficient of variation (CV) < 30%. Besides this difference, both IPA and BUME allowed for the high-throughput extraction and reproducible measurement of a large panel of complex lipids and polar metabolites, thus warranting their application in large-scale human population studies.

Keywords: sample preparation; extraction; lipidomics; metabolomics; LC-MS/MS; human plasma

---

1. Introduction

Blood plasma is one of the most commonly used biofluids for metabolic phenotyping, specifically in human population studies. This is mainly due to its easy access with minimally invasive sampling and the ability of its metabolic profile to inform about the systemic physiological status. Blood has a vital physiological role in the transport of circulating metabolites; it supplies tissues with nutrients and oxygen, and it carries away the metabolic by-products and carbon dioxide.

Human plasma contains a wide diversity of low molecular weight metabolites, including amino acids, other organic acids, fatty acids, sugars, and complex lipids [1]. Lipids represent more than 70% of plasma metabolome diversity and so far, more than 600 distinct lipid molecular species have been detected in human plasma [2]. These lipids can be classified, depending on their chemical structure, according to Lipid Maps consortium (http://www.lipidmaps.org) into sphingolipids, glycerophospholipids, fatty acids, glycerolipids, sterols, and prenols [3,4]. Their main functions comprise energy storage, signal transduction, and plasma membrane and organelles constituents, thus enabling the cell growth, proliferation, activation, and apoptosis [5]. When compared to total lipid content, the total polar metabolite content (comprising mainly carbohydrates, amino acids, and other organic acids), represents a minor part of plasma metabolome [2]. These polar metabolites are known to regulate the intracellular energy metabolism, including glucose, lipid, and amino acid oxidation [6,7]. The concentrations of endogenous, polar, and lipid metabolites in plasma are determined by cellular activity, and additional internal (e.g., microbiome, inflammation) and external factors (e.g., diet, physical activity, medication, environmental exposures, etc.). Therefore, the recorded changes in respective metabolite concentrations can serve as a readout of cellular biochemical activity and its response to changes in the environment. Hence, an approach to extract and measure the broadest possible range of lipids and polar metabolites in human plasma is fundamental for better understanding of metabolic changes in different physiological conditions [8,9].

Despite having the same analytical principle, lipidomics and metabolomicshave been considered as distinct technological approaches, which is mainly due to the application of different sample preparation and data acquisition protocols. Traditionally, polar metabolite extraction is performed in a single-step using methanol (MeOH) [10–12], and lipid extraction using biphasic Folch or Bligh and Dyer protocols with a mixture of methanol, chloroform, or dichloromethane, and water [13,14]. The main advantage of these protocols is the recovery of a wide range of lipid classes in the organic phase and polar metabolites in the aqueous phase and, therefore, their applicability for multicomponent analyses [4,15,16]. However, the main drawback constitutes the reproducible recovery of the bottom organic phase when analyzing large sample sets. Consequently, in order to overcome this issue, chloroform and dichloromethane were replaced by methyl-tert-butyl ether (MTBE) in the Matyash method, where lipids are recovered in the upper organic phase [17]. Still, the reproducible separation of organic and aqueous phase remains laborious and can introduce a bias to the measurement accuracy of some amphiphilic metabolites partitioned between these two phases [15].

Recently, the Alshehry method or 1-BUtanol:MEthanol mixture (1:1, BUME) and isopropanol (IPA, 100%) were introduced as an efficient single-step alternative for complex lipid extraction, also suitable for automation [18–20]. Single-step methods are advantageous because they require minimal sample handling to ensure reproducible sample preparation in large-scale studies. In addition, they allow for the simultaneous multicomponent extraction and sample cleaning (deproteinization). Systematic evaluations of single-step *versus* traditionally used biphasic extractions are sparse and the potential of these straightforward protocols to simultaneously extract complex lipids and polar metabolites has not yet been assessed. So far, sample preparation using the Matyash biphasic protocol was evaluated as the best compromise for the comprehensive extraction of complex lipids and an adequate yield of polar metabolites [21,22].

In this work, we demonstrate the performance of two single-step sample preparation methods, i.e., using BUME and IPA, for the simultaneous extraction of complex lipids and polar metabolites from human plasma. The extraction method evaluation was based on protein removal efficiency, relative signal intensity and signal reproducibility of the complex lipid and polar metabolite profiles acquired by high-coverage targeted HILIC-MS/MS. The complex lipidome and polar metabolome coverage was assessed for both BUME and IPA by considering the signal reproducibility (the coefficient of variation (CV) < 30%) as a prerequisite for the measurement accuracy and precision, rather than only focusing on the diversity of detectable metabolome and lipidome.

2. Results

2.1. Sample Preparation Methods and Evaluation Workflow

The ability of single-step methods to simultaneously extract complex lipids and polar metabolites was evaluated using relative abundance and repeatability of metabolite signal, against the commonly applied protocols, a biphasic extraction with MTBE for lipids, and a single-step MeOH for polar metabolites (Figure 1). The extracts were analyzed with high-coverage targeted profiling using HILIC-MS/MS in positive and in negative ionization mode (see Materials and Methods). These methods targeted a total of 1159 lipids from five different lipid classes (sphingolipids, cholesterol esters, glycerolipids, glycerophospholipids, and free fatty acids, Tables S1 and S2), and 374 polar metabolites from 12 different classes (amino acids and their derivatives, carboxylic acids, acylcarnitines, nucleotides, etc., Tables S3 and S4). The classification of polar metabolites was based on the Human Metabolome Database (HMDB) while the complex lipids were classified according to LipidMaps [4,23]. The protein precipitation efficeincy, the relative abundance and the coefficient of variation of each detected metabolite were used to evaluate the extraction performance for each lipid and polar metabolite class (Figure 1).

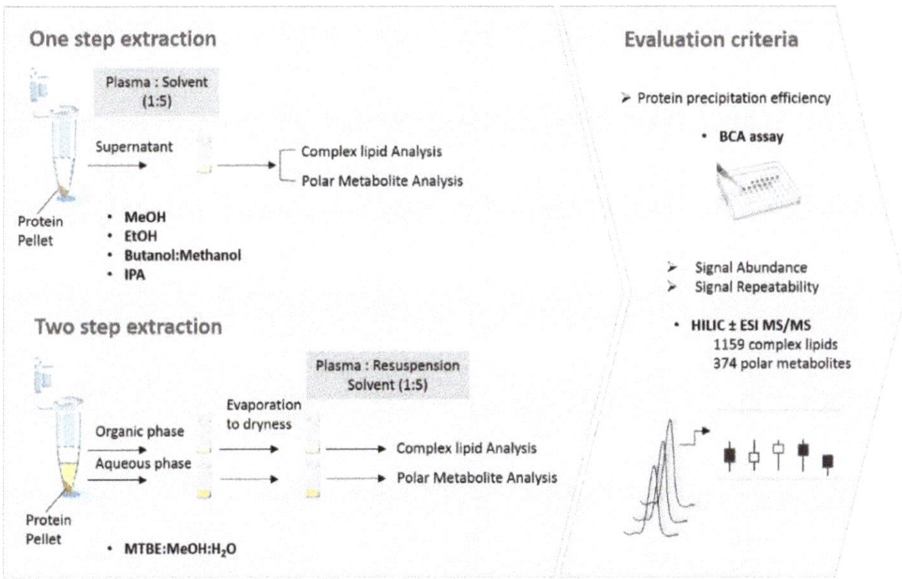

Figure 1. Sample preparation and evaluation workflow. The performance of single step extraction methods, including methanol (MeOH), ethanol (EtOH), BUtanol:MEthanol (BUME), and isopropanol (IPA), was compared to the solvent mixture considered as the best compromise, i.e., MTBE:MeOH:H_2O for complex lipid and MeOH for polar metabolite extraction. The criteria used for evaluation of solvent performance are shown on the right side of the figure. For the repeatability experiments, the extraction was done by four independent operators each performing five technical replicates per extraction protocol (n = 4 operators x 5 replicates).

Firstly, we examined the performance of single-step methods, using MeOH, ethanol (EtOH), BUME, and IPA, to extract lipids. Following the lipid extraction, the protein precipitation efficiency of these four solvents was evaluated by measuring the total protein content in the supernatant and in the pellet (see Materials and Methods section for more details). The equivalent to 95% of protein removal was achieved with all four methods (See Table S5).

Secondly, the relative signal abundance of the entire panel of detected lipid species was evaluated in all four plasma extracts (Figure S1). Methanol demonstrated the poorest performance for complex lipid extraction, with the exception of lysophosphatidylcholines (LPCs) and lysophosphatidylinositols (LPIs). A significantly lower signal for several lipid classes (i.e., sphingomyelins, fatty acids, and lysophospholipids) was also obtained from ethanol extracts, when compared to IPA and BUME. Therefore, MeOH and EtOH were excluded from further evaluation as non-suitable or less performant for complex lipid extraction, respectively. The best candidates, BUME and IPA, were further evaluated against biphasic Matyash method, for their capacity to reproducibly extract lipids and polar metabolites.

2.2. Relative Signal Abundance

BUME and IPA extractions were selected as the best single-step methods when considering the relative abundance of extracted lipids. Therefore, they were further evaluated for lipid and polar metabolite extraction against commonly used biphasic extraction MTBE:MeOH:H$_2$O in lipidomics studies, and MeOH in polar metabolome studies.

Pooled plasma samples were extracted, and the supernatant of each solvent was evaluated per metabolite and lipid class covered by targeted HILIC-MS/MS methods. To acquire the lipid and polar metabolite profiles, the metabolite separation was based on the interaction with the HILIC stationary phase: amide-based for complex lipids in positive and in negative ionization mode, and amide-based and with zwitterionic exchange for polar metabolites in positive and in negative ionization mode, respectively [11]. The extraction capacity of each solvent was evaluated by the relative abundance of each lipid and metabolite class to the average signal abundance obtained by the reference solvents (MTBE mixture for lipids and MeOH for polar metabolites, Figure 2, Figure 3 and Figures S2–S4).

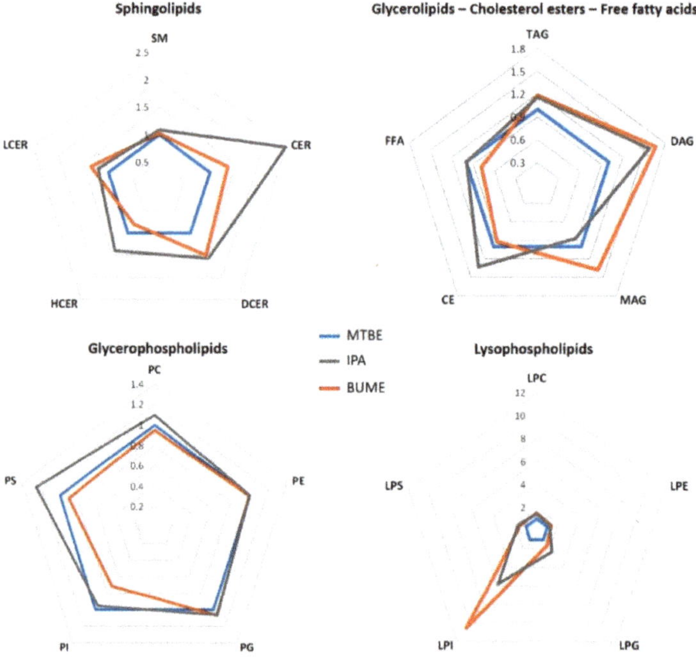

Figure 2. Relative signal abundance (per lipid class) to signal recovered with biphasic extraction with methyl-tert-butyl ether (MTBE). Values indicated on spider plots represent the relative signal

abundance (or fold change) for each single-step method, using IPA or BUME, to the reference MTBE extract. For statistical significance see Figure S2 in the Supplementary Information (data are given in Tables S7 and S8). Lipid classification abbreviations: SM—sphingomyelin, CER—ceramides, DCER—dihydroceramides, LCER—lactosylceramiedes, HCER—hexosylceramides, MAG—monoacylglycerol, DAG—diacylglycerol, TAG—triacylglycerol, CE—cholesterol esters, FFA—free fatty acids, PI—phosphatidylinositol, PC—phosphatidylcholine, PG—phosphatidylglycerol, PE—phosphatidylethanolamine, PS—phosphatidylserine, LPC—lysophospha- tidylcholine, LPG—lysophosphatidylglycerol, LPE—lysophosphatidylethanolamine, LPI—lysophosphatidylinositol, LPS—lysophosphatidylserine.

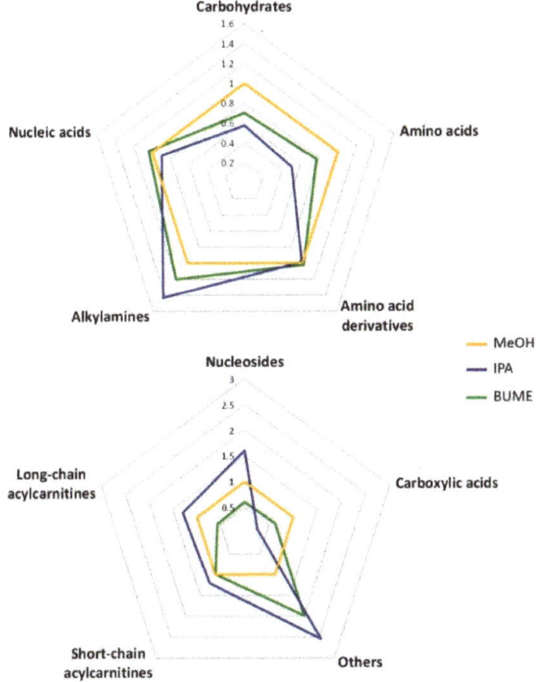

Figure 3. The relative abundance of polar metabolite classes in IPA and BUME extracts as compared to MeOH plasma extract. Values indicated on spider plots represent the relative signal abundance (or fold change) for each single-step method, using IPA or BUME, to the reference MeOH extract. For statistical significance, see Figure S7 in the Supplementary Information (data are provided in Table S9). For the comparison with aqueous phase from Matyash method see Figure S8. The class "Others", i.e., other polar metabolites comprise glycocholate, hydroquinone, hydroxyphenyllactate, pyridoxine, salsolinol, trigonelline, and tryptamine (Table S6). Carboxylic acids comprised mono-, di-, and tri-carboxylic acids.

2.2.1. Complex Lipid Profile

Initially, twenty lipid sub-classes were targeted by monitoring 1159 characteristic MRM transitions combining positive and negative ionization mode. Following the filtering of noisy signals, the detectable lipidome in human plasma was resumed to 584 lipid species. In the positive mode, 324 lipids were detected, including sphingomyelins (SM), ceramides (CER), dihydroceramides (DCer), lactosylceramides (LCer), hexosylceramides (HCer), mono, di, triacylglycerols (MAG, DAG, TAG, respectively), and cholesterol esters (CE). In negative mode, 260 lipids were detected, comprising phosphatidylinositols (PI), phosphatidylcholines (PC), phosphatidylglycerols (PG), phosphatidylethanolamines (PE), phosphatidylserines (PS), and their lysophospholipid analogues (LPC, LPG, LPE, LPI, LPS), and free fatty acids (FFA).

The signal abundance recovered (with each extraction method) was evaluated by relative comparison to the average signal of the reference extract, in this case, the MTBE for lipid extraction. Lipid species that were representative of all different lipid classes were successfully extracted by all three solvents tested, with the exception of LPIs, for which the signal in the MTBE extract was close to the limit of detection, as shown in Figure 2. Among sphingolipids, the highest signal was obtained with IPA and BUME extracts for all sub-classes, including sphingomyelins, ceramides, dihydroceramides, lactosylceramides and hexosylceramides. In the case of glycerolipids (MAGs, DAGs, and TAGs) the highest signals were obtained following single-step extraction with BUME and IPA, although the relative signal was not significantly different from the reference signal of the MTBE extract. Among glycerophospholipids, the significantly higher signal was obtained with IPA for PSs. For other subclasses the difference between three solvents was not significant. For lysophospholipids, the relative signal abundance was in general significantly higher for IPA and BUME extracts when compared to MTBE phase (Figure 2 and Figure S2, Tables S7 and S8).

To confirm these observations, six internal standards (IS), analogues for TAGs, PCs, LPCs, PEs, LPEs, and SMs, were spiked to the samples with the addition of organic solvent (see Figure S3). Based on relative signal abundance these internal standards demonstrated the similar tendency as the endogenous metabolites, depending on the lipid class. Globally, a higher signal was obtained for all different IS following IPA and BUME extraction compared to the organic MTBE phase from Matyash extraction, with the exception of TAG-d7, for which the signal was also the most variable (Figure S3).

In addition, in order to evaluate the extent of interference between lipids and polar metabolites when using one-step extraction coupled to HILIC-MS/MS analysis, the most abundant polar metabolites in human plasma (including different amino acids, organic acids, and acylcarnitines) were added to the MRM method for complex lipid analysis. We could appreciate that the selected polar metabolites, in the applied HILIC conditions, did not co-elute with complex lipids, therefore minimizing the potential effect of ion suppression on lipids (see Figures S4 and S5). Polar metabolites were effectively retained, but due to their high hydrophilicity, they eluted (with the exception of hydroquinone and hypoxanthine) after complex lipids in the chromatographic gradient.. Interestingly, similar results were found for the long chain acylcarnitines that did not co-elute with lipids, such as sphingomyelins when profiled in the positive ionization mode.

Finally, different organic solvents can lead to different contamination levels from plastic agents and, therefore, cause differences in matrix effects. To examine the potential influence of contaminants on lipid abundances, the background noise from blank extractions was examined using HRMS analysis (see Figure S6). Overall, no significant differences in the contaminant background were observed among the different solvent extracts, with the exception of MTBE extract profile in negative ionization mode, which showed several intense peaks eluting between 1 and 4 min. A few of these peaks likely represent fatty acids (stearic acid with m/z 283.26 and palmitic acid with m/z 255.23) that, in addition to being endogenous lipids, also represent the potential contaminants from plastics (confirmed with matching against MaConDa (https://www.maconda.bham.ac.uk/)).

2.2.2. Polar Metabolite Profile

The same plasma extracts were analyzed for the detection and abundance of polar metabolites (as described in Materials and Methods). The HILIC-MS/MS method initially targeted 374 polar metabolites that were classified into 12 categories: proteinogenic amino acids, amino acid derivatives, nucleosides, nucleic acids, alkylamines, short and long chain acylcarnitines (SCACs and LCACs, respectively), carbohydrate conjugates, mono-, di-, and tri-carboxylic acids (Mono-COOHs, Di-COOHs, Tri-COOHs, respectively), and others (comprising glycocholate, hydroquinone, hydroxyphenyllactate, pyridoxine, salsolinol, trigonelline, and tryptamine, see Table S6 for more information). Similar to the complex lipid evaluation, we have evaluated the capacity of the BUME and IPA extractions methods to recover different polar metabolite classes with respect to signal abundance in the MeOH, as the reference method.

Significantly higher signals for the proteinogenic amino acids, carboxylic acids and carbohydrates were observed following the extraction with MeOH, while the nucleosides, alkylamines, short and long-chain acylcarnitines, and other polar metabolites showed significantly higher signals following the extraction with IPA, compared to BUME and MeOH (Figure 3 and Figure S7, Table S9). For amino acid derivatives and nucleic acids the signal intensity did not differ significantly between the MeOH, BUME, and IPA extracts.

Finally, the aqueous phase from biphasic Matyash extraction was also analyzed in order to evaluate the polar metabolite profile when compared to the profiles that were derived from other single-step methods (rich in lipids). For the majority of polar metabolites, with the exception of nucleic acids, amines, and alkylamines, the signal from the aqueous phase (following the Matyash protocol) was significantly lower when compared to single-step methods (Figure S8).

2.3. Reproducibility and Size of Measurable Lipidome and Metabolome

The analytical variability of the selected single-step extraction methods using IPA and BUME was evaluated through independent preparation of pooled plasma samples by four different operators (n = 5 samples per operator and per solvent). The intra-batch coefficient of variation (CV) across 20 replicates, analyzed in a randomized fashion, was determined for both complex lipids and polar metabolites. Figure 4 represents the lipid and polar metabolite count, depending on the coefficient of variation across all replicates and operators (Tables S10–S12).

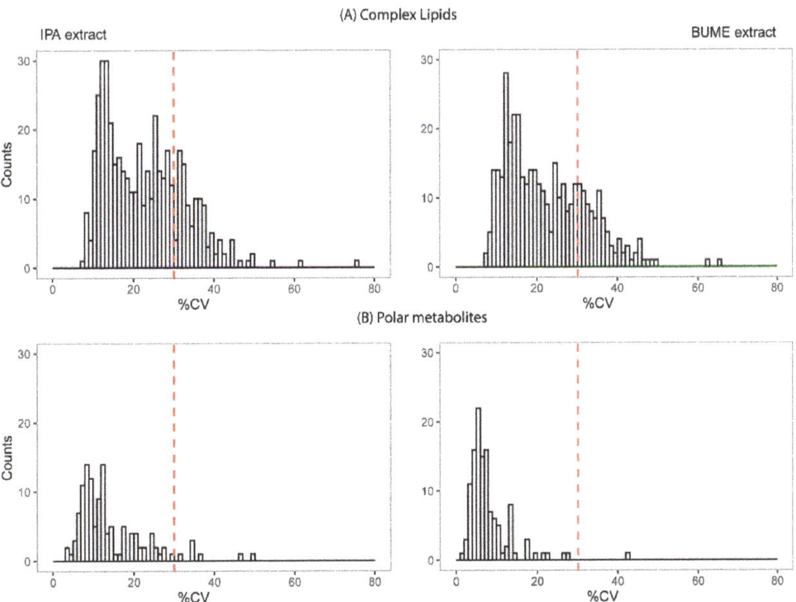

Figure 4. Signal reproducibility for complex lipids and polar metabolites. Coefficient of variation (CV) of all detected lipids (**A**) and polar metabolites (**B**) is presented in frequency histograms. The CV was evaluated across 20 independently prepared plasma samples (5 replicates × 4 different operators) using IPA and BUME. Red line indicates the threshold CV = 30% (data provided in Tables S10–S12).

Out of 584 lipid species that were detected in pooled human plasma samples, 345 from the IPA extract had CV < 30% vs. 294 lipid species from the BUME extract (Figure 4A). The lipids with CV > 30%

mainly included TAGs (169), ceramides (13, comprising DCER LCER, HCER), diacylglycerols (8), phosphatidyletanolamines (8), lysophosphatidylinositols (6), and cholesterol esters (6).

For polar metabolites, out of the 116 metabolites detected in pooled plasma samples, 109 from the IPA extract had a CV < 30% vs. 115 metabolites from the BUME extract (Figure 4B).

Following the filtering based on this analytical variability (CV < 30%), the size and diversity of the measurable lipidome and polar metabolome is shown in Figure 5 for the IPA extract, and in Figure S9 for the BUME extract.

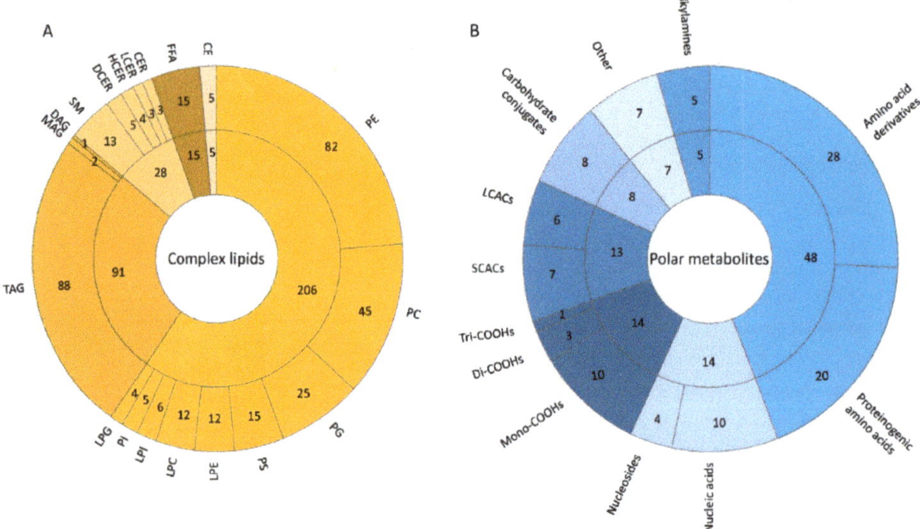

Figure 5. Diversity and size of measurable lipidome and polar metabolome. Complex lipids (**A**) and polar metabolites (**B**) that were reproducibly measured from IPA extract (CV < 30%, see Figure 4) are presented in the pie charts to show the broadest and most reproducible coverage. Lipid classification abbreviations: SM—sphingomyelin, CER—ceramides, DCER—dihydroceramides, LCER—lactosylceramiedes, HCER—hexosylceramides, MAG—monoacylglycerol, DAG—diacylglycerol, TAG—triacylglycerol, CE—cholesterol esters, FFA—free fatty acids, PI—phosphatidylinositol, PC—phosphatidylcholine, PG—phosphatidylglycerol, PE—phosphatidyletanolamine, PS—phosphatidylserine, LPC—lysophosphatidylcholine, LPG—lysophosphatidylglycerol, LPE—lysophosphatidyletanolamine, LPI—lysophosphatidylinositol, LPS—lysophosphatidylserine. Class abbreviations for polar metabolites: short (SCACs) and long chain (LCACs) acylcarnitines, mono- (Mono-COOHs)-, di- (Di-COOHs)-, and tri- (Tri-COOHs) carboxylic acids and others (comprising glycocholate, hydroquinone, hydroxyphenyllactate, pyridoxine, salsolinol, trigonelline, and tryptamine, see Table S6 for more information).

3. Discussion

In this study, we have evaluated the performance of several single-step protocols for the simultaneous extraction of complex lipids and polar metabolites from the human plasma in a high-throughput manner. The extractions were performed with solvents that cover the following range of polarity indexes H_2O > MeOH > EtOH > IPA > BuOH > MTBE, with the BUME mixture expected to be in the middle of this range [24,25]. Therefore, the extraction affinity of BUME is likely driven by the interaction of both butanol and methanol solvent, with the resulting polarity effect for the more hydrophobic (lipids) and hydrophilic (polar metabolites) compounds.

HILIC chromatography was chosen as the best compromise for performing complex lipid and polar metabolite analysis. It is important to note that different HILIC methods with different gradients and mobiles phases (see Materials and methods for details) were used for complex lipid and polar metabolite profiling, respectively. In addition to offering high chromatographic resolution for polar metabolite profiling, the HILIC approach is also advantageous for lipid analysis, since it allows for the separation of lipids classes based on their head groups. The separation by lipid class facilitates the development of quantitative methods, because each class can be covered by its corresponding internal standard. When compared to Reversed Phase (RPLC), the separation using HILIC also better corrects the matrix effects, since the endogenous lipids and their corresponding Internal Standards (IS) co-elute [26]. Importantly, HILIC separation also allows for the chromatography-assisted lipid quantification in large-scale population studies, due to the acceptable cost of a relatively low number of required internal standards.

The aim of our study was to evaluate whether the use of a single-step extraction generates the lipid and polar metabolite coverage equivalent to the commonly used biphasic protocols. To this end, we evaluated the abundance and repeatability of the MS signal of all detected endogenous lipids and polar metabolites in human plasma extracts instead of estimating the extraction efficiency of each solvent with the recovery assessment using pre- and post-extraction internal standard spike. Because complex lipids bind to protein carriers (i.e., lipoproteins), a property that cannot be reproduced by an internal standard spike, our strategy to evaluate the extraction method performance using the signal of extracted endogenous metabolites was considered as the best compromise. We argue that the best strategy to quantify the recovery of endogenous compounds is by using reference materials due to the complexity of biological matrices in general [27,28]. This approach is commonly applied for the validation of the measurement accuracy of analytical methods and was out of the scope of this study.

3.1. Complex Lipid Extraction

The extraction of lipids depends on their structure and substitute groups (i.e., headgroups that are representative of each lipid class), their polarity (determined by the head group and length of the alkyl chain) and their spatial configuration (e.g., degree of unsaturation). These structural characteristics play an important role in the interaction with the extraction solvent. As previously reported for BUME and IPA extraction, in general, more than 90% of the total lipid content of a plasma sample is extracted, and it is not deemed necessary to perform a re-extraction of the pellet [18,19].

Among different lipid classes, the relative signal abundance for sphingolipids was higher in IPA and BUME extracts when compared to MTBE. This can be explained by the polar character of the head group in the sphingolipid structure (Figure S10). The extraction of the sphingolipids with more polar character, such as lactosylceramides (LCer), was improved with more polar BUME mixture, owing to the lactosyl group. Conversely, in the case of hexosylceramides (HCER), significantly higher signal abundance was observed with IPA solvent as compared to MTBE and BUME (see Figure S11, showing the signal abundance of specific hexosylceramides and lactosylceramides with the same alkyl chain composition, across different solvents).

Glycerolipids and mono-, di-, and tri-alkyl substituted glycerols followed a similar trend of solvent affinity, MTBE < IPA < BUME, although this was observed without a statistically significant difference among the tested extraction methods. The arrangement of fatty acid chains located in any of three positions (sn-1/sn-2/sn-3) in the glycerol backbone may play an important role in the extraction affinity of this lipid class [29]. The highest affinity for BUME was the most pronounced for MAGs and DAGs having lower hydrophobicity, due to only one or two fatty acyl chains and the exposition of the hydroxyl group.

For cholesterol esters, which also have a highly lipophilic character similar to TAGs, no significant difference was observed between three extraction protocols, which could be due to the low intensity and high variability of signal due to their low electrospray ionization efficiency. The highest and the most reproducible response for these lipids was observed when using IPA as extraction solvent.

In the case of glycerophospholipids, the extraction appeared to be primarily driven by their hydrophobic character (due to their alkyl chains), since all subclasses were most efficiently recovered with the MTBE and IPA. These results, showing the strong affinity of PIs for the MTBE, and PGs, PCs and PEs for all three solvents, are in agreement with the previous reports of Lee et al. and Matyash et al. [17,30]. Importantly, the performance of the single-step extraction with IPA was at the same level as the reference method with MTBE for all glycerophospholipid classes. Even more, a significantly higher signal for PSs was obtained following the extraction with IPA (see Figure 2).

Finally, lysoglycerophospholipids, which are composed of a single fatty acyl chain, have lower hydrophobicity and, thus, their solubility mainly relies on the head group. Consequently, the extraction of lysophospholipids was more efficient with BUME and IPA. This effect was exacerbated for LPIs that were poorly recovered using the biphasic method with MTBE, thus limiting the LC-MS analysis of this class of lipids (i.e., signals close the level of detection).

3.2. Polar Metabolite Extraction

Large population studies require high-throughput methods, with minimal and highly reproducible sample preparation. Therefore, in this study, we investigated the viability of measuring polar metabolites from the same plasma extract that was obtained in single-step extraction used for the lipidome analysis. Polar metabolite analysis was performed using the optimized targeted HILIC-based methods (in acidic and basic conditions in positive and negative ionization mode, respectively) and methanol extraction was used as a reference for comparison [11,31].

Because BUME and IPA demonstrated the best performance for lipid extraction in a single step, we investigated their capacity to simultaneously extract polar metabolites implicated in central carbon metabolism (Table S9). Similar to lipid analysis, all of the polar metabolite classes were efficiently extracted by both methods. However, for the most polar classes, i.e., proteinogenic amino acids, carboxylic acids (mono-, di-, and tri-), and carbohydrate conjugates, the recovered signal was significantly lower while using IPA and BUME, as compared to methanol. Besides the hydrophilicity of these metabolites, the signal decrease following BUME and IPA extraction is also likely to be a consequence of ion suppression that is caused by co-eluting complex lipids (extracted with IPA and BUME) particularly in ESI positive mode, in the chromatographic conditions for polar metabolite analysis. Despite its decrease, the signal remained well defined and reproducible—a prerequisite for quantitative measurement.

For the majority of other polar metabolites and specifically nucleosides and alkylamines a significantly higher signal was observed with IPA when compared to BUME and MEOH. We hypothesize that this may be due to the representative heterocyclic compounds.

For acylcarnitines, the signal abundance varied, depending on the length of the acyl chain. Acylcarnitines are zwitterion molecules that are composed of a quaternary ammonium linked to a fatty acyl chain and their polarity varies, depending on the length of the acyl chain. The hydrophobic character of long-chain carnitines is evident, since the signal intensity was significantly higher in the IPA extract. For more polar short-chain acylcarnitines, the intensity of the recovered signal was also significantly higher in IPA extract but to a lesser extent. For both of these classes, methanol performance was significantly lower when compared to IPA.

It is important to note that when applying biphasic methods for lipid extraction, polar metabolites were previously reported to be successfully measured from the aqueous layer [32]. Interestingly, our results clearly showed the significantly lower signal abundance for the majority of polar metabolites following the biphasic Matyash protocol—as a result of the analysis of aqueous phase (when compared to BUME, IPA, and MEOH extracts, Figure S8). This observation is likely due to the partition of polar metabolites between the aqueous and organic phase in the Matyash extraction.

3.3. Extraction Method and Measurement Reproducibility

In terms of lipidome coverage, the single-step lipid extraction using IPA or BUME efficiently recovered all lipid sub-classes, which was already reported using untargeted approaches for the comparison of Matyash, Folch, Alshery (1:1, BUME), and IPA protocols [19,33–35]. Importantly, one-phase extraction using MMC solvent mixture (MeOH/MTBE/CHCl$_3$) showed significantly better extraction efficiencies for moderate and highly non-polar lipid species in comparison with biphasic Folch, Bligh and Dyer, and MTBE extraction systems [32]. While the size of detectable lipidome was extensively explored in several previous studies, the relative abundance and repeatability of lipid signal was rarely evaluated. When evaluating the size of measurable lipidome and polar metabolome we observed a difference in lipid signal variability between IPA and BUME extracts following the reproducibility test (CV < 30% across independently extracted replicates). This difference was mainly due to the higher variability of specific lipid signals recovered from BUME extracts, comprising TAGs (20), some phospholipids, free fatty acids, and cholesterol esters. We hypothesize that the presence of MeOH and thus the mixture of two solvents, methanol, and butanol (*vs.* pure IPA) could contribute to this difference in CVs [36]. In general, high CVs that were observed for TAGs were likely caused by ion suppression as a consequence of high degree of co-elution in the void volume (using HILIC chromatography). The difference in terms of signal reproducibility was not observed for polar metabolites, likely thanks to a significantly higher degree of separation of hydrophilic metabolites and, thereby, the lower degree of co-elution throughout the chromatographic gradient.

The robustness of IPA and BUME extractions for lipidome analysis was previously assessed by independent studies using untargeted lipidomics approaches. For example, Calderón et al. compared the extraction methods while using CHCl$_3$, MTBE, and IPA, after which IPA was reported as the most robust [37]. In another study, extraction protocols using MeOH, acetonitrile (ACN), IPA, IPA-ACN, CH$_2$Cl$_2$, CHCl$_3$, MTBE, and Hexane were compared, and IPA was revealed as the best compromise to extract lipids [19]. In an additional study, where Folch, Matyash, and Alshery methods were tested, and the Alshery method was reported with high recoveries and lowest CV values for lipids [33]. All of these previous results highlight the differences in terms of signal reproducibility, depending on the lipid class and show that the most robust methods are the least biased single-step methods.

Beside robustness, the application of single-step plasma extraction in human population studies [38] is also advantageous in terms of high throughput. For instance, BUME extraction was applied to the study of differences in lipidome composition in the Singaporean population where three main communities, Chinese, Indian, and Malays were characterized [39]. It was also recently used in large-scale plasma lipidomic profiling to reveal the associations between lipid levels and cardiometabolic risk factors [40]. When compared to BUME that has been used in lipidomics community for more than 12 years now, the potential of IPA extraction for lipidome analysis was only recently revealed and it was therefore less commonly applied in lipidomics studies. However, several recent human population phenotyping studies warrant its application [19,41,42].

4. Materials and Methods

4.1. Chemicals and Reagents

Human pooled plasma samples were purchased from Sera Laboratories International Ltd. Trading as BioIVT (West Sussex, UK). A mixture of pooled male and female (40–60 years old) plasma was prepared and aliquoted for the extraction experiments.

Internal Standards, including sphingomyelin SM-18:1(d9) (1.96 µg/mL), triacylglycerol TAG-15:0/18:1/15:0(d7) (3.5 µg/mL), phosphatidylcholine PC-15:0/18:1(d7) (9.99 µg/mL), phosphatidylethanolamine PE-15:0/18:1(d7) (0.35 µg/mL), lysophosphatidylcholine LPC-18:1(d7) (1.58 µg/mL), and lysophosphatidylethanolamine LPE-18:1(d7) (0.32 µg/mL), were purchased from Avanti (Alabaster, AL, USA).

4.2. Metabolite Extraction Protocols

Human blood plasma (25 µL) was extracted with MeOH (125 µL), EtOH (125 µL), BUtanol:MEthanol (BUME, 125 µL), or isopropanol (IPA, 125 µL) that were pre-spiked with the above indicated mixture of internal standards in order to evaluate the performance of different extraction solvents. For the biphasic MTBE extraction 10 μL plasma was extracted with MTBE:MeOH:H$_2$O (750/225/188 μL) [43]. Because plasma is homeostatically regulated no normalization to protein amount is required [44]. All of the samples were vortexed and kept at −20 °C for one hour to facilitate protein precipitation. These extracts were then centrifuged for 15 min at 20,000× *g* at 4 °C and the resulting supernatants, from MeOH, EtOH, BUME, and IPA, were collected and transferred to LC-MS vials for injection. The upper phase resulting from biphasic MTBE:MeOH:H$_2$O extraction was evaporated to dryness (in a vacuum concentrator LabConco) and re-suspended in 50 μL of IPA for lipid extraction evaluation in order to maintain the same sample to solvent ratio (1/5), as in the single step extraction. Prior to LC-MS analysis, the extracted pooled plasma samples were randomized per operator and extraction solvent.

4.3. Protein Quantification

BCA Protein Assay Kit (Thermo Scientific, Waltham, MA, USA) was used in order to measure (A562nm) total protein concentration (Hidex, Turku, Finland). For this, the protein pellets were evaporated and reconstituted in the lysis buffer (20 mM Tris-HCl (pH 7.5), 4M guanidine hydrochloride, 150 mM NaCl, 1 mM Na$_2$EDTA, 1 mM EGTA, 1% Triton, 2.5 mM sodium pyrophosphate, 1 mM beta-glycerophosphate, and 1 mM Na$_3$VO$_4$, 1 μg/mL leupeptin) using the Cryolys Precellys 24 sample Homogenizer (2 × 20 s at 10,000 rpm, Bertin Technologies, Rockville, MD, USA) with ceramic beads.

Protein quantification was performed on supernatant and plasma pellets in order to evaluate the efficiency of each solvent to precipitate proteins. The precipitation efficiency (Table S5) was calculated, as follows:

$$Protein precipitation efficiency = \frac{protein\ content\ in\ plasma\ pellet}{protein\ content\ in\ plasma\ pellet + supernatant} \times 100$$

4.4. LC-MS/MS Analysis

4.4.1. High-Coverage Targeted Lipidome Analysis

For the lipidome analysis, the extracted plasma samples were analyzed by Hydrophilic Interaction Liquid Chromatography coupled to tandem mass spectrometry (HILIC-MS/MS) in positive and negative ionization modes using a Q-TRAP 6500 plus LC-MS/MS system (Sciex, Darmstadt, Germany). In both positive and negative ionization mode, the chromatographic separation was carried out on an Acquity BEH Amide, 1.7 μm, 100 mm × 2.1 mm I.D. column (Waters, Milford, MA, USA). Mobile phase was composed of A = 10 mM ammonium acetate in Acetonitrile:H$_2$O (95:5) (pH = 8.2) and B = 10 mM ammonium acetate in Acetonitrile:H$_2$O (50:50) (pH = 7.4). The linear gradient elution from 0.1% to 20% B was applied for 2 min, from 20% to 80% B for 3 min, followed by 3 min of re-equilibration to the initial chromatographic conditions. The flow rate was 600 μL/min, column temperature 45 °C, and sample injection volume 3 μL. Optimized ESI Ion Drive Turbo V source parameters were set, as follows: Ion Spray (IS) voltage 5500 V in positive mode and −4500 V in negative mode, curtain gas 35 psi, nebulizer gas (GS1) 50 psi, auxiliary gas (GS2) 60 psi, and source temperature 550 °C. Nitrogen was used as the nebulizer and collision gas. Optimized compound-dependent parameters were used for data acquisition in scheduled multiple reaction monitoring (sMRM) mode. Transitions for the entire panel of targeted lipids were optimized by SCIEX while using the Lipidyzer™ Platform [45]. The pooled plasma sample was first analyzed by MRM (non-scheduled) in order to obtain the retention time of each lipid class in the applied chromatographic system and added later to the scheduled MRM method (Tables S1 and S2).

4.4.2. High-Coverage Targeted Metabolome Analysis

For the polar metabolome analysis, the extracted plasma samples were analyzed by HILIC-MS/MS in both positive and negative ionization modes while using a 6495 triple quadrupole system (QqQ) interfaced with 1290 UHPLC system (Agilent Technologies, Santa Clara, CA, USA). In the positive mode, the chromatographic separation was carried out in an Acquity BEH Amide, 1.7 µm, 100 mm × 2.1 mm I.D. column (Waters, MA, USA). Mobile phase was composed of A = 20 mM ammonium formate and 0.1% FA in water and B = 0.1% formic acid in ACN. The linear gradient elution from 95% B (0–1.5 min) down to 45% B was applied (from 1.5 to 17 min) and these conditions were held for 2 min. Subsequently, initial chromatographic condition were maintained as a post-run during 5 min for column re-equilibration. The flow rate was 400 µL/min, column temperature 25 °C, and sample injection volume 2 µL. In negative mode, a SeQuant ZIC-pHILIC (100 mm, 2.1 mm I.D., and 5 µm particle size (Merck, Darmstadt, Germany) column was used. The mobile phase was composed of A = 20 mM ammonium Acetate and 20 mM NH_4OH in water at pH 9.7 and B = 100% ACN. The linear gradient elution from 90% (0–1.5 min) to 50% B (8–11 min) down to 45% B (12–15 min). Finally, the initial chromatographic conditions were established as a post-run during 9 min for column re-equilibration. The flow rate was 300 µL/min, column temperature 30 °C, and sample injection volume 2 µL.

ESI source conditions were set, as follows: dry gas temperature 290 °C, nebulizer 35 psi and flow 14 L/min, sheath gas temperature 350 °C and flow 12 L/min, nozzle voltage 0 V, and capillary voltage +2000 V and −2000 V in positive and negative mode, respectively. Dynamic Multiple Reaction Monitoring (DMRM) was used as an acquisition mode with a total cycle time of 600 ms. The MRM transitions were optimized from the direct analysis of pure chemical standards that were obtained from Sigma–Aldrich (The Mass Spectrometry Metabolite Library of Standards—MSMLS) (Tables S3 and S4). These transitions are publicly available in the Metlin-MRM spectral library [46].

4.5. Data (Pre)Processing

Raw LC-MS/MS lipidome data were processed (i.e., peak extraction and alignment) while using the Multi Quant Software (version 3.0.3, Sciex, Framingham, MA, USA) and raw metabolome data was processed using the Agilent Quantitative analysis software (version B.07.00, MassHunter, Agilent technologies, Santa Clara, CA, USA). The data on peak height and peak area were extracted for each lipid and polar metabolite based on its extracted ion chromatograms (EICs) for the monitored MRM transitions. The peaks were filtered based on their presence (in 100% of replicates across all operators) and the intensity threshold (of 5×10^3 ion counts), and the obtained tables (containing peak areas of detected metabolites across all replicates) were exported to R version 3.5.1 software http://cran.r-project.org/ and RStudio version 1.1.463. For a quality control, the peaks were filtered based on their coefficient of variation calculated per solvent, while using the independent replicates analyzed across the entire run, when considering all four operators. Peaks with CV > 30% were removed from further analysis with the aim to determine the size of measurable and not only detectable lipidome and metabolome.

4.6. Statistical Data Analysis

R packages "tidyverse" and "ggplot2" were used to format the data and plot the figures, respectively. Fold changes were calculated while using the peak area values (no transformation and/or scaling was applied). GraphPad Prism 6 (GraphPad Software Inc., La Jolla, CA, USA) was used for statistical data analysis. One-way ANOVA was used in order to test the significance of signal abundance between replicates that were extracted with different solvents with an arbitrary level of significance p-value = 0.05.

5. Conclusions

Two single-step sample preparation methods, using IPA and BUME, were evaluated as the best compromise for the simultaneous and reproducible extraction of complex lipids and polar metabolites from human plasma. The relative signal abundance of complex lipids in IPA and BUME extracts was greater or equivalent to the signal that was recovered with the Matyash method using MTBE (commonly applied in lipidomics). Importantly, the MTBE showed limited performance for the extraction of lysophospholipids, and particularly lysophosphatidylinositols (LPIs). Although the breadth of coverage was the same for both single-step methods, the most robust lipid profiling was achieved following IPA extraction with the greatest number of profiled lipids with CV < 30%, and specifically TAGs. In addition to complex lipids, both IPA and BUME method extracted polar metabolites successfully, but less efficiently than methanol. These polar metabolites included proteinogenic amino acids and acylcarnitines, di- and tri-carboxylic acids, carbohydrates, and nucleosides. Based on the examined lipidome and polar metabolome coverage, extraction efficiencies, effectiveness of protein precipitation, and reproducibility, we conclude that both methods, IPA and BUME, are suitable for merged lipid and polar metabolite analysis in large-scale human population studies.

Supplementary Materials: The following supplementary figures and tables are available online at http://www.mdpi.com/2218-1989/10/12/495/s1: Figure S1. Relative signal abundance of different lipid classes in MeOH, EtOH, IPA and BUME extracts of human plasma, Figure S2. Relative signal abundance (per lipid class) to signal recovered with biphasic extraction using MTBE, Figure S3. Signal abundance of internal standards (representing six lipid classes) spiked into plasma samples during single step IPA, BUME and biphasic Matyash extraction (organic MTBE phase), Figure S4. Retention of polar metabolites and complex lipids throughout the chromatographic gradient applied for complex lipid analysis using HILIC-MS/MS in positive ionization mode, Figure S5. Retention of polar metabolites and complex lipids throughout the chromatographic gradient applied for complex lipid analysis using HILIC-MS/MS in negative ionization mode, Figure S6. Background noise from blank extractions performed with (A) methanol, (B) Matyash extraction (organic MTBE phase), (C) IPA and (D) BUME, Figure S7. Relative signal abundance (per polar metabolite class) to signal recovered with MeOH extract, Figure S8. Relative signal abundance of polar metabolites detected in the aqueous phase of Matyash extraction compared to other extraction protocols (including methanol extract as a reference), Figure S9. Diversity and size of measurable lipidome and polar metabolome from BUME extract, Figure S10. Signal intensity of the sphingomyelins detected in the method, Figure S11. Signal intensity of specific hexosylceramides and lactosylceramides with the same alkyl chain composition in IPA, BUME and MTBE extracts, Table S1. MRM transitions of lipids in positive mode, Table S2. MRM transitions of lipids in negative mode, Table S3. MRM transitions of polar metabolites in positive mode, Table S4. MRM transitions of polar metabolites in negative mode, Table S5. Protein content per each solvent or solvent mixture, Table S6. Classification of polar metabolites, Table S7. Abundances (peak areas) of lipid species detected by the HILIC-MS/MS analysis of each solvent extract (organic MTBE phase from Matyash protocol, IPA and BUME) in positive ESI mode, Table S8. Abundances of lipid species detected by the HILIC-MS/MS analysis of each solvent extract (organic MTBE phase from Matyash protocol, IPA and BUME) in negative ESI mode, Table S9. Abundances (peak areas) of polar metabolites detected by the HILIC-MS/MS analysis of each solvent extract (methanol, IPA and BUME) in positive and negative ESI modes, Table S10. Abundances and CVs of lipid species detected by the HILIC-MS/MS analysis of selected single-step extractions (IPA and BUME) in positive ESI mode, Table S11. Abundances and CVs of lipid species detected by the HILIC-MS/MS analysis of selected single-step extractions (IPA and BUME) in negative ESI mode, Table S12. Abundances and CVs of polar metabolites detected by the HILIC-MS/MS analysis of selected single-step extractions (IPA and BUME) in positive and negative ESI mode.

Author Contributions: Conceptualization: J.M., H.G.-A., J.I.; Data curation: J.M., T.T., V.v.d.V.; Formal analysis: J.M.; Visualization: J.M., V.v.d.V.; Investigation, J.M., T.T., Y.G., H.G.-A.; Writing—original draft: J.M. Writing—review and editing: H.G.-A., J.I.; Supervision: H.G.-A., J.I. All authors have read and agreed to the published version of the manuscript.

Funding: This work was supported by funds from Faculty of Biology and Medicine (FBM) at the University of Lausanne (UNIL), Fondation Pierre-Mercier pour la science and Swiss National Science Foundation (SNF) grant 316030_183377.

Conflicts of Interest: The authors declare no conflict of interest.

References

1. Psychogios, N.; Hau, D.D.; Peng, J.; Guo, A.C.; Mandal, R.; Bouatra, S.; Sinelnikov, I.; Krishnamurthy, R.; Eisner, R.; Gautam, B.; et al. The human serum metabolome. *PLoS ONE* **2011**, *6*, e16957. [CrossRef] [PubMed]

2. Quehenberger, O.; Dennis, E.A. The Human Plasma Lipidome. *N. Engl. J. Med.* **2011**, *365*, 1812–1823. [CrossRef] [PubMed]

3. Fahy, E.; Subramaniam, S.; Murphy, R.C.; Nishijima, M.; Raetz, C.R.H.; Shimizu, T.; Spener, F.; Van Meer, G.; Wakelam, M.J.O.; Dennis, E.A. Update of the LIPID MAPS comprehensive classification system for lipids. *J. Lipid Res.* **2009**, *50*, 9–14. [CrossRef] [PubMed]

4. Quehenberger, O.; Armando, A.M.; Brown, A.H.; Milne, S.B.; Myers, D.S.; Merrill, A.H.; Bandyopadhyay, S.; Jones, K.N.; Kelly, S.; Shaner, R.L.; et al. Lipidomics reveals a remarkable diversity of lipids in human plasma. *J. Lipid Res.* **2010**, *51*, 3299–3305. [CrossRef]

5. Holčapek, M.; Liebisch, G.; Ekroos, K. Lipidomic Analysis. *Anal. Chem.* **2018**, *90*, 4249–4257. [CrossRef]

6. Qu, Q.; Zeng, F.; Liu, X.; Wang, Q.J.; Deng, F. Fatty acid oxidation and carnitine palmitoyltransferase I: Emerging therapeutic targets in cancer. *Cell Death Dis.* **2016**, *7*, 1–9. [CrossRef]

7. Schooneman, M.G.; Vaz, F.M.; Houten, S.M.; Soeters, M.R. Acylcarnitines: Reflecting or inflicting insulin resistance? *Diabetes* **2013**, *62*, 1–8. [CrossRef]

8. Bagheri, M.; Djazayery, A.; Farzadfar, F.; Qi, L.; Yekaninejad, M.S.; Aslibekyan, S.; Chamari, M.; Hassani, H.; Koletzko, B.; Uhl, O. Plasma metabolomic profiling of amino acids and polar lipids in Iranian obese adults. *Lipids Health Dis.* **2019**, *18*, 1–9. [CrossRef]

9. Liu, J.; Liu, X.; Xiao, Z.; Locasale, J.W. Quantitative evaluation of a high resolution lipidomics platform. *bioRxiv* **2019**, 627687. [CrossRef]

10. Ivanisevic, J.; Want, E.J. From samples to insights into metabolism: Uncovering biologically relevant information in LC- HRMS metabolomics data. *Metabolites* **2019**, *9*, 308. [CrossRef]

11. Gallart-Ayala, H.; Konz, I.; Mehl, F.; Teav, T.; Oikonomidi, A.; Peyratout, G.; van der Velpen, V.; Popp, J.; Ivanisevic, J. A global HILIC-MS approach to measure polar human cerebrospinal fluid metabolome: Exploring gender-associated variation in a cohort of elderly cognitively healthy subjects. *Anal. Chim. Acta* **2018**, *1037*, 327–337. [CrossRef] [PubMed]

12. Lu, W.; Su, X.; Klein, M.S.; Lewis, I.A.; Fiehn, O.; Rabinowitz, J.D. Metabolite measurement: Pitfalls to avoid and practices to follow. *Annu. Rev. Biochem.* **2017**, *86*, 277–304. [CrossRef] [PubMed]

13. Folch, J.; Ascoli, I.; Lees, M.; Meath, J.A.; LeBaron, F.N. Preparation of Lipide Extracts From Brain Tisuue. *J. Biol. Chem.* **1951**, *191*, 833–841. [PubMed]

14. Bligh, E.G.; Dyer, W.J. A rapid method of total lipid extraction and purification. *Can. J. Biochem. Physiol.* **1959**, *37*, 911–917. [CrossRef]

15. Ulmer, C.Z.; Jones, C.M.; Yost, R.A.; Garrett, T.J.; Bowden, J.A. Optimization of Folch, Bligh-Dyer, and Matyash sample-to-extraction solvent ratios for human plasma-based lipidomics studies. *Anal. Chim. Acta* **2018**, *1037*, 351–357. [CrossRef]

16. Vvedenskaya, O.; Wang, Y.; Ackerman, J.M.; Knittelfelder, O.; Shevchenko, A. Analytical challenges in human plasma lipidomics: A winding path towards the truth. *TrAC Trends Anal. Chem.* **2019**, *120*, 115277. [CrossRef]

17. Matyash, V.; Liebisch, G.; Kurzchalia, T.V.; Shevchenko, A.; Schwudke, D. Lipid extraction by methyl- tert -butyl ether for high-throughput lipidomics. *J. Lipid Res.* **2008**, *49*, 1137–1146. [CrossRef]

18. Alshehry, Z.H.; Barlow, C.K.; Weir, J.M.; Zhou, Y.; McConville, M.J.; Meikle, P.J. An efficient single phase method for the extraction of plasma lipids. *Metabolites* **2015**, *5*, 389–403. [CrossRef]

19. Sarafian, M.H.; Gaudin, M.; Lewis, M.R.; Martin, F.P.; Holmes, E.; Nicholson, J.K.; Dumas, M.E. Objective set of criteria for optimization of sample preparation procedures for ultra-high throughput untargeted blood plasma lipid profiling by ultra performance liquid chromatography-mass spectrometry. *Anal. Chem.* **2014**, *86*, 5766–5774. [CrossRef]

20. Löfgren, L.; Ståhlman, M.; Forsberg, G.B.; Saarinen, S.; Nilsson, R.; Hansson, G.I. The BUME method: A novel automated chloroform-free 96-well total lipid extraction method for blood plasma. *J. Lipid Res.* **2012**, *53*, 1690–1700. [CrossRef]

21. Cai, X.; Li, R. Concurrent profiling of polar metabolites and lipids in human plasma using HILIC-FTMS. *Sci. Rep.* **2016**, *6*, 36490. [CrossRef]

22. Patterson, R.E.; Ducrocq, A.J.; McDougall, D.J.; Garrett, T.J.; Yost, R.A. Comparison of blood plasma sample preparation methods for combined LC-MS lipidomics and metabolomics. *J. Chromatogr. B Anal. Technol. Biomed. Life Sci.* **2015**, *1002*, 260–266. [CrossRef] [PubMed]

23. Sud, M.; Fahy, E.; Cotter, D.; Brown, A.; Dennis, E.A.; Glass, C.K.; Merrill, A.H.; Murphy, R.C.; Raetz, C.R.H.; Russell, D.W.; et al. LMSD: LIPID MAPS structure database. *Nucleic Acids Res.* **2007**, *35*, 527–532. [CrossRef]

24. Cajka, T.; Fiehn, O. Toward Merging Untargeted and Targeted Methods in Mass Spectrometry-Based Metabolomics and Lipidomics. *Anal. Chem.* **2016**, *88*, 524–545. [CrossRef]

25. Snyder, L.R. Classification off the solvent properties of common liquids. *J. Chromatogr. Sci.* **1978**, *16*, 223–234. [CrossRef]

26. Lange, M.; Fedorova, M. Evaluation of lipid quantification accuracy using HILIC and RPLC MS on the example of NIST® SRM® 1950 metabolites in human plasma. *Anal. Bioanal. Chem.* **2020**, *412*, 3573–3584. [CrossRef] [PubMed]

27. Bowden, J.A.; Heckert, A.; Ulmer, C.Z.; Jones, C.M.; Koelmel, J.P.; Abdullah, L.; Ahonen, L.; Alnouti, Y.; Armando, A.M.; Asara, J.M.; et al. Harmonizing lipidomics: NIST interlaboratory comparison exercise for lipidomics using SRM 1950–Metabolites in Frozen Human Plasma. *J. Lipid Res.* **2017**, *58*, 2275–2288. [CrossRef]

28. Simón-Manso, Y.; Lowenthal, M.S.; Kilpatrick, L.E.; Sampson, M.L.; Telu, K.H.; Rudnick, P.A.; Mallard, W.G.; Bearden, D.W.; Schock, T.B.; Tchekhovskoi, D.V.; et al. Metabolite profiling of a NIST standard reference material for human plasma (SRM 1950): GC-MS, LC-MS, NMR, and clinical laboratory analyses, libraries, and web-based resources. *Anal. Chem.* **2013**, *85*, 11725–11731. [CrossRef]

29. Lee, J.W.; Nagai, T.; Gotoh, N.; Fukusaki, E.; Bamba, T. Profiling of regioisomeric triacylglycerols in edible oils by supercritical fluid chromatography/tandem mass spectrometry. *J. Chromatogr. B Anal. Technol. Biomed. Life Sci.* **2014**, *966*, 193–199. [CrossRef]

30. Lee, D.Y.; Kind, T.; Yoon, Y.R.; Fiehn, O.; Liu, K.H. Comparative evaluation of extraction methods for simultaneous mass-spectrometric analysis of complex lipids and primary metabolites from human blood plasma. *Anal. Bioanal. Chem.* **2014**, *406*, 7275–7286. [CrossRef]

31. Ivanisevic, J.; Zhu, Z.-J.; Plate, L.; Tautenhahn, R.; Chen, S.; O'Brien, P.J.; Johnson, C.H.; Marletta, M.A.; Patti, G.J.; Siuzdak, G. Toward '{Omic} {Scale} {Metabolite} {Profiling}: {A} {Dual} {Separation}–{Mass} {Spectrometry} {Approach} for {Coverage} of {Lipid} and {Central} {Carbon} {Metabolism}. *Anal. Chem.* **2013**, *85*, 6876–6884. [CrossRef] [PubMed]

32. Gil, A.; Zhang, W.; Wolters, J.C.; Permentier, H.; Boer, T.; Horvatovich, P.; Heiner-Fokkema, M.R.; Reijngoud, D.J.; Bischoff, R. One- vs two-phase extraction: Re-evaluation of sample preparation procedures for untargeted lipidomics in plasma samples. *Anal. Bioanal. Chem.* **2018**, *410*, 5859–5870. [CrossRef] [PubMed]

33. Wong, M.W.K.; Braidy, N.; Pickford, R.; Sachdev, P.S.; Poljak, A. Comparison of single phase and biphasic extraction protocols for lipidomic studies using human plasma. *Front. Neurol.* **2019**, *10*, 1–11. [CrossRef] [PubMed]

34. Cruz, M.; Wang, M.; Frisch-Daiello, J.; Han, X. Improved Butanol–Methanol (BUME) Method by Replacing Acetic Acid for Lipid Extraction of Biological Samples. *Lipids* **2016**, *51*, 887–896. [CrossRef]

35. Sitnikov, D.G.; Monnin, C.S.; Vuckovic, D. Systematic Assessment of Seven Solvent and Solid-Phase Extraction Methods for Metabolomics Analysis of Human Plasma by LC-MS. *Sci. Rep.* **2016**, *6*, 38885. [CrossRef]

36. Satomi, Y.; Hirayama, M.; Kobayashi, H. One-step lipid extraction for plasma lipidomics analysis by liquid chromatography mass spectrometry. *J. Chromatogr. B Anal. Technol. Biomed. Life Sci.* **2017**, *1063*, 93–100. [CrossRef]

37. Calderón, C.; Sanwald, C.; Schlotterbeck, J.; Drotleff, B.; Lämmerhofer, M. Comparison of simple monophasic versus classical biphasic extraction protocols for comprehensive UHPLC-MS/MS lipidomic analysis of Hela cells. *Anal. Chim. Acta* **2019**, *1048*, 66–74. [CrossRef]

38. Alshehry, Z.H.; Mundra, P.A.; Barlow, C.K.; Mellett, N.A.; Wong, G.; McConville, M.J.; Simes, J.; Tonkin, A.M.; Sullivan, D.R.; Barnes, E.H.; et al. Plasma Lipidomic Profiles Improve on Traditional Risk Factors for the Prediction of Cardiovascular Events in Type 2 Diabetes Mellitus. *Circulation* **2016**, *134*, 1637–1650. [CrossRef]

39. Saw, W.Y.; Tantoso, E.; Begum, H.; Zhou, L.; Zou, R.; He, C.; Chan, S.L.; Tan, L.W.L.; Wong, L.P.; Xu, W.; et al. Establishing multiple omics baselines for three Southeast Asian populations in the Singapore Integrative Omics Study. *Nat. Commun.* **2017**, *8*, 1–11. [CrossRef]

40. Huynh, K.; Barlow, C.K.; Jayawardana, K.S.; Weir, J.M.; Mellett, N.A.; Cinel, M.; Magliano, D.J.; Shaw, J.E.; Drew, B.G.; Meikle, P.J. High-Throughput Plasma Lipidomics: Detailed Mapping of the Associations with Cardiometabolic Risk Factors. *Cell Chem. Biol.* **2019**, *26*, 71–84.e4. [CrossRef]

41. Lamour, S.D.; Gomez-romero, M.; Vorkas, P.A.; Alibu, V.P.; Saric, J.; Holmes, E.; Sternberg, J.M. Discovery of Infection Associated Metabolic Markers in Human African Trypanosomiasis. *PLoS Negl. Trop. Dis.* **2015**, *9*, e0004200. [CrossRef] [PubMed]

42. Wang, X.; Nijman, R.; Camuzeaux, S.; Sands, C.; Jackson, H.; Kaforou, M.; Emonts, M.; Herberg, J.A.; Maconochie, I.; Carrol, E.D.; et al. Plasma lipid profiles discriminate bacterial from viral infection in febrile children. *Sci. Rep.* **2019**, *9*, 17714. [CrossRef]

43. Bhattacharya, S.K. Lipidomics: Methods and Protocols, Methods in Molecular Biology. *Methods Mol. Biol.* **2017**, *1609*, 91–106. [CrossRef]

44. Cannon, W.B. Organization for Physiological Homeostasis. *Physiol. Rev.* **1929**, *9*, 399–431. [CrossRef]

45. Cao, Z.; Schmitt, T.C.; Varma, V.; Sloper, D.; Beger, R.D.; Sun, J. Evaluation of the Performance of Lipidyzer Platform and Its Application in the Lipidomics Analysis in Mouse Heart and Liver. *J. Proteome Res.* **2020**, *19*, 2742–2749. [CrossRef] [PubMed]

46. Domingo-Almenara, X.; Montenegro-Burke, J.R.; Ivanisevic, J.; Thomas, A.; Sidibé, J.; Teav, T.; Guijas, C.; Aisporna, A.E.; Rinehart, D.; Hoang, L.; et al. XCMS-MRM and METLIN-MRM: A cloud library and public resource for targeted analysis of small molecules. *Nat. Methods* **2018**, *15*, 681–684. [CrossRef] [PubMed]

Publisher's Note: MDPI stays neutral with regard to jurisdictional claims in published maps and institutional affiliations.

 *metabolites*

Article

Electromembrane Extraction of Highly Polar Compounds: Analysis of Cardiovascular Biomarkers in Plasma

Nicolas Drouin [1,2], Tim Kloots [2], Julie Schappler [1], Serge Rudaz [1], Isabelle Kohler [2], Amy Harms [2], Petrus Wilhelmus Lindenburg [2,3,]* and Thomas Hankemeier [2]

[1]   Institute of Pharmaceutical Sciences of Western Switzerland (ISPSO), University Medical Centre, 1206 Geneva, Switzerland; n.f.p.drouin@lacdr.leidenuniv.nl (N.D.); julie.schappler@icloud.com (J.S.); serge.rudaz@unige.ch (S.R.)

[2]   Division of Systems Biomedicine and Pharmacology, Leiden Academic Center for Drug Research, Leiden University, Einsteinweg 55, 2333 CC Leiden, The Netherlands; t.n.kloots@lacdr.leidenuniv.nl (T.K.); i.kohler@lacdr.leidenuniv.nl (I.K.); a.c.harms@lacdr.leidenuniv.nl (A.H.); hankemeier@lacdr.leidenuniv.nl (T.H.)

[3]   Research Group Metabolomics, Leiden Centre for Applied Bioscience, University of Applied Sciences Leiden, Zernikedreef 11, 2333 CK Leiden, The Netherlands

*   Correspondence: p.w.lindenburg@lacdr.leidenuniv.nl; Tel.: +31-(0)715274220

Received: 15 November 2019; Accepted: 11 December 2019; Published: 18 December 2019

Abstract: Cardiovascular diseases (CVDs) represent a major concern in today's society, with more than 17.5 million deaths reported annually worldwide. Recently, five metabolites related to the gut metabolism of phospholipids were identified as promising predictive biomarker candidates for CVD. Validation of those biomarker candidates is crucial for applications to the clinic, showing the need for high-throughput analysis of large numbers of samples. These five compounds, trimethylamine N-oxide (TMAO), choline, betaine, L-carnitine, and deoxy-L-carnitine (4-trimethylammoniobutanoic acid), are highly polar compounds and show poor retention on conventional reversed phase chromatography, which can lead to strong matrix effects when using mass spectrometry detection, especially when high-throughput analysis approaches are used with limited separation of analytes from interferences. In order to reduce the potential matrix effects, we propose a novel fast parallel electromembrane extraction (Pa-EME) method for the analysis of these metabolites in plasma samples. The evaluation of Pa-EME parameters was performed using multi segment injection–capillary electrophoresis–mass spectrometry (MSI-CE-MS). Recoveries up to 100% were achieved, with variability as low as 2%. Overall, this study highlights the necessity of protein precipitation prior to EME for the extraction of highly polar compounds. The developed Pa-EME method was evaluated in terms of concentration range and response function, as well as matrix effects using fast-LC-MS/MS. Finally, the developed workflow was compared to conventional sample pre-treatment, i.e., protein precipitation using methanol, and fast-LC-MS/MS. Data show very strong correlations between both workflows, highlighting the great potential of Pa-EME for high-throughput biological applications.

Keywords: electromembrane extraction; cardiovascular disease; multi-segment injection; capillary electrophoresis–mass spectrometry; liquid chromatography–mass spectrometry

---

1. Introduction

With 17.5 million deaths per year, cardiovascular disease (CVD) represents the leading cause of death worldwide and is thus considered a major public health issue [1]. Gut flora-dependent metabolism has been recently reported as a risk factor for CVD, most notably the metabolic pathway for dietary phosphatidylcholine which includes trimethylamine oxide (TMAO), choline, betaine,

L-carnitine, and deoxy-L-carnitine. These metabolites are related to the development of various CVDs including stroke and myocardial dysfunction [2–5].

Mass spectrometry (MS) coupled with separation techniques is a widely used tool in metabolomics, enabling the identification and quantitation of metabolites via the analysis of thousands of samples. Hydrophilic interaction chromatography (HILIC)-based methods allow for an efficient separation of polar compounds such as the ones involved in the dietary phosphatidylcholine metabolic pathway, but they require long re-equilibration times, resulting in low throughput [6]. Fast chromatography is often preferred in combination with fast sample pre-treatment. However, this approach may lead to significant matrix effects (MEs), i.e., the alteration of the analyte signal due to the presence of co-eluting matrix interference. Deuterated internal standards (dISTDs) can be added to correct for the deleterious effect of ME on quantitation. However, even though quantitative corrections can be applied, ion suppression will still lead to decreased signal intensity and lower sensitivity, showing the need for alternative approaches.

In order to reduce MEs, a sample preparation step is highly recommended. The most common practice in metabolomics is to remove proteins by protein precipitation [7] or ultrafiltration [8]. However, these two approaches cannot separate the compounds of interest from interferences such as salts or phospholipids [9,10]. To do this, more complex sample preparation approaches, e.g., liquid–liquid extraction or reversed phase solid phase extraction are often considered. However, in the present case the metabolites of interest are quaternary ammoniums. Therefore, only weak cation exchanger resins are conceivable. However, this extraction mode is complicated and expensive.

Electromembrane extraction (EME) is a recently developed electro-driven sample preparation method developed for the extraction of charged compounds [11]. In this approach, two compartments, namely, the donor and acceptor compartments, are separated by a supported liquid membrane (SLM) containing an organic solvent. When applying an electric field between the two compartments, the ions present in the donor compartment are selectively extracted to the acceptor compartment [12,13]. With this approach, it is possible to reach very high recovery (up to 100%) and enrichment factor (up to 50-fold) in a few minutes [14,15]. In addition, this method allows for an efficient sample clean-up by separation of metabolites of interest from phospholipids and salts, as well as proteins [16–18]. EME has extensively been applied to the extraction of non-polar and moderately polar drugs (1 < logP < 5). Recent work has also demonstrated the great potential of EME for the extraction of highly polar compounds (logP < −1) [19–22]. TMAO, choline betaine, L-carnitine, and deoxy-L-carnitine, as quaternary ammonium compounds, are excellent candidates for EME due to their permanent positive charge. Indeed, acylcarnitines have been already successfully extracted by electroextraction [23–25]. However, the EME set-ups described are not suited for high throughput workflows necessary for large cohort studies.

In this study, we developed and optimized a parallel EME (Pa-EME) extraction method for the these five metabolites from human plasma samples. This approach enabled the extraction from several samples in parallel, dramatically increasing the extraction throughput. A non-polar drug, bupivacaine, was also added to the set of selected compounds to monitor the extraction process. The influence of the applied voltage as well as the composition of the donor compartment were investigated. For method optimization, the extracted compounds were analyzed using a multi segment-capillary electrophoresis-mass spectrometry (MSI-CE-MS) method to both improve the throughput of the analysis and reduce the analytical variability. The optimized method was then combined with fast-liquid chromatography–tandem mass spectrometry (Fast-LC-MS/MS) for further evaluation of response function and dynamic range. ME was evaluated by comparing the developed extraction method to a sample pre-treatment method conventionally used in metabolomics, i.e., protein precipitation using methanol (PP-MeOH) [26,27]. Finally, the developed EME-LC-MS/MS method was applied to the analysis of 40 plasma samples and compared to PP-MeOH to show the strength of the method. This highlighted the relevance of such approaches for large-scale metabolomics studies, where the analysis of highly polar metabolites in complex matrices remains challenging.

2. Experimental

2.1. Chemical and Reagents

MS-grade acetic acid and sodium hydroxide ($\geq$ 99%) were purchased from VWR International BV (Amsterdam, The Netherlands). MS-grade methanol (MeOH), isopropanol (iPrOH), and acetonitrile (MeCN) were supplied from Actu-All Chemicals (Oss, the Netherlands). Model compounds (purity $\geq$ 95%) as well as 6.1 N trichloroacetic acid (TCA) and 2-nitrophenyl pentyl ether (NPPE, $\geq$ 99%) were purchased from Sigma-Aldrich Chemie NV (Zwijndrecht, The Netherlands). Deuterated internal standards (dISTDs), isotopic purity $\geq$ 99%, were purchased from Cambridge Isotopes Laboratories, Inc. (Tewksbury, MA, USA), except TMAO-d9 (CDM Isotopes, Pointe-Claire, Quebec, Canada). MS-grade formic acid (FA) and 37% (*v/v*) hydrochloric acid were obtained from Thermo Fisher Acros Organics (Geel, Belgium)

2.2. Preparation of Standard Solutions

Individual stock solutions of both standard and deuterated internal standard (dISTD) were prepared in water/MeCN/FA (94.9:5:0.1, *v/v*) to reach a final concentration of 5 mg/mL. The method development was performed using neat standards of carnitine-d_3, deoxycarnitine-d_9, choline-d_4, betaine-d_9, TMAO-d_9, and bupivacaine prepared in 5% MeCN and 0.1% FA at 100 µg/mL. This mixture was diluted in the donor solution to a final concentration of 2.5 µg/mL before use. The acceptor solution consisted of a solution of 1% acetic acid.

In addition, for MeOH for protein precipitation, dISTD solutions were also prepared in MeOH at the following concentrations: 4 µg/mL for carnitine-d3, 0.2 µg/mL for deoxycarnitine-d9, 2 µg/mL for choline-d4, 4 µg/mL for betaine-d9, and 0.3 µg/mL for TMAO-d9.

2.3. Plasma Samples

The method development was carried out using citrate plasma samples collected from eight healthy volunteers obtained from Sanquin (Amsterdam, The Netherlands) and pooled together.

The method comparison was performed with 40 plasma samples collected from healthy volunteers from the Growing Old Together (GOTO) study [28].

2.4. Sample Preparation: Protein Precipitation

In order to enhance extraction performance of Pa-EME, three different PP methods were investigated during this study, namely, (1) PP using 6.1 N TCA (referred to as PP-TCA), (2) adjusted PP-TCA, and (3) PP using MeOH with a ratio MeOH/sample of 9:1 ratio (referred to as PP-MeOH). The resulting solutions were used as donor solution during the Pa-EME procedure.

The PP-TCA was performed using a solution of 6.1 N TCA with a ratio of 0.05:1 (TCA/sample, *v/v*). Briefly, 25 µL of 6.1 N TCA were added to 475 µL of plasma. After 30 min of agitation at 1400 rpm at 23 °C using a Thermomixer (Vaudaux-Eppendorf AG, Bale, Switzerland), the sample was centrifuged at 15,000 rpm for 15 min at 4 °C. Supernatant was then collected. In order to avoid potential variability, multiple PPs were performed and their supernatants were mixed together prior to further division into aliquots. Aliquots were then kept at −20 °C until analysis. As a donor solution for Pa-EME, the aliquots were then diluted with adequate volume of water to reach an equivalent concentration of 10, 20, and 50% of untreated plasma, respectively.

This PP-TCA method was then adapted for the analysis of the 40 different plasma samples due to the small volume of sample available, i.e., ca. 25 µL. Briefly, 20 µL of plasma were mixed with 20 µL of dISTD solution. Then, 2 µL of 6.1 N TCA were added to obtain a final ratio of TCA/sample of 0.05:1 (*v/v*). After agitation at 1400 rpm for 30 min at 23 °C, 358 µL of water was added prior to centrifugation at 24,400× *g* during 20 min at 4 °C. The supernatant was then collected and stored at −20 °C prior to electroextraction. For EME, 300 µL of the supernatant was used, leading to an equivalent untreated plasma content of 5% in the donor compartment.

The third method, namely, PP-MeOH, was performed using 90 μL of MeOH containing dISTD added to 10 μL of plasma. After vortex agitation for 1 min, samples were centrifuged for 5 min at 18,300× *g*. The supernatant was then collected, leading to a corresponding concentration of 10% of untreated plasma in the analyzed sample.

2.5. Pa-EME Set-Up and Procedure

The electroextraction procedure and the Pa-EME device used in this study have been already described elsewhere [28]. Briefly, the Pa-EME device consisted of a donor and an acceptor 96 well-plate, as is illustrated in Figure 1. The acceptor plate consists of a custom made conductive 96 well-plate in polyether ether ketone (PEEK) polymer. Conductivity was ensured by a piece of aluminum foil with a thickness of 0.14 mm placed on the bottom of the well plate. In order to limit carryover and cross contamination, the donor plate used was a disposable MultiScreen-IP Filter Plate of 300 μL with a polyvinyldifluoride (PVDF) membrane with a thickness of 100 to 145 μm and a pore size of 0.45 μm, and purchased from Millipore (Milford, MA, USA).

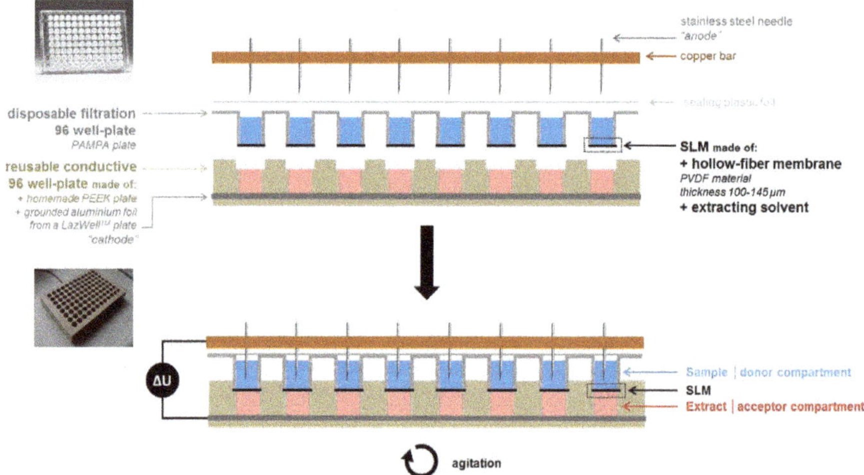

Figure 1. Schematic representation of the parallel electromembrane extraction (Pa-EME) device. Reprinted with permission [29]. SLM: supported liquid membrane; PAMPA: Parallel artificial membrane permeability assay; PVDF: polyvinyldifluoride.

Impregnation of the PVDF membrane was performed by pipetting 3 μL of 2-nitrophenylpentyl ether (NPPE) on the external face of the PVDF. Then, the excess of SLM is removed by placing the plate on a tissue and by application of a 2.5 psi for 5s in each well using a positive pressure manifold (Biotage, Uppsala, Sweden). This cleaning step was repeated until the tissue appeared dry. The donor compartment was then filled with 300 μL of sample and was sealed using an adhesive sealing film (PCR-TS, Axygen, MA, USA).

After filling of the acceptor plate with 300 μL of acceptor solution made of 1% acetic acid, the donor plate was inserted into the acceptor one and the Pa-EME system was placed on a thermomixer for agitation. The electrode needles were then inserted in the donor compartment. Extraction took place for 15 min at 1400 rpm, with application of a constant voltage or current between the needles and the aluminum foil of the conductive well-plate using a Power Supply ES 0300–0.45 from Delta Elektronica (Zierikzee, the Netherlands). After 15 min, the power supply was turned off, needles were removed from the acceptor plate, and both plates were separated. Finally, the extracts present in the acceptor compartment are collected and were ready to be analyzed.

In order to avoid potential carry over, the donor plate was discarded after use and the acceptor plate was rinsed with a mixture of iPrOH/H2O 50:50 (*v/v*) and dried under nitrogen after each experiment.

2.6. Capillary Electrophoresis–Mass Spectrometry

CE separations were carried out using a G7100 capillary electrophoresis (CE) system from Agilent (Waldbronn, Germany) using a fused silica capillary (BGB Analytik Benelux B.V, Harderwijk, The Netherlands) with a length of 90 cm and an internal diameter of 50 μm. Separation was carried out using a background electrolyte (BGE) composed of 10% acetic acid (*v/v*). Prior to first use, the capillary was conditioned with MeOH, H_2O, 1 M NaOH, H_2O, 1 M HCl, H_2O, 0.1 M HCl, H_2O, and BGE at 5 bar. Between each run, the capillary was rinsed with MeOH and BGE at 5 bar for 130 s. CE-MS analyses were performed using a MSI-CE-MS approach with electrokinetic plugs between samples which were hydrodynamically injected [20]. Briefly, the first sample was hydrodynamically injected at 100 mbar for 20 s (corresponding to 1.6% of total length of the capillary). Prior to the second sample injection, a voltage of +30 kV was applied during 60 s. This process was repeated until 7 samples were injected. Typically, the first sample injected consisted of the neat standard at 2.5 μg/mL, followed by six injections of other samples consisting of the replicates of a specific extraction condition. The samples were kept at 7–8 °C in the auto-sampler using an external water cooling system.

The CE system was hyphenated with an Agilent 6230 TOF mass spectrometer (Santa Clara, CA, USA) via an electrospray ionization (ESI) source and a coaxial sheath-flow ESI interface equipped with a stainless-steel triple-tube sprayer (P/N G1607A) from Agilent Technologies. The sheath liquid was composed of a mixture of H_2O/iPrOH/acetic acid 50:50:1 (*v/v/v*) and delivered at a flow rate of 3 μL/min using a 2300 Series isocratic pump (Agilent Technologies) equipped with a 1:100 split ratio. MS experiments were acquired in positive mode between 50 and 1000 *m/z* with an acquisition rate of 1.5 spectrum/s. The nebulizer gas was set to 0 psi, while the sheath gas flow rate and temperature were set at 11 L/min and 100 °C, respectively. The ESI capillary voltage was adjusted to 5500 V. Fragmentor and skimmer voltages were set at 150 V and 50 V, respectively. MassHunter version B.06.00 (Agilent, Santa Clara, CA, USA) was used for data acquisition, instrument control, and data treatment. Isopropanol acetate adduct $[CH_3COOH\text{-}C_3H_8O\text{+}H]^+$ (*m/z* 121.08592) was used as reference mass for TOF calibration of each spectrum.

2.7. Fast-Liquid Chromatography–Mass Spectrometry

A previously developed fast LC-MS/MS method for analysis of metabolites linked to gut metabolism was used in this study [10]. Briefly, a 1290 Infinity II ultra-high pressure liquid chromatography (UHPLC) system from Agilent Technologies (Waldbronn, Germany) was used for fast LC-MS/MS experiments. The instrument was equipped with an autosampler, a column oven and a binary pump with a maximum delivery flow rate of 5 mL/min. Separations were performed with a Waters AccQ-TagTM Ultra column (2.1 mm × 100 mm, 1.7 μm) maintained at 60 °C. The mobile phases consisted of 0.1% formic acid (A) and MeCN (B). The injection volume was 1 μL and the flow rate of the mobile phase was set at 0.7 mL/min.

The separation was carried out using the following gradient: (B) maintained at 5% for 0.8 min, further increased to 50% over 0.05 min, followed by an increase to 100% in 0.1 min. These conditions were kept during 1.25 min before returning to initial conditions in 0.02 min and re-equilibration over 0.8 min. The total analysis time was 3 min.

The UHPLC system was hyphenated with an AB Sciex 6500 Q-Trap MS (AB Sciex, Concord, ON, Canada) equipped with a Turbo Spray ionDrive source. MS experiments were performed in the positive ionization mode using the selected reaction monitoring (SRM) acquisition mode. The precursor and product ions that were monitored for each compound, as well as the respective collision energies, are reported in Table S1. The SRM experiments were acquired with a chromatographic time window of 60 s and a cycle time of 0.2 s. The drying gas temperature and flow rate were set at 220 °C and 14 L/min, respectively. The ion source gas 1 and 2 pressures were fixed at 80 and 70 psi, respectively, with a temperature of 350 °C for both.

The ion spray voltage, declustering potential, and collision cell exit potential were adjusted to 2500 V, 70 V, and 10 V, respectively. The curtain and collision gas were set at 20 psi and "medium", respectively.

Data acquisition and instrument control were monitored using AB Sciex Analyst version 1.6.2 (AB Sciex, Concord ON, Canada). Data treatment was performed using Skyline-daily version 4.1 (MacCoss Lab, Seattle, WA, USA).

2.8. Calculation of Extraction Yield, Process Efficiency, and Matrix Effect

The extraction yield (EY) is described as the recovery in absence of matrix. In this study, EY was determined by comparing a neat standard solution (2.5 µg/mL) with a neat spiked solution (2.5 µg/mL in 50 mM FA) extracted with Pa-EME [30], according to Equation (1).

$$EY = \frac{AUC_{extract}}{AUC_{neat\ standard}} \times \frac{V_{donor}}{V_{acceptor}} \tag{1}$$

where $AUC_{extract}$ is the peak area of the compound measured in the acceptor solution, $AUC_{neat\ standard}$ the peak area of the compound in the neat standard solution, V_{donor} the volume of the donor compartment, and $V_{acceptor}$ the theoretical volume recovered in the acceptor compartment.

Process efficiency (PE) describes the extraction performance in presence of matrix and is determined by comparing a neat standard solution (2.5 µg/mL) to a spiked biological sample (2.5 µg/mL) extracted with Pa-EME [31]. The PE was calculated according to Equation (2).

$$PE = \frac{AUC_{extracted\ plasma}}{AUC_{neat\ standard}} \times \frac{V_{donor}}{V_{acceptor}} \tag{2}$$

where $AUC_{extracted\ plasma}$ is the peak area of the compound measured in the acceptor solution.

The ME is defined as the difference in signal due to ion suppression or signal enhancement. The ME was evaluated using a method described by Matuszewski et al. [31]. In this case, the sample was first extracted and then spiked with compounds of interest to a known final concentration. This post-extraction spiked sample was then compared to a neat standard at the same concentration.

The ME was calculated according to Equation (3).

$$ME = 1 - \frac{AUC_{post\ extraction\ spiked\ matrix}}{AUC_{neat\ standard}} \tag{3}$$

where $AUC_{post\ extraction\ spiked\ matrix}$ corresponds to the peak area of compounds detected in a biological matrix spiked after extraction with a known analyte concentration and $AUC_{neat\ standard}$ the peak area of the compound at known concentration measured in neat standard.

3. Results and Discussion

TMAO, choline, betaine, L-carnitine, and deoxy-L-carnitine, known predictive biomarkers for CVD [2], were used as model compounds. All these metabolites are very polar compounds with logP between −4.49 and −0.93 (Table 1). Due to their very low lipophilicity, these highly polar molecules are difficult to extract using conventional sample preparation methods such as solid-phase extraction or reversed-phase SPE [9], which are based on the partition coefficients of analytes in two-phase systems. Indeed, only PP has been reported as efficient sample pre-treatment for these class metabolites so far [31,32]. PP is typically used in metabolomics, especially in large-scale studies where high throughput is essential. Fast analytical techniques are also required, such as fast LC-MS, but may lead to strong ME, typically for poorly-retained compounds and particularly in combination with straightforward sample pre-treatments [9,33]. This ME issue highlights the needs for novel sample preparation approaches adapted to the extraction of highly polar compounds.

Table 1. Physicochemical properties of the compounds of interest.

	Molecular Weight (Da)	LogP *	pKa *	Structure	ChEBI
TMAO	75.11	−0.9	4.7		15724
Choline	104.17	−4.7	14.1		15354
Betaine	117.15	−4.5	2.3		17750
Deoxy-ʟ-carnitine (4-trimethylammoniobutanoic acid)	145.20	−4.0	4.5		16244
ʟ-carnitine	161.20	−4.9	4.2		17126
Bupivacaine	288.43	4.5	13.6		3215

*: calculated using Chemaxon, www.chemicalize.org. TMAO: trimethylamine N-oxide.

3.1. Optimization of the Parallel Electromembrane Extraction Set-Up

The EME experimental conditions, i.e., applied voltage and sample composition, were first optimized to reach the highest EY and PE while lowering ME. Because most of the compounds of interest are endogenously present in human plasma, dISTDs were used during the Pa-EME optimization step, except for bupivacaine, which is a xenobiotic compound. Based on previous work [20], 2-nitrophenylpentyl ether (NPPE) as SLM and 1% acetic acid (pH 2.8) as both acceptor and donor solutions were used as starting conditions. NPPE was selected due the expected good extraction recovery and low extraction variability for the selected compounds, while 1% acetic acid allowed for both protonation of the basic moiety and neutralization of carboxylic group of ʟ-carnitine, deoxy-ʟ-carnitine and betaine. Moreover, 1% acetic acid generated a relatively low current which allowed for the application of higher voltages without generating excessive Joule heating. The obtained extracts were then analyzed using MSI-CE-MS. MSI-CE-MS consists of consecutive injections of up to seven different samples within the same analytical run. This leads to a significant increase in analysis throughput as well as decrease of analytical variability for sample injected in the same run [34–36].

First, the extraction voltage was investigated, since EME recoveries are known to be directly correlated to the electric field applied during the electroextraction process [14,29]. Figures 2 and 3 illustrate the results obtained for three compounds, i.e., choline (positively charged and polar), ʟ-carnitine (partially charged and polar), and bupivacaine (positively charged and non-polar). As shown in Figure 2, the EY (calculated according to Equation (1)) increased for all compounds with an increased extraction voltage. The gain in EY was especially important for ʟ-carnitine, where the EY showed a 5-fold enhancement when increasing the voltage from 75 V to 100 V. At 100 V, EYs up to 92%

were obtained, with relative standard deviations (RSDs) as low as 4%. Choline and L-carnitine are close compounds with logD values in the same range, i.e., −4.6 and −4.8 at pH 2.8, respectively. The difference observed in EY between both compounds might be explained by the higher molecular charge of L-carnitine and the partial deprotonation of its carboxylic group (pKa 4.2) at pH 2.8, leading to a decrease of its net charge [37,38] and thereby its susceptibility to electromigration. As expected, bupivacaine was easily extracted with EY above 85% with all tested voltages, showing that the Pa-EME setup was functioning well.

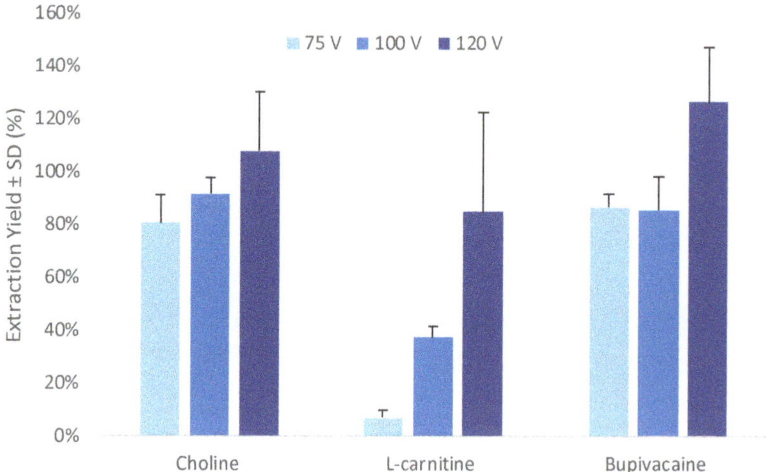

Figure 2. Effect of applied voltage on extraction yield (*n* = 3). Error bars are expressed as the standard deviations.

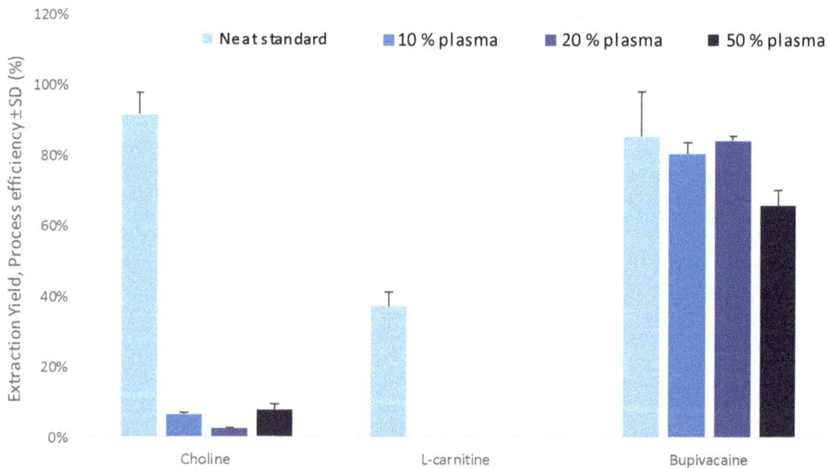

Figure 3. Effect of untreated plasma content in the acceptor compartment on process efficiency (*n* = 3).

The highest voltage tested, i.e., 120 V, led to the highest EYs for all compounds. However, at this voltage, a significant fluid leakage between the acceptor and donor plate was observed. This was explained by gas production caused by electrolysis in the acceptor compartment, leading to an overpressure in this closed compartment and, ultimately, to the loss of the acceptor phase.

This supports the apparent higher EY that were observed due to an overestimation of the acceptor compartment volume according to Equation (1).

Therefore, an applied voltage value of 100 V was selected for further experiments, leading to the highest EYs without any volume loss observed.

The influence of the concentration of untreated plasma in the donor compartment on the extraction was then investigated. As shown in Figure 3, a significant drop of PE was observed for highly polar compounds in the presence of 10% untreated plasma or higher. The strong decrease in PE for betaine, L-carnitine, and deoxy-L-carnitine might be explained by partial deprotonation of their carboxylic acid group due to a pH increase in the compartment caused by addition of plasma (up to pH 4.5 with 50% of plasma) and the poor buffer capacity of acetic acid 1%. Therefore, the pH increase of the donor compartment led to a decrease of metabolite net charge. In these conditions, TMAO and choline remained both fully ionized, irrespective of the pH.

The observed decrease in PE for the two polar compounds might be further explained by the drastic reduction of the electric field in the system due to the plasma ionic strength, leading to lower logD values of polar compounds and slower migration into the SLM [39]. On the other hand, bupivacaine was more slightly affected by this phenomenon when using up to 20% of plasma content, due to the known high logD of non-polar compounds in the SLM. However, when using 50% of plasma content, the important increase of ionic strength decreased the electric field, leading to a significantly lower PE for this analyte.

Another hypothesis, i.e., the disturbance of the interface between the organic layer and the plasma sample due to a superficial protein precipitation in this region, seems unlikely since a vortex is created in the donor compartment thanks to the very high agitation rate (i.e., 1400 rpm). In addition, since bupivacaine is a drug known to be 95% linked to plasmatic proteins but showed high PE values, the protein-binding hypothesis was discarded. However, PVDF material is well-known for its very high affinity and protein binding capacity. Therefore, this might lead to perturbation of the organic layer by competition between the organic solvent and proteins.

Finally, as high sensitivity is essential in metabolomics, an untreated plasma content of 10% was selected for further experiments.

In order to test our two hypotheses, we evaluated two approaches to modify the sample composition, namely, (1) addition of an organic solvent to the sample (e.g., MeOH) to enhance the electric field, and (2) PP prior to extraction to remove proteins.

Various proportions of MeOH were added to the donor compartment, i.e., 10%, 20%, and 50%, but no significant difference in PE was observed (data not shown). Higher concentrations of organic modifier were not tested to avoid possible SLM dissolution [40].

PP using TCA is known to allow for an efficient protein removal (i.e., above 99%) with a limited dilution factor [41,42]. The PP-TCA method takes place in pure aqueous phase, showing the benefit of avoiding the evaporation step that is necessary when using organic solvents for PP, which is favorable for compatibility with the EME approach. Moreover, the low pH that occurs when using PP-TCA is suitable for the extraction of cationic compounds. Therefore, PP-TCA was selected for further investigations, in a 0.05:1 ratio (TCA:sample). After a 10-fold dilution of the precipitated plasma, the TCA concentration was close to 30 mM, which was sufficient to obtain a pH of 2.0 and ensure protonation of all the compounds of interest. With an applied voltage of 100 V, a high current (more than 1–2 mA/well) was observed. This relatively high current could be explained by higher conductivity of the solution of 30 mM TCA compared to 1% acetic acid. In order to avoid the potential issues generated by a high current, the extraction current was set to 400 µA/well to minimize electrolysis and gas production, while maximizing both EY and PE. As presented in Figure 4, good EY (up to 75% for polar compounds) and low variability (as low as 7%) were obtained using the optimal conditions.

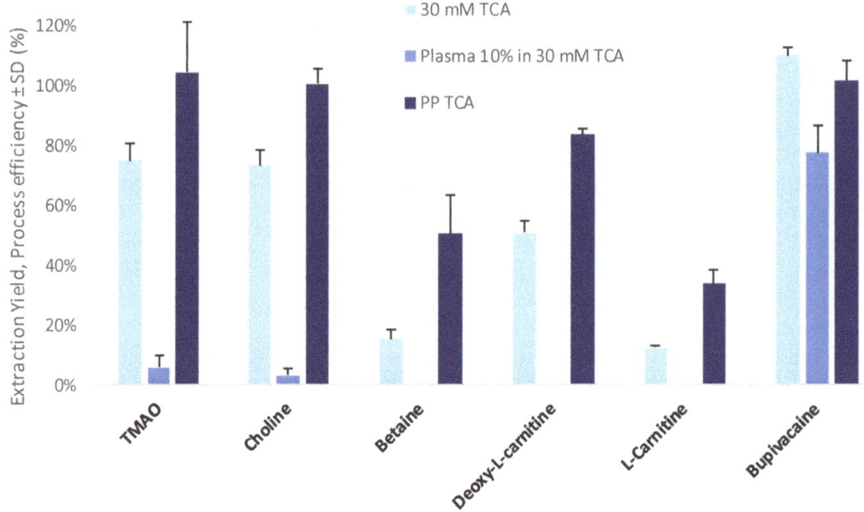

Figure 4. Influence of protein precipitation (PP) on process efficiency (PE) ($n = 6$). Experimental conditions: current, 400 µA/well; extraction time, 15 min; agitation, 1400 rpm, 1% acetic acid as acceptor compartment. TCA: trichloroacetic acid; PP-TCA: protein precipitation using trichloroacetic acid, ratio trichloroacetic acid/sample 0.05:1 (*v/v*).

For all the selected compounds, similar or higher PE (up to 100%) and low variability (as low as 2%) were obtained on protein precipitated plasma samples compared to neat solutions, especially for betaine and L-carnitine where PE were 3.3 and 2.7-fold higher, respectively, when PP-TCA was used. This increase in PE remains unexplained and requires further investigations. Nevertheless, a PP step appears essential prior to EME of highly polar compounds to reduce both ionic strength and buffer capacity of biological fluids. To assess the maximal precipitated plasma volume extractable with the developed approach, different plasma contents, i.e., up to 50%, were evaluated. Using the same PP-TCA method, an increase of the precipitated plasma content into the donor compartment involved an increase of TCA concentration, up to 150 mM. The results are summarized in Table 2.

Table 2. PE (RSD) in % ($n = 6$) according to the precipitated plasma amount in the sample using the PP-TCA method and linearity ranges obtained using 30 mM TCA in water. RSD: relative standard deviations.

	PE (RSD) in %			Dynamic Range (µM)	R^2
	10% Plasma 30 mM TCA	20% Plasma 60 mM TCA	50% Plasma 150 mM TCA		
TMAO	104 (16)	96 (13)	94 (13)	0.27–43.19	0.997
Choline	100 (5)	84 (4)	89 (4)	1.8–286.5	0.996
Betaine	55 (12)	66 (14)	71 (9)	4.6–716.1	0.994
Deoxy-L-carnitine	84 (2)	60 (14)	60 (10)	0.1–17.6	0.997
L-carnitine	34 (14)	18 (54)	12 (6)	3.5–556.5	0.995
Bupivacaine	102 (7)	101 (3)	105 (3)	n.d.	

Except for L-carnitine, good PE (between 50% and 100%) values were obtained for all tested precipitated plasma amounts. No change in PE was observed for TMAO, choline, and bupivacaine. An increase of PE for betaine with an increased plasma content and TCA concentration was obtained, explained by the lower pH observed, i.e., a pH value of 1.0–1.5 with 50% of plasma and 150 mM of TCA versus pH of 2.0 with 30 mM of TCA and 10% of plasma content. This decrease of pH induces a higher positive net charge on betaine, leading to an increase of its electrophoretic mobility. Nevertheless, the decrease in PE observed for deoxy-L-carnitine and carnitine remain unexplained but might be the consequence of stability issues of these metabolites in highly acidic conditions (pH ~1).

Finally, the linear response function of the developed method was evaluated. For this purpose, a fast LC-MS/MS method was used to be in similar conditions as what is observed in the context of large cohort studies with thousands of samples. The calibration curve was plotted using ratios of non-deuterated compounds and dISTDs in a neat solution. Calibration samples with increasing concentrations of TMAO, choline, betaine, L-carnitine and deoxy-L-carnitine (Table 2) were made and mixed to constant concentrations of their dISTD in 30 mM TCA, to mimic EME conditions of the donor compartment composition. These calibration samples were extracted using the optimized Pa-EME setup and the selected experimental parameters (i.e., 400 μA/well, 15 min, 1400 rpm). As shown in Table 2, a linear response function ($R^2 > 0.995$) was obtained on concentration ranges of two orders of magnitude for all compounds of interest (Figure S1).

3.2. Evaluation of Matrix Effects

The ME were evaluated for the developed Pa-EME set-up in combination with fast-LC-MS/MS.

The observed ME when using a conventional PP-MeOH were compared to the method combining PP-TCA and Pa-EME. For all these experiments, the same amount of untreated plasma concentration was used, i.e., 10 μL of untreated plasma for 100 μL of precipitated plasma for PP-MeOH and 30 μL of untreated plasma for 300 μL of donor solution for EME, respectively, leading in both case to an equivalent of 10% of untreated plasma after PP. As shown in Figure 5, lower MEs were observed using the combination of PP-TCA and Pa-EME. Compared to the conventional PP-MeOH approach, a noticeable decreased ME was observed, especially in the case of TMAO and betaine where the observed ME was 2-fold lower with the combined approach versus MeOH-PP alone. This decrease might be explained by the cationic selectivity of EME and efficient salt removal.

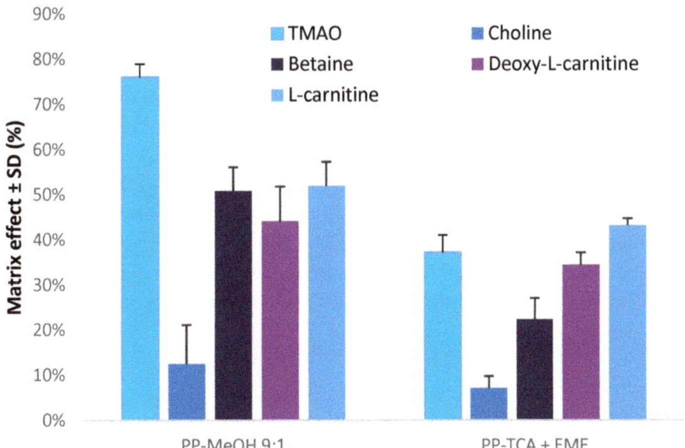

Figure 5. Comparison of matrix effect between conventional PP-MeOH and combination of PP-TCA and Pa-EME using optimal conditions. PP-MeOH: protein precipitation using methanol, ratio methanol/sample 9:1 (*v/v*).

Along with the decrease in ME and high PE, a 2.9-fold increase of peak intensity was observed with EME compared to PP-MeOH for TMAO. The sensitivity observed for choline and deoxy-L-carnitine was not significantly impacted, with an increased factor of 1.2 and 1.1, respectively. With same signal intensity, the lower ME observed for betaine compared to PP-MEOH compensated for the low PE of this compound (i.e., 55%). Finally, L-carnitine, which was the compound with the lower PE (i.e., 34%), showed reduced ME but was detected with a signal intensity 0.6-fold lower than with PP-MeOH.

3.3. Application to Metabolomics Studies

The potential of the developed Pa-EME set-up for large scale metabolomics studies has been investigated by correlating the data obtained with this optimized extraction method (combination of PP-TCA+EME) versus a typically used sample clean-up, i.e., PP-MeOH, on a set of 40 human plasma samples.

Due to limited volume of plasma available (ca. 25 µL), the developed PP-TCA was first downscaled. This adjusted PP-TCA method led to an equivalent of 5% of untreated plasma after PP. The downscaled method led to similar extraction performance compared to the conventional PP-TCA (data not shown), and was therefore used for subsequent experiments.

As shown in Figure 6, excellent correlations were obtained between the two evaluated sample preparation approaches for TMAO, choline, L-carnitine, and deoxy-L-carnitine, with correlation coefficients (R) between 0.88 and 0.98. In addition, the linear models are highly significant, with p-values between 1.4×10^{-14} and 1.3×10^{-32}. These results highlight the relevance of the developed Pa-EME method in comparison with gold standard methods used for sample preparation in metabolomics. However, a poor correlation was observed for betaine (R = 0.46), probably due to a contamination of the milliQ water used for preparation of acceptor and donor solution with a compound detected in the same SRM transition as betaine.

During the analysis of these clinical samples, an unexpected boiling and loss of acceptor compartment was observed for many samples during the EME process. This unexpected phenomenon was likely explained by the differences in plasma composition. Indeed, the optimization phase was carried out using the same sample split in six aliquots which were simultaneously extracted using Pa-EME. Therefore, the total applied current (2.4 mA) was equally distributed over the six wells, leading to a current of 400 µA/well. No boiling or loss of acceptor phase was observed during the optimization process. However, with a parallel extraction of six different plasma samples, the applied current was not uniformly distributed over the wells. Indeed, due to small differences of sample composition (e.g., ionic strength, residual proteins, etc.), plasma samples possess different conductivities. According to Kirchhoff's current law, these different conductivities lead to different current in every parallel circuit. Consequently, several samples were subjected to currents lower than 400 µA and other samples to a current above this limit, the latter leading to boiling and loss of acceptor volume. However, this variability was successfully corrected using a dISTD for each compounds of interest, as shown by the good correlation obtained with PP-MeOH (Figure 6). A solution to circumvent this issue could be to use lower currents to stay below the limit of 400 µA/well, but this solution requires enhanced extraction times or would lead to lower PE.

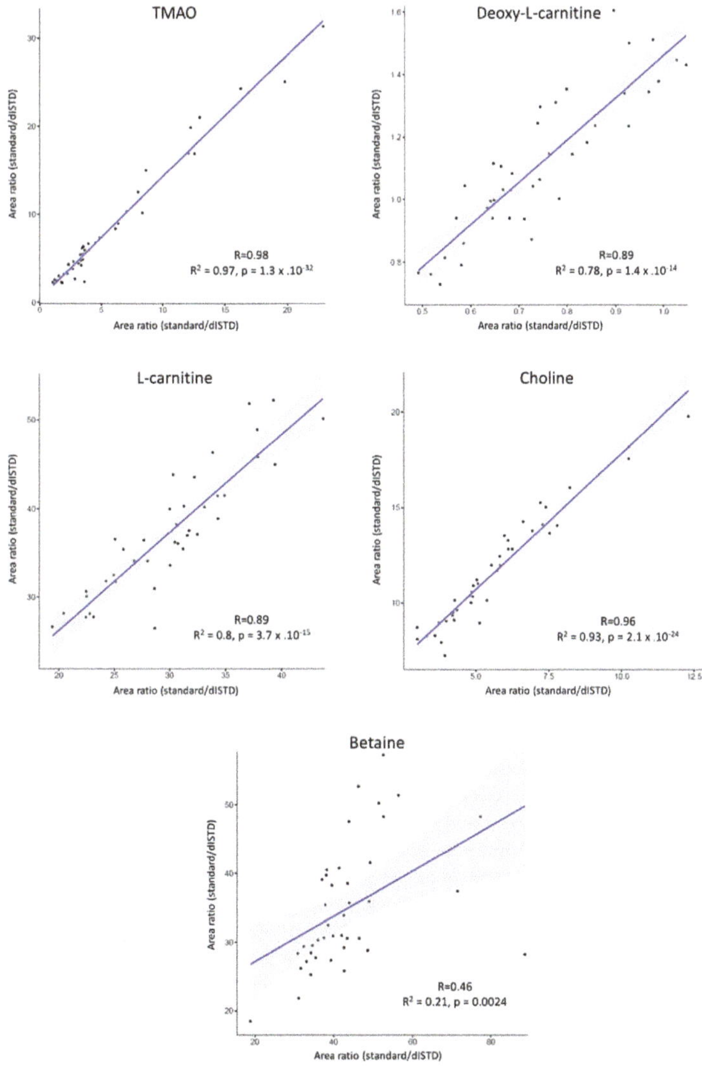

Figure 6. Area ratio correlations observed for analyte response between PP-TCA + EME (*x*-axis) and PP-MeOH (*y*-axis). 95 % confidence intervals were calculated using Pearson correlation. dISTD: deuterated internal standard.

4. Conclusions

This study demonstrated the power of EME approaches for the extraction of highly polar compounds from biological fluids and its potential for metabolomics studies. In this study, we presented an optimized Pa-EME method for the efficient extraction of highly polar compounds from plasma. The developed Pa-EME method involves a PP step using TCA before the actual EME process; this is an essential step to reduce the buffer capacity of plasma, and additionally to avoid possible interferences with the PVDF membrane. We demonstrated that the combination of PP-TCA with Pa-EME allowed for high PE (up to 100%) as well as low variabilities (RSD as low as 7%) for the extraction of selected highly polar compounds from plasma samples. Moreover, the PP-TCA-Pa-EME set-up led to decreased ME in comparison to conventional PP-MeOH when using fast chromatography (up to 2-fold matrix effect reduction for TMAO and betaine). A 3-fold gain in sensitivity was observed for TMAO with PP-TCA-Pa-EME compared to PP-MeOH. Similar sensitivity was obtained for choline, betaine, and deoxy-L-carnitine. However, using PP-TCA-PaEME, L-carnitine presented a decrease of sensitivity in comparison with PP-MeOH due to incomplete extraction. This poor sensitivity could be nevertheless improved with further experiments and the development of a new Pa-EME set-up for higher enrichment with favorable acceptor/donor volume ratio. The developed method showed a linear response function (R^2 between 0.994 and 0.997) of more than two orders of concentration magnitude for all the metabolites of interest.

Finally, the developed PP-TCA-Pa-EME method was compared to PP-MeOH using 40 different plasma samples. The influence of sample conductivity, which is a common concern in electromigration-based sample pre-treatment, was highlighted but was fully compensated using dISTDs for each compound. Overall, the combination of PP-TCA and EME showed excellent correlation with the conventional PP-MeOH.

The great potential of electromembrane extraction in bioanalysis is highlighted by its analytical merit in terms of high recovery (up to 100%) and low variability (down to 7%) of highly polar metabolites from a complex matrix such as plasma, a significant reduction of the matrix effect, and the strong correlation to gold standard sample preparation practices in metabolomics.

The development of a new Pa-EME device to further reach higher enrichment factors for such metabolites will represent the next logical step for application of this method to state-of-the-art metabolomics-based analysis.

Supplementary Materials: The following are available online at http://www.mdpi.com/2218-1989/10/1/4/s1. Table S1: SRM transitions and collision energies used in the developed Fast-LC-MS/MS method. Figure S1: Calibration curves built for TMAO, choline, betaine, L-carnitine and deoxy-L-carnitine in water.

Author Contributions: Conceptualization, N.D. and A.H.; methodology, N.D. and A.H.; validation, N.D., A.H. and P.W.L.; formal analysis, N.D.; investigation, N.D. and T.K.; resources, T.H. and S.R.; data curation, N.D. and T.K.; writing—original draft preparation, N.D.; writing—review and editing, N.D., T.K., J.S., S.R., I.K., A.H., P.W.L. and T.H.; visualization, N.D.; supervision, A.H., P.W.L., J.S. and S.R.; project administration, N.D., I.K., P.W.L., A.H.; funding acquisition, N.D., J.S., S.R., I.K., P.W.L. and T.H. All authors have read and agreed to the published version of the manuscript.

Funding: Financial support from the Swiss National Science Foundation (Doc.Mobility Grant No. P1GEP3_174791) and from Horizon 2020 Marie Sklodowska-Curie CO-FUND (Grant Agreement No: 707404) are gratefully acknowledged.

Acknowledgments: Authors would like to acknowledge Eline Slagboom and Marian Beekman for providing the plasma samples from the Growing Old Together (GOTO) study used in this work.

Conflicts of Interest: Authors declare no conflict of interest.

Abbreviations

CVD	cardiovascular disease
CE-MS	capillary electrophoresis–mass spectrometry
dISTD	deuterated internal standard
EME	electromembrane extraction
EY	extraction yield
FA	formic acid
Fast-LC-MS/MS	fast liquid chromatography–tandem mass spectrometry
HILIC	hydrophilic interaction chromatography
ME	matrix effect
MSI-CE-MS	multisegment injection–capillary electrophoresis–mass spectrometry
NPPE	2-nitrophenylpentyl ether

MeCN	acetonitrile
MeOH	methanol
Pa-EME	parallel electromembrane extraction
PE	process efficiency
PEEK	polyether ether ketone
PP	protein precipitation
PP-MeOH	protein precipitation using methanol, ratio methanol/sample 9:1 (*v/v*)
PP-TCA	protein precipitation using trichloroacetic acid, ratio trichloroacetic acid/sample 0.05:1 (*v/v*)
PVDF	polyvinyldifluoride
SLM	supported liquid membrane
SRM	selected reaction monitoring
TCA	trichloroacetic acid
TMAO	trimethylamine N-oxide
UHPLC	ultra-high pressure liquid chromatography

References

1. Mortality, G.B.D.; Causes of Death, C. Global, regional, and national life expectancy, all-cause mortality, and cause-specific mortality for 249 causes of death, 1980–2015: A systematic analysis for the Global Burden of Disease Study 2015. *Lancet* **2016**, *388*, 1459–1544. [CrossRef]

2. Wang, Z.; Klipfell, E.; Bennett, B.J.; Koeth, R.; Levison, B.S.; Dugar, B.; Feldstein, A.E.; Britt, E.B.; Fu, X.; Chung, Y.M.; et al. Gut flora metabolism of phosphatidylcholine promotes cardiovascular disease. *Nature* **2011**, *472*, 57–63. [CrossRef]

3. Tang, W.H.; Wang, Z.; Shrestha, K.; Borowski, A.G.; Wu, Y.; Troughton, R.W.; Klein, A.L.; Hazen, S.L. Intestinal microbiota-dependent phosphatidylcholine metabolites, diastolic dysfunction, and adverse clinical outcomes in chronic systolic heart failure. *J. Card. Fail.* **2015**, *21*, 91–96. [CrossRef] [PubMed]

4. Koeth, R.A.; Wang, Z.; Levison, B.S.; Buffa, J.A.; Org, E.; Sheehy, B.T.; Britt, E.B.; Fu, X.; Wu, Y.; Li, L.; et al. Intestinal microbiota metabolism of L-carnitine, a nutrient in red meat, promotes atherosclerosis. *Nat. Med.* **2013**, *19*, 576–585. [CrossRef] [PubMed]

5. Dambrova, M.; Skapare-Makarova, E.; Konrade, I.; Pugovics, O.; Grinberga, S.; Tirzite, D.; Petrovska, R.; Kalvins, I.; Liepins, E. Meldonium decreases the diet-increased plasma levels of trimethylamine N-oxide, a metabolite associated with atherosclerosis. *J. Clin. Pharmacol.* **2013**, *53*, 1095–1098. [CrossRef] [PubMed]

6. Steuer, C.; Schutz, P.; Bernasconi, L.; Huber, A.R. Simultaneous determination of phosphatidylcholine-derived quaternary ammonium compounds by a LC-MS/MS method in human blood plasma, serum and urine samples. *J. Chromatogr. B Anal. Technol. Biomed. Life Sci.* **2016**, *1008*, 206–211. [CrossRef]

7. Gagnebin, Y.; Pezzatti, J.; Lescuyer, P.; Boccard, J.; Ponte, B.; Rudaz, S. Toward a better understanding of chronic kidney disease with complementary chromatographic methods hyphenated with mass spectrometry for improved polar metabolome coverage. *J. Chromatogr. B* **2019**. [CrossRef] [PubMed]

8. Lomonaco, T.; Ghimenti, S.; Piga, I.; Onor, M.; Melai, B.; Fuoco, R.; Di Francesco, F. Determination of total and unbound warfarin and warfarin alcohols in human plasma by high performance liquid chromatography with fluorescence detection. *J. Chromatogr. A* **2013**, *1314*, 54–62. [CrossRef]

9. Drouin, N.; Rudaz, S.; Schappler, J. Sample preparation for polar metabolites in bioanalysis. *Analyst* **2017**, *143*, 16–20. [CrossRef]

10. van der Laan, T.; Kloots, T.; Beekman, M.; Kindt, A.; Dubbelman, A.C.; Harms, A.; van Duijn, C.M.; Slagboom, P.E.; Hankemeier, T. Fast LC-ESI-MS/MS analysis and influence of sampling conditions for gut metabolites in plasma and serum. *Sci. Rep.* **2019**, *9*, 12370. [CrossRef] [PubMed]

11. Pedersen-Bjergaard, S.; Rasmussen, K.E. Electrokinetic migration across artificial liquid membranes. New concept for rapid sample preparation of biological fluids. *J. Chromatogr. A* **2006**, *1109*, 183–190. [CrossRef]

12. Drouin, N.; Kubáň, P.; Rudaz, S.; Pedersen-Bjergaard, S.; Schappler, J. Electromembrane extraction: Overview of the last decade. *TrAC Trends Anal. Chem.* **2018**. [CrossRef]

13. Oedit, A.; Ramautar, R.; Hankemeier, T.; Lindenburg, P.W. Electroextraction and electromembrane extraction: Advances in hyphenation to analytical techniques. *Electrophoresis* **2016**, *37*, 1170–1186. [CrossRef]

14. Drouin, N.; Rudaz, S.; Schappler, J. Dynamic-Electromembrane Extraction: A Technical Development for the Extraction of Neuropeptides. *Anal. Chem.* **2016**, *88*, 5308–5315. [CrossRef]

15. Balchen, M.; Gjelstad, A.; Rasmussen, K.E.; Pedersen-Bjergaard, S. Electrokinetic migration of acidic drugs across a supported liquid membrane. *J. Chromatogr. A* **2007**, *1152*, 220–225. [CrossRef]

16. Gjelstad, A.; Rasmussen, K.E.; Pedersen-Bjergaard, S. Electromembrane extraction of basic drugs from untreated human plasma and whole blood under physiological pH conditions. *Anal. Bioanal. Chem.* **2009**, *393*, 921–928. [CrossRef]

17. Vardal, L.; Gjelstad, A.; Huang, C.; Oiestad, E.L.; Pedersen-Bjergaard, S. Efficient discrimination and removal of phospholipids during electromembrane extraction from human plasma samples. *Bioanalysis* **2017**, *9*, 631–641. [CrossRef]

18. Seip, K.F.; Jensen, H.; Kieu, T.E.; Gjelstad, A.; Pedersen-Bjergaard, S. Salt effects in electromembrane extraction. *J. Chromatogr. A* **2014**, *1347*, 1–7. [CrossRef]

19. Huang, C.; Seip, K.F.; Gjelstad, A.; Pedersen-Bjergaard, S. Electromembrane extraction of polar basic drugs from plasma with pure bis(2-ethylhexyl) phosphite as supported liquid membrane. *Anal. Chim. Acta* **2016**, *934*, 80–87. [CrossRef]

20. Drouin, N.; Rudaz, S.; Schappler, J. New supported liquid membrane for electromembrane extraction of polar basic endogenous metabolites. *J. Pharm. Biomed. Anal.* **2018**, *159*, 53–59. [CrossRef]

21. Fernandez, E.; Vardal, L.; Vidal, L.; Canals, A.; Gjelstad, A.; Pedersen-Bjergaard, S. Complexation-mediated electromembrane extraction of highly polar basic drugs-a fundamental study with catecholamines in urine as model system. *Anal. Bioanal. Chem.* **2017**, *409*, 4215–4223. [CrossRef]

22. Huang, C.X.; Shen, X.T.; Gjelstad, A.; Pedersen-Bjergaard, S. Investigation of alternative supported liquid membranes in electromembrane extraction of basic drugs from human plasma. *J. Membr. Sci.* **2018**, *548*, 176–183. [CrossRef]

23. Lindenburg, P.W.; Tjaden, U.R.; van der Greef, J.; Hankemeier, T. Feasibility of electroextraction as versatile sample preconcentration for fast and sensitive analysis of urine metabolites, demonstrated on acylcarnitines. *Electrophoresis* **2012**, *33*, 2987–2995. [CrossRef]

24. Raterink, R.J.; Lindenburg, P.W.; Vreeken, R.J.; Hankemeier, T. Three-phase electroextraction: A new (online) sample purification and enrichment method for bioanalysis. *Anal. Chem.* **2013**, *85*, 7762–7768. [CrossRef]

25. Schoonen, J.W.; van Duinen, V.; Oedit, A.; Vulto, P.; Hankemeier, T.; Lindenburg, P.W. Continuous-flow microelectroextraction for enrichment of low abundant compounds. *Anal. Chem.* **2014**, *86*, 8048–8056. [CrossRef]

26. Schoeman, J.C.; Hou, J.; Harms, A.C.; Vreeken, R.J.; Berger, R.; Hankemeier, T.; Boonstra, A. Metabolic characterization of the natural progression of chronic hepatitis B. *Genome Med.* **2016**, *8*, 64. [CrossRef]

27. Noga, M.J.; Dane, A.; Shi, S.; Attali, A.; van Aken, H.; Suidgeest, E.; Tuinstra, T.; Muilwijk, B.; Coulier, L.; Luider, T.; et al. Metabolomics of cerebrospinal fluid reveals changes in the central nervous system metabolism in a rat model of multiple sclerosis. *Metab. Off. J. Metab. Soc.* **2012**, *8*, 253–263. [CrossRef] [PubMed]

28. van de Rest, O.; Schutte, B.A.; Deelen, J.; Stassen, S.A.; van den Akker, E.B.; van Heemst, D.; Dibbets-Schneider, P.; van Dipten-van der Veen, R.A.; Kelderman, M.; Hankemeier, T.; et al. Metabolic effects of a 13-weeks lifestyle intervention in older adults: The Growing Old Together Study. *Aging* **2016**, *8*, 111–126. [CrossRef]

29. Drouin, N.; Mandscheff, J.F.; Rudaz, S.; Schappler, J. Development of a New Extraction Device Based on Parallel-Electromembrane Extraction. *Anal. Chem.* **2017**, *89*, 6346–6350. [CrossRef]

30. Marchi, I.; Viette, V.; Badoud, F.; Fathi, M.; Saugy, M.; Rudaz, S.; Veuthey, J.L. Characterization and classification of matrix effects in biological samples analyses. *J. Chromatogr. A* **2010**, *1217*, 4071–4078. [CrossRef]

31. Matuszewski, B.K.; Constanzer, M.L.; Chavez-Eng, C.M. Strategies for the assessment of matrix effect in quantitative bioanalytical methods based on HPLC-MS/MS. *Anal. Chem.* **2003**, *75*, 3019–3030. [CrossRef] [PubMed]

32. Beale, R.; Airs, R. Quantification of glycine betaine, choline and trimethylamine N-oxide in seawater particulates: Minimisation of seawater associated ion suppression. *Anal. Chim. Acta* **2016**, *938*, 114–122. [CrossRef] [PubMed]

33. Drouin, N.; Rudaz, S.; Schappler, J. New Trends in Sample Preparation for Bioanalysis. *Am. Pharm. Rev.* **2016**, *16*, 62–66.

34. Geiser, L.; Rudaz, S.; Veuthey, J.L. Decreasing analysis time in capillary electrophoresis: Validation and comparison of quantitative performances in several approaches. *Electrophoresis* **2005**, *26*, 2293–2302. [CrossRef]

35. Kuehnbaum, N.L.; Kormendi, A.; Britz-McKibbin, P. Multisegment injection-capillary electrophoresis-mass spectrometry: A high-throughput platform for metabolomics with high data fidelity. *Anal. Chem.* **2013**, *85*, 10664–10669. [CrossRef]

36. DiBattista, A.; Rampersaud, D.; Lee, H.; Kim, M.; Britz-McKibbin, P. High Throughput Screening Method for Systematic Surveillance of Drugs of Abuse by Multisegment Injection-Capillary Electrophoresis-Mass Spectrometry. *Anal. Chem.* **2017**, *89*, 11853–11861. [CrossRef]

37. Gonzalez-Ruiz, V.; Gagnebin, Y.; Drouin, N.; Codesido, S.; Rudaz, S.; Schappler, J. ROMANCE: A new software tool to improve data robustness and feature identification in CE-MS metabolomics. *Electrophoresis* **2018**. [CrossRef]

38. Drouin, N.; Pezzatti, J.; Gagnebin, Y.; Gonzalez-Ruiz, V.; Schappler, J.; Rudaz, S. Effective mobility as a robust criterion for compound annotation and identification in metabolomics: Toward a mobility-based library. *Anal. Chim. Acta* **2018**, *1032*, 178–187. [CrossRef]

39. Seip, K.F.; Jensen, H.; Sonsteby, M.H.; Gjelstad, A.; Pedersen-Bjergaard, S. Electromembrane extraction: Distribution or electrophoresis? *Electrophoresis* **2013**, *34*, 792–799. [CrossRef]

40. Seip, K.F.; Gjelstad, A.; Pedersen-Bjergaard, S. Electromembrane extraction from aqueous samples containing polar organic solvents. *J. Chromatogr. A* **2013**, *1308*, 37–44. [CrossRef]

41. Blanchard, J. Evaluation of the Relative Efficacy of Various Techniques for Deproteinizing Plasma Samples Prior to High-Performance Liquid-Chromatographic Analysis. *J. Chromatogr.* **1981**, *226*, 455–460. [CrossRef]

42. Polson, C.; Sarkar, P.; Incledon, B.; Raguvaran, V.; Grant, R. Optimization of protein precipitation based upon effectiveness of protein removal and ionization effect in liquid chromatography–tandem mass spectrometry. *J. Chromatogr. B* **2003**, *785*, 263–275. [CrossRef]

 *metabolites*

Article

General Guidelines for Sample Preparation Strategies in HR-*μ*MAS NMR-based Metabolomics of Microscopic Specimens

Covadonga Lucas-Torres [1],*, Thierry Bernard [1], Gaspard Huber [1], Patrick Berthault [1], Yusuke Nishiyama [2,3], Pancham S. Kandiyal [4], Bénédicte Elena-Herrmann [4], Laurent Molin [5], Florence Solari [5], Anne-Karine Bouzier-Sore [6] and Alan Wong [1],*

[1] NIMBE, CEA, CNRS, Université Paris-Saclay, CEA Saclay, 91191 Gif-sur-Yvette, France; thierry.bernard@cea.fr (T.B.); gaspard.huber@cea.fr (G.H.); patrick.berthault@cea.fr (P.B.)
[2] JEOL RESONANCE Inc., Musashino, Akishima, Tokyo 196-8558, Japan; yunishiy@jeol.co.jp
[3] RIKEN-JEOL Collaboration Center, Yokohama, Kanagawa 230-0045, Japan
[4] Univ Grenoble Alpes, CNRS, INSERM, IAB, Allée des Alpes, 38000 Grenoble, France; pancham-singh.kandiyal@univ-grenoble-alpes.fr (P.S.K.); benedicte.elena@univ-grenoble-alpes.fr (B.E.-H.)
[5] Univ Lyon, Université Claude Bernard Lyon 1, CNRS UMR 5310, INSERM U 1217, Institut NeuroMyoGène, 69008 Lyon, France; laurent.molin@univ-lyon1.fr (L.M.); florence.solari@univ-lyon1.fr (F.S.)
[6] Centre de Résonance Magnétique des Systèmes Biologiques, CNRS-Université de Bordeaux, UMR5536 Bordeaux, France; anne-karine.bouzier-sore@rmsb.u-bordeaux.fr
*   Correspondence: Covadonga.Lperez@gmail.com (C.L.-T.); alan.wong@cea.fr (A.W.)

Received: 16 December 2019; Accepted: 28 January 2020; Published: 30 January 2020

Abstract: The study of the metabolome within tissues, organisms, cells or biofluids can be carried out by several bioanalytical techniques. Among them, nuclear magnetic resonance (NMR) is one of the principal spectroscopic methods. This is due to a sample rotation technique, high-resolution magic angle spinning (HR-MAS), which targets the analysis of heterogeneous specimens with a bulk sample mass from 5 to 10 mg. Recently, a new approach, high-resolution micro-magic angle spinning (HR-μMAS), has been introduced. It opens, for the first time, the possibility of investigating microscopic specimens (<500 μg) with NMR spectroscopy, strengthening the concept of homogeneous sampling in a heterogeneous specimen. As in all bioanalytical approaches, a clean and reliable sample preparation strategy is a significant component in designing metabolomics (or -omics, in general) studies. The sample preparation for HR-μMAS is consequentially complicated by the μg-scale specimen and has yet to be addressed. This report details the strategies for three specimen types: biofluids, fluid matrices and tissues. It also provides the basis for designing future μMAS NMR studies of microscopic specimens.

Keywords: high-resolution magic angle spinning; NMR; microscopic samples; metabolomics; sample preparation

---

1. Introduction

Sample preparation is an essential component in metabolomics [1,2]. It requires dedication in designing protocols for precise and reliable acquisition of data with the appropriate analytical platform (i.e., gas (GC) [3] or liquid (LC) chromatography [4], mass spectrometry (MS) [5,6] and nuclear magnetic resonance (NMR) [7,8]). In any case, incorporating a convenient and reliable strategy is of underlying importance for avoiding any metabolic loss, for ensuring reproducibility and for the valid biological interpretation of the data [9,10].

NMR spectroscopy has been present in the development of metabolomics for decades. Strategies for sample preparations for both the standard high-resolution liquid [7] and high-resolution magic angle spinning (HR-MAS) [11] NMR are well documented and established. The fundamental basis is to minimize the sample exposure to unfavorable conditions (such as contamination, temperature, time, etc.) during the preparation so as to preserve sample integrity and experimental reproducibility. In general, the procedure is straightforward. For example, in HR-MAS with a Bruker 4 mm rotor, it consists of introducing the mg (or µL) level of a sample into a Kel-F bio-insert (Figure 1a) by either a pipette or a biopsy punch, depending on the sample morphology (liquid or semi-solid); if necessary, this is followed by adding sufficient D_2O (or buffer) to homogenize the content for high-quality data acquisition; and lastly, the insert and the rotor are sealed with the corresponding use of dedicated toolsets (Figure 1). The entire procedure takes about 5–10 min [11–15].

Recently, a new NMR technology, high-resolution micro-magic angle spinning (HR-µMAS), was introduced [16]. It targets the microgram (µg) level of specimens and has shown promising results towards metabolomics [17,18]. HR-µMAS could open a new and unexplored NMR platform [19], such as the possibility of carrying out longitudinal metabolic investigation on living animals. This contributes to the fact that µg scale analysis permits (1) minimal surgical tissue excision and (2) homogenous sampling from a heterogeneous specimen.

However, the sample preparation for HR-µMAS NMR spectroscopy (i.e., with a 1 mm µ-rotor) is a strenuous task [19] compared to the mg sampling with HR-MAS (i.e., 4 mm rotor). As trivial as it may seem at first glance, handling delicate specimens at the µg level requires high precision skills and tools. One has to consider that displacing such a minuscule sample mass (<500 µg) in a confined volume (<500 nL), in a clean and efficient manner, creates an inevitable difference with HR-MAS. Moreover, unlike HR-MAS, different sample morphologies also require different strategies with different toolsets, constituting additional steps for HR-µMAS.

Presently, the sample preparation for HR-µMAS has yet to be fully described and documented. The main bottleneck lies in the necessary requirements, namely methodical sample collection, feasible sample filling and quick µ-rotor sealing. Few attempts have been made to simplify the preparation of µMAS for metabolomics [20–22]. One example is the use of a 1 mm disposable Kel-F µ-rotor along with the concept of sampling insert (i.e. a 4 mm Kel-F bioinsert) using a glass capillary [20]. The intent is to facilitate the sample filling and eliminate the rotor sealing. However, the overall procedure is not efficient due to the limitations of handling a non-rigid Kel-F rotor. Moreover, the use of an insert lowers the filling factor and, hence, lowers the detection sensitivity by, in this case, nearly one-third. In addition, this approach presents a high risk of damaging the MAS stator due to the spinning of a fragile glass capillary insert. As such, new strategies must be explored and attuned to metabolomics studies.

After over 500 sample preparations in a span of two years with µ-rotor, this report summarizes these experiences and outlines the general preparation strategies for different specimens targeting to HR-µMAS NMR metabolomics. The strategies adopt three principal criteria: (1) rapid and clean sampling, (2) direct, clean and consistent sample filling and (3) quick µ-rotor sealing. Although the guidelines herein are based on a JEOL 1 mm µ-rotor, the strategies could provide the basis for designing new procedures for metabolomics studies including with µ-rotors from other manufacturers.

2. HR-µMAS Sample Preparation

As an initial and major modification from HR-MAS, all manipulations with both the sample and the µ-rotor are carried out under a stereomicroscope (Figure 1b) with a large set of high-precision tools (Figure 1a), each with a specific function. For example, a holder (Figure 1a, iii) facilitates all manipulations of the µ-rotor and is an essential tool in all sample filling procedures described below. The use of a holder also prevents a constant contamination on the rotor surface. Specific toolsets are dedicated for closing (Figure 1a, iv and 1a, v) and opening (Figure 1a, vi and 1a, vii) the µ-rotor caps. Since there are different toolset designs for different µ-rotors (e.g., Bruker, JEOL, etc.), the report will omit the descriptions of operating the tools. Formal training from the vendor is recommended.

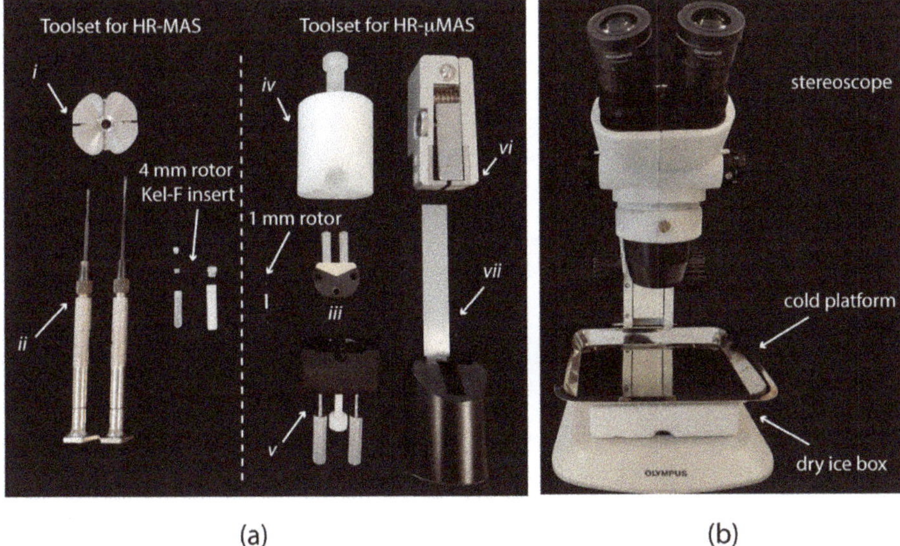

Figure 1. (**a**) Toolsets for (left) high-resolution magic angle spinning (HR-MAS) sample preparation compared to (right) high-resolution micro-magic angle spinning (HR-μMAS) with details of ZrO_2 rotor of different sizes and packing tools. (i) Clamp tool for opening the rotor cap, (ii) screw drivers for handling Kel-F insert, (iii) μ-rotor holder, (iv) and (v) toolset for closing the μ-rotor caps, (vi) and (vii) toolset for opening the μ-rotor caps. (**b**) Cold workstation and stereomicroscope. A dry ice bucket is placed under a metallic platform, which will be consequently cooled down for allocating the sample manipulations with the μ-rotor under the stereomicroscope.

Due to the small sample volume, <500 nL, even the tiniest contamination is in line with the sample metabolic content. As a result, the unwanted signals will inevitably disrupt the metabolic spectral profile. Therefore, prior to the sample preparation, special attention to the cleanness of the rotor, the toolsets (i.e., contact with the rotor) and even the working space entirely is essential. Supplementary Materials Figure S1 shows an example of a contamination from a cleaning solvent, ethanol, even after a long drying period. Tips: (1) It is strongly recommended to avoid using solvent other than water for cleaning (i.e., rotor, toolsets, workspace); (2) sonicating the μ-rotor in a warm water bath is recommended to assist in eliminating the tiniest residues.

2.1. μg Sampling

As aforementioned, ensuring the sample integrity is crucial [10], and therefore preconditions must be regulated. For example, the time for the samples to be exposed to an unfavorable temperature must be short. Therefore, the availability of dry ice during all procedures is important. While working with the stereomicroscope, one of the complications, the use of a cold platform (Figure 1b) is highly recommended for carrying out the entire sample preparation for keeping the specimens under a favorable environment. Ideally, the entire procedure from sampling to filling would be better performed inside a walk-in cold room facility.

Prior to μg sampling, specific preconditions must be considered for different specimens. For example, intact cells are generally susceptible to their surroundings, and the permeability of the membrane should be considered. In that sense, the pH should be controlled using a buffer [23], as well as the cell concentration in the suspensions; it should be efficient for spectral sensitivity but avoiding a crowding effect causing cell asphyxiation or rupture [24]. In addition, cells are particularly sensitive to the changes in temperature, so the freeze/thawing cycles have to be avoided as the formation of ice crystals will damage the cell membrane [25].

For animal tissues, the presence of excessive blood content distorts spectral resolution due to the existence of paramagnetic species [11]. Unlike in HR-MAS, even a very small content of blood in a 500 µg sample could render a significant paramagnetic effect on the spectrum. Therefore, it is advised to wash the tissue with D_2O or saline solution (0.9% NaCl w/v) prior to filling. However, caution must be applied to prevent removal of the metabolic content. We recommend a swift immersion into a deuterated saline solution.

Depending on the sample morphologies, the µg sampling procedure is performed either by a pipette for biofluids or fluid matrices (i.e., intact specimens suspended in a liquid) or by a microsized biopsy punch for tissues.

Tips: (1) With a pipette, it is recommended to use a micropipette tip with hydrophilic surface (ideally with glass) to avoid surface tensions with the individual specimens; (2) with a microbiopsy punch, it is recommended to cool down the tip to prevent a complete thawing of the tissue during the collection and transferring.

2.2. Sample Filling

The sample filling procedure can be considered the most significant step in the sample preparation. This is because of its extensive manipulation of the sample. Unfortunately, it is not straightforward to fill µg specimens into a tiny rotor (with a 0.5 mm inner diameter). It should comply with the following criteria: (1) a good sample homogeneity inside the µ-rotor to achieve high spectral resolution data (i.e., avoiding the presence of air bubbles). For example, the tiniest air pocket can worsen the spectral resolution; (2) a correct sample displacement inside the µ-rotor for maximum sensitivity detection; (3) a sufficient sample mass to achieve a good sensitivity; (4) a good weight balance of the µ-rotor to avoid spinning deficiency; and (5) a repeatable sampling procedure for data reproducibility. Subtle deviations in all these criteria could affect the individual spectral data and diminish both the data repeatability and reproducibility. What follows are the details of three strategies, each with different toolsets targeting different specimens.

2.2.1. Micropipette or Microsyringe

Target samples: biofluids such as serum, plasma, urine and tissue extract.

Pipette tip/fine-needle characteristics: both must be narrow in diameter (i.e., < 0.5 mm) to be able to traverse through the entire µ-rotor length (e.g., Eppendorf GELoader®tips, Hilgenberg glass needles).

µ-rotor requirements: one closed end (i.e., one end is open while the other is closed with a µ-rotor cap).

Guidelines:

○ Convey 1–2 µL of fluid inside the µ-rotor by placing the tip (or needle) at the bottom. Tip: the µ-rotor is placed in the holder (Figure 1a, iii) to facilitate the handling.
○ Release the fluid slowly while moving upwards to avoid air bubbles.
○ Centrifuge (~3000 rpm, ~30 s; recommended at 4 °C) the filled µ-rotor to ensure the exclusion of air bubbles.
○ Seal the µ-rotor with a designated µ-rotor cap using the dedicated toolset (e.g., Figure 1a). Caution: Ensure a sufficient space for the sealing; if not, the sealing would be impossible.

Estimated time: 5–10 min.

2.2.2. Centrifugal Microfunnel

Target samples: fluid matrix samples such as cells or whole organisms in biofluids (e.g., blood, nematode and microbe).

Funnel characteristics: biocompatible material (ideally with glass or parylene coating, etc.). Caution: The surface must be smooth and hydrophilic to prevent sample shearing, and to assist the transference into the μ-rotor.

μ-rotor requirement: one closed end.

Guidelines (based on a 3D-printed microfunnel shown in Figure 2a):

- ○ Place the μ-rotor inside the designated space in the funnel.
- ○ Convey (by either pipette or glass syringe) the matrix into the funnel reservoir.
- ○ Centrifuge at 4 °C. The speed and time depend on the funnel materials (polymers such as Kel-F and Teflon allow for faster centrifugation, while glass will only tolerate a gentle centrifugation).
- ○ Close the μ-rotor.

Estimated time: 10–15 min.

2.2.3. Microbiopsy Punch

Target samples: semi-solids such as animal and food tissues, or cell pellets.

Punch characteristics: sharp edges for clean-cut and reproducibility. The outer diameter must be smaller than or equal to the inner diameter of the μ-rotor (i.e., <0.5 mm, examples in Figure 2b).

μ-rotor requirements: Both ends must be opened to prevent the effect of the air pressure during the filling.

Guidelines:

- ○ Extract μg sample by punching. Tip: Frozen samples facilitate a clean excision.
- ○ Fast transfer of the excised sample into the μ-rotor (placed in a holder).
- ○ Follow by adding a drop of D_2O (or buffer) into the sample to homogenize and to avoid dehydration. Note: Water content in the sample can have a large effect on the spectral quality (Supplementary Materials Figure S2).
- ○ Close one end of the μ-rotor followed by a gentle centrifugation (~1500 rpm, ~30 s, at 4 °C) for sample positioning and releasing air bubbles.
- ○ Fill the remaining μ-rotor volume with D_2O or buffer. Tip: Use the holder.
- ○ A second centrifugation should be applied to further homogenize the sample.
- ○ Close the μ-rotor.

Estimated time: 15–20 min.

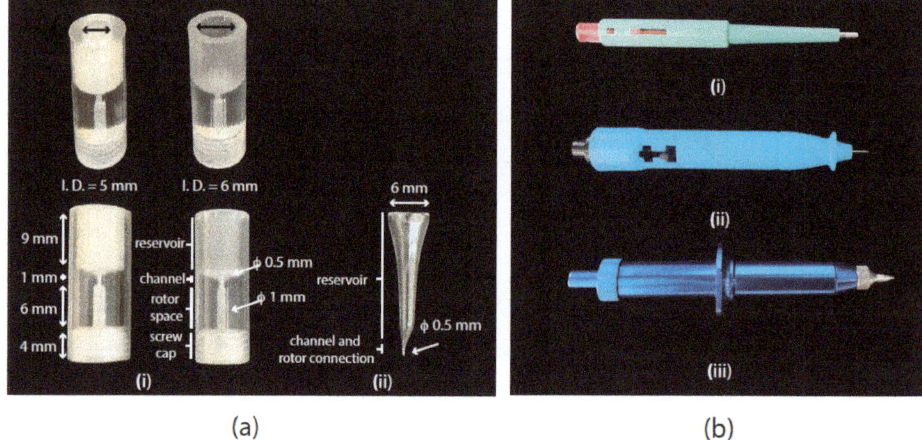

Figure 2. (**a**) (i) 3D-printed funnel. It consists of a bulk polymeric funnel, which leaves a space for the μ-rotor, and a μ-channel connecting the rotor volume with the sample reservoir. After printing, the funnel should be submitted to a coating process with deposited poly(p-xylylene) (i.e., parylene), which adds a layer of 0.5 μm and generates a biocompatible and smoother surface. (ii) Custom-made glass funnel. It connects the funnel reservoir with the μ-rotor through a short channel. Suggested convenient dimensions are shown in the picture for both types of funnel. (**b**) Different biopsy punch models. (i) 2 mm biopsy punch used for sample collection and filling process inside the standard HR-MAS Kel-F insert. (ii) Disposable and (iii) reusable 0.5 mm biopsy punch fitting the inner diameter of the μ-rotor for HR-μMAS to facilitate the filling process.

2.3. Pre-Acquisition Considerations

After the sample filling, a few critical precautions must be applied prior to inserting the μ-rotor into the probe. Under the stereomicroscope, one must carefully inspect the μ-rotor caps to see if they are in good condition (i.e., no sign of damage) to ensure a good and stable sample spinning, and that the caps are tightly fit and secure in the μ-rotor to prevent sample leakage. In addition, the cleanliness on the rotor surface is absolute; any tiny particles (i.e., dried sample residue, dust, etc.) could damage the stator. Hence, for the same reason, it is advised to clean the entire μ-rotor surface, including the caps, with high-quality tissue (e.g., Kimtech wipe) and/or with sticky pens (e.g., standard JEOL RESONANCE Inc. preparation tool set, Supplementary Materials Figure S3) prior to displacing the μ-rotor into the stator.

Once the μ-rotor is introduced in the stator, sample spinning must proceed with caution. A manual adjustment to the desired spinning frequency is recommended. Typical spinning rates ranging from 4 to 6 kHz are sufficient to suppress the susceptibility broadening in MAS NMR spectra of semi-solids such as tissues and cells [11]. Such moderate rate prevents the sample temperature from increasing and provides adequate conditions during the data acquisition. As an example, Figure 3 shows excellent spectral quality from a spinning rate of 4 kHz. The spinning side-bands are displaced outside the metabolites' chemical shift range, and the metabolic isotropic signals are with good resolution.

^{2}H-field locking can be difficult due to the low ^{2}H content in the sample volume; the signal is often weak and unreliable for field shimming. Consequently, the strategy of field shimming with HR-μMAS is different to that with HR-MAS. Despite the applied MAS shim sets being the same [26], the metabolic signals are generally weak with HR-μMAS to shim with a continuous acquisition; therefore, it is recommended to shim first on the water signal with continuous mode, followed by an incremental acquisition on metabolic signals. Consequently, the shimming can be a long and demanding process; hence, the use of sacrificed sample(s) is strongly recommended. One can also consider adding a small content of a known metabolite (e.g., <1 mM alanine, sucrose) in the sacrificed sample rendering a shimming process with a continuous acquisition. Supplementary Materials Figure S4 shows the resultant signals of the metabolic additive used for shimming.

Provided the sample filling into a μ-rotor is identical, then the same shim set of the sacrificed sample can be applied, with minor adjustments. However, one should note that a slight deviation in the sample filling, resulting in an air pocket or insufficient water content, can deviate the shims from a sacrificed sample.

The experimental parameters used in the spectral acquisitions are no different with the standard NMR experiments, except with one exception: the power level for a 90° pulse is low owing to the high B_1 efficiency with a μ-size coil.

Figure 3 shows the resultant NMR spectra of different specimens (biofluid, cell matrix, small organism and animal tissue), each prepared by the different strategies described above. The details of the preparations are summarized in the Supplementary Materials (Protocol S1) and should provide the basis for future metabolomic studies with HR-μMAS or with μMAS in general.

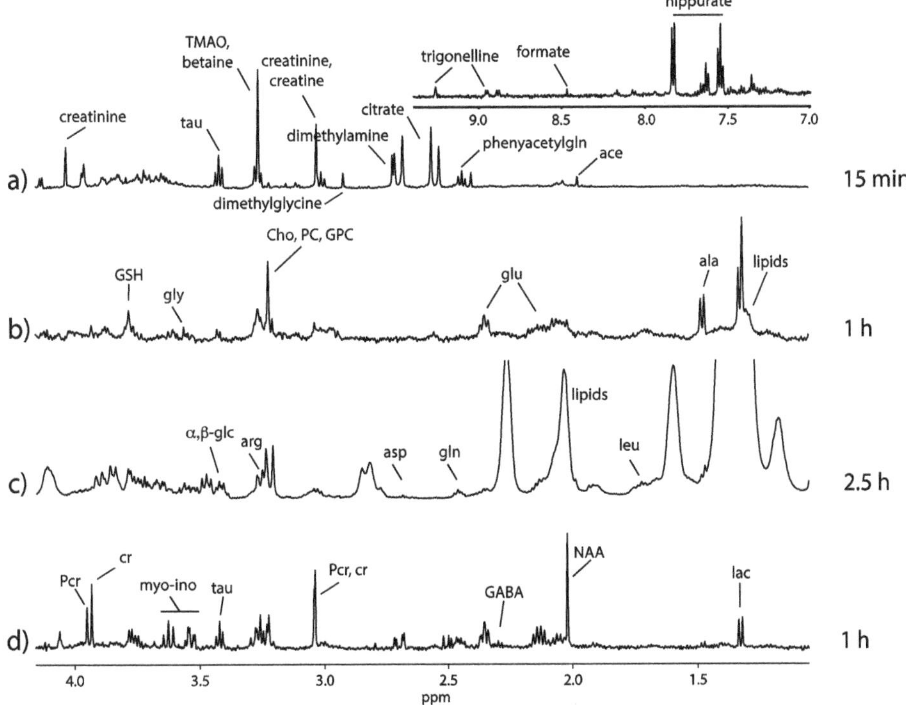

Figure 3. ^{1}H HR-μMAS nuclear magnetic resonance (NMR) spectra spinning at 4 kHz from (**a**) rat urine in PBS/D$_2$O prepared with automatic pipette, (**b**) 400 nL K562 cell suspension in PBS/D$_2$O buffer (pH = 7.4) prepared with a 3D-printed funnel, (**c**) 400 nL of *C. elegans* (*n* = 30) suspension in D$_2$O prepared with a custom-made glass funnel and (**d**) 500 μg brain tissue prepared with a disposable 0.5 mm biopsy punch. Total acquisition times are indicated for each spectrum. Spectra (**b**) and (**d**) were acquired using the Carr-Purcell-Meiboom-Gill (CPMG) pulse sequence (d20 = 0.2 ms, loop = 200), and spectra (**a**) and (**c**) were acquired using the NOESY pulse sequence (mixing time 0.1 s). The main metabolic signatures are identified on the spectra. The preparation for each specimen is detailed in the Supplementary Protocol S1.

3. Final Remarks

Acknowledging the difficulties for preparing µg-scale specimens in a specific tiny sampling volume for HR-µMAS NMR spectroscopy, this report presents general guidelines of the sample preparation strategies for the different type of specimens (biofluid, fluid matrix and tissue). Although the basis of these preparations is similar to those for HR-MAS, they are considerably complicated by the fact that the manipulation of the minuscule specimens along with a tiny rotor must be performed in a repeatable, clean and timely manner. A slight deviation could affect the overall data, resulting in non-reproducible data acquisition and consequently in variable or misinterpreted analysis.

The guidelines herein can provide a good basis for designing NMR-based metabolomics studies of µg-scale heterogeneous specimens with HR-µMAS or with µMAS in general. An example is shown in Figure 4; over 100 sampling data on tissue (control and disease) were acquired adopting the guidelines stated in this report. The results offer good data reproducibility for reliable multivariate data analysis. The Principal Component Analysis (PCA) score plots (Figure 4a) clearly display two groups within the data set. Its quality parameters ($R^2X = 0.85$, $Q^2 = 0.63$) demonstrate the data acquisition is trustworthy where the sample preparation has an important contribution. Figure 4b exhibits good regularity of the NMR spectral profiles within the groups. Acceptable values of the relative standard deviation (%RSD) of the individual spectral bins are found with a median of 30.7% for control and 37.6% for disease (Figure 4c). The slightly higher value of the latter is attributed to the accentuated heterogeneity of the tissue itself.

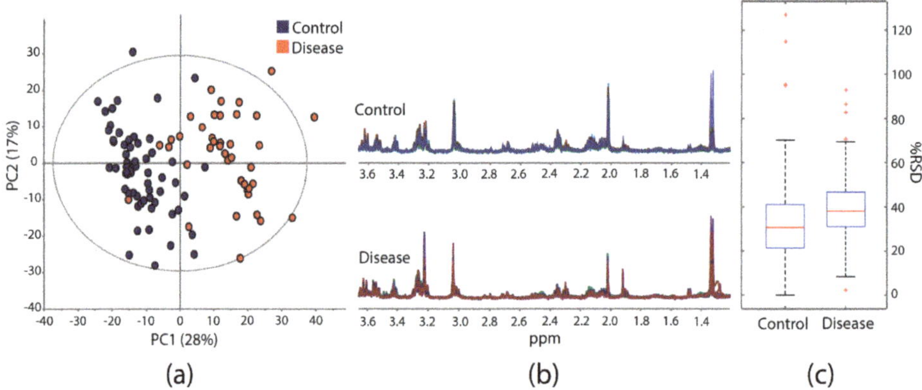

Figure 4. (a) PCA scores plot obtained from a model study containing 102 data on brain tissue (control and disease) from an initial 112 samples, where 10 (9%) were discarded due to either extra peaks from contamination or poor spectral quality from air pockets or dryness. Quality parameters: 14 components, $R^2X = 0.85$, $Q^2 = 0.63$. (b) Overlaid ^{1}H-HR-µMAS NMR spectra of (blue) 62 control and (red) 40 diseased tissue samples. (c) Boxplots of the relative standard deviation (%RSD) values calculated from the individual bucket intensity ($\Delta = 0.04$ ppm) of both groups across the spectral region (0.76–5.28 ppm). It summarizes the lower, median and upper quartiles, with the black whiskers displaying the range of data, and the red cross indicating the outlier data points.

It should be noted that among the different types of specimens, fluid matrices are the most challenging of all. This is due to the difficulty in achieving a required quantity of intact specimens (e.g., tiny organisms and cells) in a nL scale volume. Although the strategy (centrifugal microfunnel) described above improves the sample filling, it is still a primitive and tedious approach. One should consider redesigning the µ-rotor to incorporate a filtration system that can improve sample filling for fluid matrices, which would be a great advantage. For example, an internal filter inside the rotor would retain the matrix (i.e., specimens), while the fluid would be guided out. This method of collecting and filling the sample would assist in concentrating the specimens inside the µ-rotor and render an increase in the sensitivity of the experiments. In addition, this potential methodology would benefit from the use of microfluidic technology for gently guiding the susceptible specimens inside the μ-rotor.

Supplementary Materials: The following are available online at http://www.mdpi.com/2218-1989/10/2/54/s1, Protocol S1: Detailed preparation for the samples generating the spectra in Figure 3, Figure S1: Garlic and brain tissue displaying ethanol impurity signals. Figure S2: Differences in the spectral resolution depending on the water content shown in different specimens. Figure S3: μ-rotor handling and cleaning tools. Figure S4: Example of spectrum obtained by adding a known amount of sucrose embedded in a brain tissue sample.

Author Contributions: C.L.-T. and P.S.K. carried out experimental work. C.L.-T. planned and drafted the manuscript; A.W., Y.N., F.S., B.E.-H. and A.-K.B.-S. assisted in editing and reviewing; T.B., G.H., P.B. and L.M. provided supports in designing the sample filling and the tools (e.g. microfunnels). All authors have read and agreed to the published version of the manuscript.

Funding: C.L.T. and P.S.K. the APC were funded by Agence Nationale de la Recherche in France, grant number ANR-16-CE11-0023. A.K.B.-S was also supported by LabEx TRAIL, grant number ANR-10-LABX-57.

Acknowledgments: Christina Sizun (Institut de Chimie des Substances Naturelles, CNRS) for offering samples; Guillaume Carret and Melanie Moskura (CEA Saclay) for their assistance in fabricating the 3D-printed microfunnel.

Conflicts of Interest: The authors declare no conflict of interest.

References

1. Dunn, W.B.; Ellis, D.I. Metabolomics: Current analytical platforms and methodologies. *Trends Anal. Chem.* **2005**, *24*, 285–294. [CrossRef]

2. Moco, S.; Bino, R.J.; de Vos, R.C.H.; Vervoort, J. Metabolomics technologies and metabolite identification. *Trends Anal. Chem.* **2007**, *26*, 855–866. [CrossRef]

3. Dunn, W.B.; Broadhurst, D.; Begley, P.; Zelena, E.; Francis-Mcintyre, S.; Anderson, N.; Brown, M.; Knowles, J.D.; Halsall, A.; Haselden, J.N.; et al. Procedures for large-scale metabolic profiling of serum and plasma using gas chromatography and liquid chromatography coupled to mass spectrometry. *Nat. Protoc.* **2011**, *6*, 1060–1083. [CrossRef] [PubMed]

4. Martano, G.; Delmotte, N.; Kiefer, P.; Christen, P.; Kentner, D.; Bumann, D.; Vorholt, J.A. Fast sampling method for mammalian cell metabolic analyses using liquid chromatography-mass spectrometry. *Nat. Protoc.* **2015**, *10*, 1–11. [CrossRef]

5. Rubakhin, S.S.; Sweedler, J.V. Characterizing peptides in individual mammalian cells using mass spectrometry. *Nat. Protoc.* **2007**, *2*, 1987–1997. [CrossRef]

6. Vuckovic, D. Current trends and challenges in sample preparation for global metabolomics using liquid chromatography-mass spectrometry. *Anal. Bioanal. Chem.* **2012**, *403*, 1523–1548. [CrossRef]

7. Beckonert, O.; Keun, H.C.; Ebbels, T.M.D.; Bundy, J.; Holmes, E.; Lindon, J.C.; Nicholson, J.K. Metabolic profiling, metabolomic and metabonomic procedures for NMR spectroscopy of urine, plasma, serum and tissue extracts. *Nat. Protoc.* **2007**, *2*, 2692–2703. [CrossRef]

8. Pontes, J.G.M.; Brasil, A.J.M.; Cruz, G.C.F.; de Souza, R.N.; Tasic, L. NMR-based metabolomics strategies: Plants, animals and humans. *Anal. Methods* **2017**, *9*, 1078–1096. [CrossRef]

9. Álvarez-Sánchez, B.; Priego-Capote, F.; Luque de Castro, M.D. Metabolomics analysis I. Selection of biological samples and practical aspects preceding sample preparation. *Trends Anal. Chem.* **2010**, *29*, 111–119. [CrossRef]

10. Álvarez-Sánchez, B.; Priego-Capote, F.; Luque de Castro, M.D. de Metabolomics analysis II. Preparation of biological samples prior to detection. *Trends Anal. Chem.* **2010**, *29*, 120–127. [CrossRef]

11. Beckonert, O.; Coen, M.; Keun, H.C.; Wang, Y.; Ebbels, T.M.D.; Holmes, E.; Lindon, J.C.; Nicholson, J.K. High-resolution magic-angle-spinning NMR spectroscopy for metabolic profiling of intact tissues. *Nat. Protoc.* **2010**, *5*, 1019–1032. [CrossRef]

12. Mirnezami, R.; Jiménez, B.; Li, J.V.; Kinross, J.M.; Veselkov, K.; Goldin, R.D.; Holmes, E.; Nicholson, J.K.; Darzi, A. Rapid diagnosis and staging of colorectal cancer via high-resolution magic angle spinning nuclear magnetic resonance (HR-MAS NMR) spectroscopy of intact tissue biopsies. *Ann. Surg.* **2014**, *259*, 1138–1149. [CrossRef]

13. Choi, J.S.; Baek, H.-M.; Kim, S.; Kim, M.J.; Youk, J.H.; Moon, H.J.; Kim, E.-K.; Han, K.H.; Kim, D.; Kim, S.I.; et al. HR-MAS MR Spectroscopy of Breast Cancer Tissue Obtained with Core Needle Biopsy: Correlation with Prognostic Factors. *PLoS ONE* **2012**, *7*, e51712. [CrossRef] [PubMed]

14. Farooq, H.; Courtier-Murias, D.; Soong, R.; Bermel, W.; Kingery, W.; Simpson, A. HR-MAS NMR Spectroscopy: A Practical Guide for Natural Samples. *Curr. Org. Chem.* **2013**, *17*, 3013–3031. [CrossRef]

15. Blaise, B.J.; Giacomotto, J.; Triba, M.N.; Toulhoat, P.; Piotto, M.; Emsley, L.; Ségalat, L.; Dumas, M.E.; Elena, B. Metabolic profiling strategy of caenorhabditis elegans by whole-organism nuclear magnetic resonance. *J. Proteome Res.* **2009**, *8*, 2542–2550. [CrossRef] [PubMed]

16. Nishiyama, Y.; Endo, Y.; Nemoto, T.; Bouzier-Sore, A.-K.; Wong, A. High-resolution NMR-based metabolic detection of microgram biopsies using a 1 mm HRμMAS probe. *Analyst* **2015**, *140*, 8097–8100. [CrossRef]

17. Duong, N.T.; Endo, Y.; Nemoto, T.; Kato, H.; Bouzier-Sore, A.-K.; Nishiyama, Y.; Wong, A. Evaluation of a high-resolution micro-sized magic angle spinning (HRμMAS) probe for NMR-based metabolomic studies of nanoliter samples. *Anal. Methods* **2016**, *8*, 6815–6820. [CrossRef]

18. Lucas-Torres, C.; Huber, G.; Ichikawa, A.; Nishiyama, Y.; Wong, A. HR-μMAS NMR-Based Metabolomics: Localized Metabolic Profiling of a Garlic Clove with μg Tissues. *Anal. Chem.* **2018**, *90*, 13736–13743. [CrossRef]

19. Lucas-Torres, C.; Wong, A. Current developments in μMAS NMR analysis for metabolomics. *Metabolites* **2019**, *9*, 29. [CrossRef]

20. Duong, N.T.; Yamato, M.; Nakano, M.; Kume, S.; Tamura, Y.; Kataoka, Y.; Wong, A.; Nishiyama, Y. Capillary-inserted rotor design for HRμMAS NMR-based metabolomics on mass-limited neurospheres. *Molecules* **2017**, *22*, 1289. [CrossRef]

21. Feng, J.; Hu, J.; Burton, S.D.; Hoyt, D.W. High Resolution Magic Angle Spinning ^{1}H NMR metabolic profiling of nanoliter biological tissues at high magnetic field. *Chinese J. Magn. Reson.* **2013**, *30*, 1–11.

22. Wong, A.; Li, X.; Molin, L.; Solari, F.; Elena-Herrmann, B.; Sakellariou, D. μHR-MAS NMR Spectroscopy for Metabolic Phenotyping of Caenorhabditis Elegans. *Anal. Chem.* **2014**, *86*, 6064–6070. [CrossRef] [PubMed]

23. León, Z.; García-Cañaveras, J.C.; Donato, M.T.; Lahoz, A. Mammalian cell metabolomics: Experimental design and sample preparation. *Electrophoresis* **2013**, *34*, 2762–2775. [CrossRef] [PubMed]

24. Lu, W.; Su, X.; Klein, M.S.; Lewis, I.A.; Fiehn, O.; Rabinowitz, J.D. Metabolite Measurement: Pitfalls to Avoid and Practices to Follow. *Annu. Rev. Biochem.* **2017**, *86*, 277–304. [CrossRef] [PubMed]

25. Mashego, M.R.; Rumbold, K.; de Mey, M.; Vandamme, E.; Soetaert, W.; Heijnen, J.J. Microbial metabolomics: Past, present and future methodologies. *Biotechnol. Lett.* **2007**, *29*, 1–16. [CrossRef] [PubMed]

26. Piotto, M.; Elbayed, K.; Wieruszeski, J.M.; Lippens, G. Practical aspects of shimming a high resolution magic angle spinning probe. *J. Magn. Reson.* **2005**, *173*, 84–89. [CrossRef]

 *metabolites*

Article

Choosing an Optimal Sample Preparation in *Caulobacter crescentus* for Untargeted Metabolomics Approaches

Julian Pezzatti [1], Matthieu Bergé [2], Julien Boccard [1,3], Santiago Codesido [1], Yoric Gagnebin [1], Patrick H. Viollier [2], Víctor González-Ruiz [1,3] and Serge Rudaz [1,3,*]

[1] Institute of Pharmaceutical Sciences of Western Switzerland (ISPSO), University Medical Centre, 1206 Geneva, Switzerland; julian.pezzatti@unige.ch (J.P.); julien.boccard@unige.ch (J.B.); santiago.codesido@unige.ch (S.C.); yoric.gagnebin@unige.ch (Y.G.); victor.gonzalez@unige.ch (V.G.-R.)

[2] Department of Microbiology and Molecular Medicine, Faculty of Medicine, University of Geneva, University Medical Centre, 1206 Geneva, Switzerland; matthieu.berge@unige.ch (M.B.); patrick.viollier@unige.ch (P.H.V.)

[3] Swiss Centre for Applied Human Toxicology, 4055 Basel, Switzerland

* Correspondence: serge.rudaz@unige.ch; Tel.: +41-22-379-6572

Received: 27 August 2019; Accepted: 17 September 2019; Published: 20 September 2019

Abstract: Untargeted metabolomics aims to provide a global picture of the metabolites present in the system under study. To this end, making a careful choice of sample preparation is mandatory to obtain reliable and reproducible biological information. In this study, eight different sample preparation techniques were evaluated using *Caulobacter crescentus* as a model for Gram-negative bacteria. Two cell retrieval systems, two quenching and extraction solvents, and two cell disruption procedures were combined in a full factorial experimental design. To fully exploit the multivariate structure of the generated data, the ANOVA multiblock orthogonal partial least squares (AMOPLS) algorithm was employed to decompose the contribution of each factor studied and their potential interactions for a set of annotated metabolites. All main effects of the factors studied were found to have a significant contribution on the total observed variability. Cell retrieval, quenching and extraction solvent, and cell disrupting mechanism accounted respectively for 27.6%, 8.4%, and 7.0% of the total variability. The reproducibility and metabolome coverage of the sample preparation procedures were then compared and evaluated in terms of relative standard deviation (RSD) on the area for the detected metabolites. The protocol showing the best performance in terms of recovery, versatility, and variability was centrifugation for cell retrieval, using MeOH:H$_2$O (8:2) as quenching and extraction solvent, and freeze-thaw cycles as the cell disrupting mechanism.

Keywords: metabolomics; sample preparation; hydrophilic interaction liquid chromatography; ion mobility spectrometry; high resolution mass spectrometry; design of experiments; AMOPLS

---

1. Introduction

Untargeted metabolomic approaches aim to analyze the metabolic composition of biological samples in a holistic manner, with the goal of retrieving and storing as much chemical information as possible. To this end, adequate sample preparation is one of the most critical steps to obtain reliable biological insights [1]. Moreover, for an optimal interpretation of the information gathered from such systems, the annotation of metabolites of interest is a requirement. Indeed, the recent advances in liquid chromatography (LC) coupled with high resolution mass spectrometry (LC-HRMS) and the development of in-house and external databases have made possible the identification and relative quantification of an increasing number of metabolites in many types of biological samples [2,3].

When analyzing polar metabolites, techniques with separation mechanisms that are orthogonal to those of the classical reversed phase LC (RPLC) are needed [4]. Among these, hydrophilic liquid interaction chromatography (HILIC) has proven to be an attractive strategy for such classes of compounds and is well referenced in the literature [5–11]. Several groups have recently demonstrated that different HILIC stationary phases enable the analysis of many chemical classes of metabolites [12–14]. It has been shown that polymeric columns with zwitterionic stationary phases are interesting for the analysis of compounds related to the tricarboxylic acid cycle (TCA) [14].

To perform metabolite annotation in untargeted metabolomics, several criteria must be fulfilled to provide a certain level of confidence to the putatively identified compounds, as defined by the metabolomics standards initiative (MSI) [15]. To maximize the reliability of the identification of features, the retention time, exact mass, isotopic pattern, and MS/MS pattern must be used to match compound properties to those of standards measured under identical experimental conditions [16,17]. With the advent of the latest generation of quadrupole time-of-flight instruments (QTOF), coupled with ion mobility (IM), peak capacity and dynamic range have been increased, enabling to perform all ion fragmentation (AIF) approaches and including the measurement of collision cross section (CCS), which offers another structural information for identification [18,19]. Indeed, CCS values provide valuable physicochemical information about the ions such as three-dimensional conformations [20,21]. Furthermore, the IM technology allows to selectively isolate coeluting peaks resulting in cleaner MS and MS/MS spectra [22].

In untargeted metabolomics experiments, it has already been demonstrated that various sample preparation procedures result in very different metabolite recovery rates, regardless of the biological matrix studied, e.g., human plasma [23–27], mammalian cell cultures [28–31], and microbial cell cultures [32–35]. During sample preparation, subtle modifications of the conditions, such as slight changes in the composition of the extraction solvents, can have remarkable effects on the recovery of some metabolites [36]. In vitro metabolomic studies on cell cultures offer many advantages over more complex biological systems including an easily controlled environment, greater reproducibility, lower cost, and easier method transfer to external laboratories for validation [30]. However, when it comes to microbial cell cultures, sample preparation faces challenges such as fast metabolite turnover, degradation, leakage, and poor extraction reproducibility [30,37,38]. These cell cultures require harsher sample processing procedures than body fluids. Metabolite leakage must be prevented during the cell retrieval step [38]. While metabolite losses can be minimized by reducing the number of steps and simplifying the protocol, reproducible sample preparation procedures are necessary to ensure an accurate determination of small changes in metabolite levels [39]. As a rule of thumb, cell metabolism must be quenched quickly, and metabolites extracted rapidly, non-selectively, and reproducibly upon cell sampling [39]. Quenching and extraction steps are of upmost importance in the sample preparation process to obtain metabolic profiles which are representative of the physiological status at the time of sampling [30].

In this study, we aimed to perform a systematic evaluation of the influence of different sample preparation conditions on the nature and variability of the recovered metabolomic profile in Gram-negative bacterium *Caulobacter crescentus*. The latter was chosen as a model for Gram-negative bacteria because its central carbon metabolism, closely linked to cell cycle regulation and cell division, remains to be fully understood [40]. After a comprehensive evaluation of the literature on sample preparation for bacterial cell cultures [32–35], it was found that no systematic evaluation of different protocols supported by a design of experiment (DOE) approach was available. In the present study, two techniques of cell retrieval, two quenching and extraction solvents, and two cell disrupting mechanisms were investigated. Considering a full factorial design (FFD), eight different sample preparation protocols were tested for the relative quantification of polar metabolites from diverse chemical classes. The simultaneous investigation of several experimental factors constitutes a powerful approach to assess the impact of the metabolites extraction capabilities of each studied combination. Even if principal component analysis (PCA) constitutes often a potent tool to detect relevant variables and patterns [41], an approach dedicated to fully exploit the data structure of complex omics data generated from DOE, i.e., analysis of variance multiblock orthogonal partial least squares (AMOPLS) [42–44], was used to assess the contribution of each studied factor on the annotated metabolites.

2. Materials and Methods

2.1. Chemicals

UPLC-MS grade acetonitrile (ACN), methanol (MeOH), and water (H_2O) were obtained from Fisher Scientific (Loughborough, UK). UPLC-MS grade formic acid (FA) was supplied by Biosolve (Valkenswaard, Netherlands). Chloroform ($CHCl_3$) was provided by Acros Organics (Geel, Belgium). Ammonium hydroxide and ethylenediaminetetraacetic acid disodium salt dehydrate (EDTA) were obtained from Sigma-Aldrich (Buchs, Switzerland). The major mix IMS/T of calibration kit and leucine-encephalin were acquired from Waters (Milford, MA, USA).

2.2. Bacterial Sample Preparation

The eight evaluated sample preparation protocols are summarized in Table 1. Two cell retrieval procedures, two systems of quenching and extraction solvents, and two cell disrupting mechanisms were compared.

Table 1. Design of experiment (DOE) of the eight sample preparations, each investigated in triplicate, with differing cell retrieval systems (filter or centrifugation), quenching and extraction solvents (MeOH:H_2O 8:2 or MeOH:H_2O:$CHCl_3$ 7:2:1 + EDTA 1 mM), and cell disruption mechanisms (bead beating or freeze-thaw cycles).

Filter/Centrifugation	Solvent MeOH/Solvent CHCl$_3$	F/T Cycles/Beadbeating	Combination
+	+	+	Filter MeOH F/T
-	+	+	Centri MeOH F/T
+	-	+	Filter CHCl$_3$ F/T
-	-	+	Centri CHCl$_3$ F/T
+	+	-	Filter MeOH Beads
-	+	-	Centri MeOH Beads
+	-	-	Filter CHCl$_3$ Beads
-	-	-	Centri CHCl$_3$ Beads

2.2.1. Bacterial Cell Cultures and Cell Retrieval

C. crescentus cells were grown overnight at 30 °C in a PYE rich medium (2 g/L bactopeptone, 1 g/L yeast extract, 1 mM $MgSO_4$, and 0.5 mM $CaCl_2$). Two cell retrieval methods were tested, the first one in a liquid medium, and the second as an adaptation of a filter culture-based method [40].

For the liquid culture, overnight cultures were diluted to reach OD_{600nm} ~ 0.4 and 10 mL of cells were centrifuged at 8000 g for 5 min at 4 °C, then cells were resuspended in the desired precooled quenching and extracting solution. For metabolite extraction from filter cultures, overnight cultures were diluted at 30 °C in PYE medium until reaching OD_{600nm} ~ 0.2. Then, 10 mL of culture were transferred onto 0.22 µm mixed cellulose ester (MCE) membrane filter (Millipore) by vacuum filtration. The filters were then deposited on the surface of PYE-agar made with the same growth medium, and the cells were allowed to continue growing at 30 °C for 3 h. Finally, filters were dropped directly into desired precooled quenching and extraction solution.

2.2.2. Quenching and Extraction Methods

The two cold solvents (−20 °C) for quenching and extraction were respectively MeOH:H_2O 8:2 and MeOH:H_2O:$CHCl_3$ 7:2:1 + EDTA 1 mM. For each bacterial culture sample, 1 mL of solvent was added.

2.2.3. Cell Disruption

Two lysis methods were compared, one by freeze and thaw (F/T) and another by bead beating. Cells were subjected either to lysis by 5 cycles of F/T (−80 °C/40 °C) or to mechanical lysis using 0.1 mm zirconia beads in a Fast prep-24 (MP) disruption instrument (5 cycles of 15″ on, 1′ off) maintained at 4 °C. In both cases, cellular debris were removed by centrifugation at 14,000 g, 20 min at 4 °C. Metabolite extracts were kept at −80 °C.

2.2.4. Extract Preparation for LC-MS Analysis

The supernatants were collected, evaporated to dryness using a SpeedVac (ThermoFisher, Langenselbold, Germany) and reconstituted in 100 µL ACN:H_2O 50:50. Quality control (QC) and diluted QC (dQC) samples were respectively prepared by pooling equivalent volumes of all reconstituted samples and by 1:1 dilution of a certain amount of the QC pool with ACN:H_2O 50:50. QC and dQC were injected at regular intervals throughout the LC-MS analyses to assess analytical variability.

2.3. Liquid Chromatography Conditions

For the LC experiments, a Waters H-Class Acquity UPLC system composed of a quaternary pump, an autosampler including a 15 µL flow-through-needle injector for which temperature was set at 7 °C, and a two-way column manager (Waters, Milford, MA, USA) were used. The injected volume was 10 µL. HILIC separations were performed on a Merck SeQuant ZIC-pHILIC column (150 × 2.1 mm, 5 µm) with the appropriate guard kit (Merck KGaA, Darmstadt, Germany). Solvent A was ACN and solvent B was H_2O containing 2.8 mM ammonium formate adjusted at pH 9.00. The pH of the solution was checked and found to be stable during one week at room temperature. Column temperature and flow rate were set at 40 °C and 300 µL min^{-1}, respectively. The gradient elution was as follows: 5% B for 1 min, increasing to 51% B during 9 min, holding for 3 min at 51% B, and then returning back to 5% B in 0.1 min and re-equilibrating the column for 6.9 min.

2.4. Mass Spectrometric Conditions

HRMS was carried out on a Vion TWIMS-QTOF (Waters, Manchester, UK) equipped with an ESI source. Analyses were performed in negative ESI mode to acquire continuum data in the range of 50–1000 *m/z* with a scan time of 0.2 s. The source parameters were set as follows: capillary voltage was −2.0 kV, source and desolvation temperatures were set at 120 and 500 °C, respectively, cone and desolvation gas flow were 50 and 800 L/h, respectively. Velocity and height of StepWave1 and StepWave2 were set to 300 m/s and 5 V and to 200 m/s and 30 V, respectively. The high definition MS^E (HDMS^E, using ion mobility) settings consisted of trap wave velocity at 100 m/s; trap pulse height A at 10 V; trap pulse height B at 5 V; ion mobility spectrometry (IMS) wave velocity at 250 m/s; IMS pulse height at 45 V; wave delay set at 20 pushes, and gate delay at 0 ms. Gas flows of ion mobility instrument were set to 1.60 L/min for trap gas, and 25 mL/min for IMS gas. Buffer gas was nitrogen.

Fragmentation was performed in HDMS^E mode. For the collision energy, 6.0 eV was used for low energy and high energy was a ramp from 10 to 60 eV. Nitrogen was used as collision gas. Leucine-encephalin served as a lock mass (554.2615 *m/z* for ESI-) infused at 5 min intervals. The CCS and mass calibration of the instrument were done with the calibration mix "Major mix IMS-TOF calibration" (Waters, Manchester, UK).

2.5. Analysis of Raw Data

Chromatogram alignment, peak picking, adduct deconvolution, and feature annotation were sequentially performed on Progenesis QI v2.3 (Nonlinear Dynamics, Waters, Newcastle upon Tyne, UK). The following tolerances were used for feature annotation with regard to a set of pure reference standards (MSMLS Library of Standards, Sigma-Aldrich) measured in the same instrument: 2.5 ppm for precursor and fragment mass, 10% for Rt, and 5% in the case of CCS. Data pretreatment was performed with SUPreMe, an in-house software with capabilities for drift correction, noise filtering, and sample normalization. Finally, data were transferred to SIMCA-P 15.0 software (Umetrics, Umea, Sweden) to perform Principal Component Analysis (PCA). AMOPLS analysis was conducted after unit variance scaling as previously described [42] under the MATLAB® 8 environment (The MathWorks, Natick, MA, USA). A series of 10^4 random permutations was performed to validate the AMOPLS model and assess the statistical significance of the effects.

3. Results and Discussion

Bacterial cell cultures are an interesting resource to evaluate how a given system reacts upon the deletion of certain genes, modifications in the conditions of the growing media, or changes happening in the cell morphogenesis during the cell growth and proliferation processes. To increase our understanding of the endogenous metabolic changes appearing in such biological systems, untargeted metabolomics has already proven to be a valuable approach [39,40,45].

To study the effects of each sample preparation procedure on the extraction of polar metabolites, three parameters, namely cell retrieval, quenching and extraction solvent, and cell disruption, were systematically investigated by two methods. Medium filtering and centrifugation were compared for cell retrieval, cold MeOH:H_2O 8:2 and cold MeOH:H_2O:$CHCl_3$ 7:3:1 + EDTA 1 mM were compared as two quenching and extraction solvents and, finally, two cell disruption mechanisms (bead beating and freeze-thaw cycles) were tested. In total, the overall performance of eight different procedures was evaluated and compared against the others for the metabolomic analysis of *C. crescentus* cell cultures (Table 1).

The cell retrieval procedures were chosen on the basis of (i) their ability to separate the cells from media, (ii) their ease of application, and (iii) their recognized suitability for the determination of intracellular metabolites [46]. The quenching and extraction solvent mixtures were chosen due to their aptitude to (i) freeze the metabolic status after cell harvesting, (ii) inactivate intracellular enzymes and extract at the same time metabolites, and (iii) extract hydrophilic compounds as widely as possible with the best recoveries [46]. $CHCl_3$ was studied because of its potential usefulness to enhance cell wall disruption and to inactivate enzymes, and EDTA due to its ability to chelate metal ions preventing metabolite oxidation and degradation [30,35,37]. Cell disrupting mechanisms were selected due to their capacity to (i) disrupt cell walls and facilitate metabolite extraction [30], (ii) complete the denaturation of all enzymes to avoid further metabolite interconversions, and (iii) prevent significant degradation and chemical conversion of the extracted metabolites [47]. Finally, metabolomic analyses were performed in HILIC mode coupled to a Vion TWIMS-QTOF operating in negative ESI polarity. The choice of this UHPLC-HRMS platform enables the investigation of highly polar metabolites such as those related to the energy metabolism with excellent performance, as described in our previous paper [14]. Of course, the use of other LC-MS platforms to analyze samples issued from the same experimental design would provide different results, since the performance of each method to retain and ionize different classes of metabolites will be different.

3.1. Evaluation of the Extraction

The first goal of the investigation was to assess the recovery and the overall variability on annotated metabolites. Analyses were performed in triplicate for each combination of conditions and evaluated with the help of various procedures such as relative standard deviation (RSD) and multivariate analysis (MVA). PCA was used to evaluate the similarity and dissimilarity of the different extraction conditions, while AMOPLS was used to estimate the contribution of experimental factors to variations annotated metabolites abundances.

The performance reached by each sample preparation protocol was first assessed on the overall set of features remaining after data pretreatment. From the raw dataset, more than 8500 features were found in the initial peak-picking step. Data pretreatment included filtering of the features based on QC-to-dQC peak area ratio and RSD in QC samples, LOESS drift correction by interpolation with the QC samples, and probabilistic quotient normalization (PQN) methods [48]. These respectively eliminate noisy features, alleviate the changes in the per-metabolite variations of the instrument's response factor, and fix any sample normalization issues remaining from the sample preparation. In particular, LOESS has the advantage of being nonparametric, so no external information needs to be introduced, which would unnecessarily increase variability. The only requirement is the presence of QC samples. PQN was chosen for similar reasons, since it is highly robust and does not depend either on external information that would introduce uncertainty into the correction. Its only drawback is the need for enough variables to derive the ratio distribution, but that need is more than fulfilled in untargeted analyses such as the present one. The whole pretreatment procedure reduces the analytical variability in the data, so as to present a dataset to the statistical analyses, where as much variability as possible is of biological origin. Altogether, a total of 904 filtered and corrected features were found to be analytically reliable across all the samples in the experimental design, and they were retained for the posterior analysis (Supplementary Materials, Table S1).

3.2. Metabolites Annotation

For the purpose of completing the LC-MS untargeted workflow of such experiments, feature annotation must be used to assess which metabolites are preferentially retrieved by each sample preparation procedure, thus enabling the study of certain metabolic pathways in future experiments with *C. crescentus*. In the present study, features were first annotated by using their *m/z* and retention time [15] and their identities were then confirmed by MS/MS or CCS data. MS/MS acquisitions were carried out in all ion fragmentation (AIF) mode by using an energy ramp (10–60 eV). The AIF mode presents the potential that a simultaneous acquisition of both precursor and product ions can be considered in the same analytical run. However, AIF suffers from inaccuracies, which can happen when LC coeluting peaks with similar peak shapes possess similar MS/MS spectra (common product ions). This limitation can be overcome by (i) computational deconvolution algorithms that link the precursor ions (at low energy) and the product ions (at high energy) [49] and (ii) the IMS technology, i.e., HDMS^E, which allows the separation of LC coeluting precursor ions on the drift time dimension before fragmentation. Therefore precursor and product ions can be IMS-aligned, allowing to filter out fragments that do not match the precursor's drift time [18]. Moreover, IMS offers the potential to gain further structural information by providing CCS values, and therefore a gain in confidence in the metabolites annotation. An example of how both precursor and product ions can be cleared with the help of the IMS technology is presented in Figure 1 with the example of acetyl-CoA. Cleaner MS/MS also reduces the false positives rate during the annotation step. Among the 904 kept features, 133 of them were annotated with the help of *m/z* and retention time and 106 were confirmed by MS/MS or CCS values (Table S2). This result highlights the fact that false positive annotations (27 unconfirmed annotations) can occur when they are not validated by other orthogonal molecular descriptors. Most of the 106 identified metabolites (almost 55% of the confirmed compounds) belong to one of the following chemical groups: amino acids, nucleotides, and organic acids [50], as presented in Figure 2. This shows the ability of the developed analytical method to detect compounds related to fundamental metabolic pathways in biological samples, such as the energy metabolism.

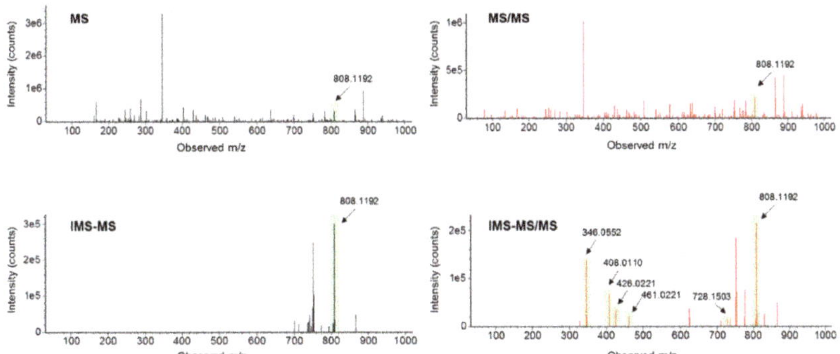

Figure 1. Example of MS and MS/MS spectra without and with IMS filtration for the precursor ion of acetyl-CoA (M-H). Precursor ion and fragment ions are highlighted in filled bars. Cleaner MS and MS/MS spectra are obtained thanks to the IMS separation. Fragment ions were matched to the MS/MS spectra of the chemical standard analyzed under the same analytical conditions.

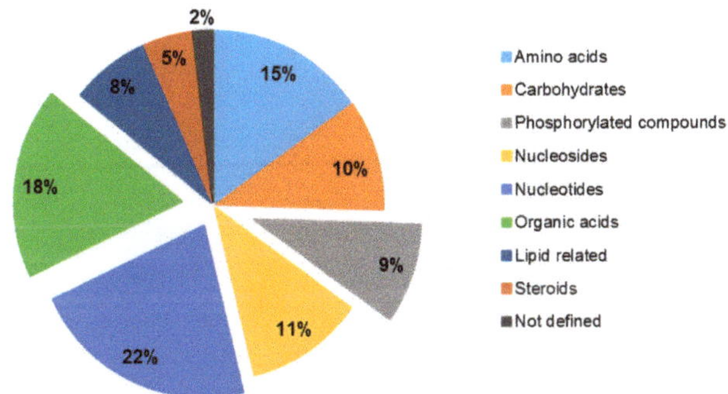

Figure 2. Annotated metabolites chemical groups. Percentage based on the total number of annotated metabolites (106).

3.3. AMOPLS for Assessing the Contribution of Each Studied Factor to the Overall Variability

To obtain an overview of the major sources of variability in the dataset, unsupervised data analysis was first performed on the 106 annotated metabolites, as measured from the eight sample preparation procedures. According to the PCA score plot presented in Figure 3, the sample preparation approaches that appear well clustered, indicating low intragroup variability, are: centrifugation-CHCl$_3$-F/T, centrifugation-MeOH-F/T, and filter-CHCl$_3$-F/T. An important difference in the sample preparation between centrifugation and filtering can also be highlighted. Both centrifugation-CHCl$_3$-F/T and centrifugation-MeOH-F/T are separated, while filter procedures are regrouped at the top-left region of the PCA.

As a second step of data analysis, AMOPLS was used to get a deeper insight into the specific contribution of each experimental factor (i.e. main effects), and their potential interactions, to the total observed variability. This approach allows us to decompose, quantify, and evaluate the significance of each effect in large datasets issued from full factorial designs. The proposed experimental setup allows the simultaneous evaluation of the three experimental factors under study, i.e., the cell retrieval procedure, the quenching and extraction solvent, and cell disrupting mechanisms, as well as their interactions. The interpretation of the signal variations is performed using effect-specific scores and loadings.

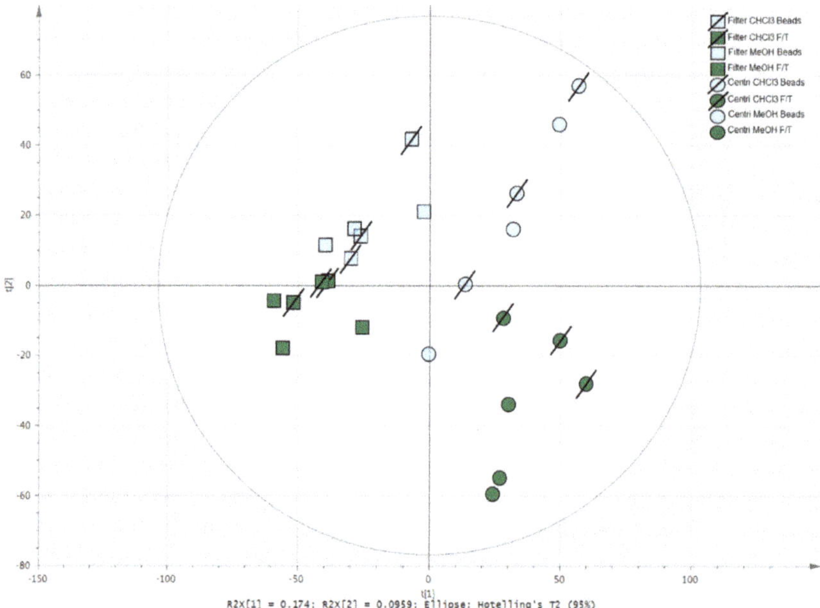

Figure 3. Principal component analysis (PCA) score plot of the eight sample preparation procedures evaluated in triplicates.

The results of the AMOPLS model are presented in Table 2. All main effects had a significant contribution to the total observed variability, but no interaction between them was found to be statistically significant. The cell retrieval procedure accounted for 27.6% of the total variability, the quenching and extraction solvent for 8.4%, and the cell disrupting mechanism for 7.0%. Remarkably, the cell retrieval procedure (filter/centrifugation) is the most influential parameter. The residuals accounted for 57.0% of the total observed variability, suggesting an important contribution of other sources of variation not formally considered in the experimental design and possibly coming from the biological variability or the sample preparation itself.

Table 2. Relative variability and block contributions of the AMOPLS model of the data acquired from the investigated biological samples. RSR: residual structure ratio, tp1-3: predictive components, to: orthogonal component.

Effect	Contribution	RSR	*p*-Value	tp1	tp2	tp3	to
Cell retrieval	27.6%	1.92	0.2%	96.7% [1]	3.9%	1.8%	15.9%
Quenching/ extraction solvent	8.4%	1.17	0.1%	1.0%	81.8%	3.1%	26.7%
Cell disrupting mechanisms	7.0%	1.14	0.4%	1.0%	6.7%	91.6%	26.9%
Residuals	57.0%	1.00	N/A	1.2%	7.6%	3.5%	30.5%

[1] The highest contribution for each component is reported in bold.

The model was then interpreted with the help of the distribution of the observations (groupings) and the contribution of the variables to the specific predictive component associated with each main effect. The score plots are presented in Figure S1, generated with data coming from Table S3. Squared variable importance in the projection (VIP^2) values were then calculated (Table S3) and taken as the quantitative measure of the contribution of all modelled effects coming from a single variable [44].

VIP^2 values are helpful to rank compounds and spot the ones that play a major role in the model for each studied effect. Figure 4 presents the VIP^2 values for each studied effect, for the 50 most relevant metabolites out of the 106 annotated ones. Larger VIP^2 values indicate which effect(s) are the most important ones for each annotated metabolite. In Figure 4, metabolites are ranked according to the importance of the cell retrieval effect and compounds on the left side are those for which this procedure had the strongest impact (largest VIP^2 values). On the contrary, this effect exerted little influence on metabolites appearing towards the right side of the plot (Figure S2). Some metabolites belonging to the groups of sugars (2-deoxy-glucose, glucose, and xylose) and nucleosides (cytidine, guanine, and xanthine) are among the most influenced ones. Besides the scores and corresponding loading plots (Figures S1 and S3, representing the data from Tables S4 and S5, respectively), these metabolites were found in higher abundance in the filtering procedure. Secondly, highly polar metabolites related to the amino acid family (L-histidine, L-glutamine, and L-asparagine), phosphorylated sugars (ribose-5-phosphate, glucose-6-phosphate, fructose-1, and 5-bisphosphate) and nucleotides (5′-CMP, IDP, IMP, and GTP) were highly influenced by the quenching and extraction solvent. Probably, the polarity of the used solvents accounts for these observed results, since MeOH:H$_2$O 8:2 is more polar than the second system used, thus enhancing the recovery of more polar molecules. Indeed, by looking at the scores and corresponding loading plots (Figures S1 and S3), almost all metabolites given as examples were found in higher abundance in MeOH:H$_2$O extracting solvent, except IDP, GTP, and fructose-1,5-bisphosphate which were found in higher abundance when CHCl$_3$ was present, an observation that can be explained by an enhanced enzyme-denaturing process expected when this solvent was added. Finally, the effect of cells lysis was studied. Metabolites related to lipids (petroselinic acid, palmitic acid, and stearic acid) and nucleotides (ADP, CTP, GDP, CDP, GTP, and ATP) were the most influenced chemical classes. All of these metabolites were found in higher abundance in the F/T procedure except for palmitic acid and petroselinic acid (Figures S1 and S3). One hypothesis to explain these results would be the fact that both cell disrupting mechanisms are quite different in terms of the amount of energy applied to the sample. Indeed, beads beating is a harsher procedure compared to F/T cycles and metabolites amenable to degradation via hydrolysis, decarboxylation or oxidation could be more easily impacted.

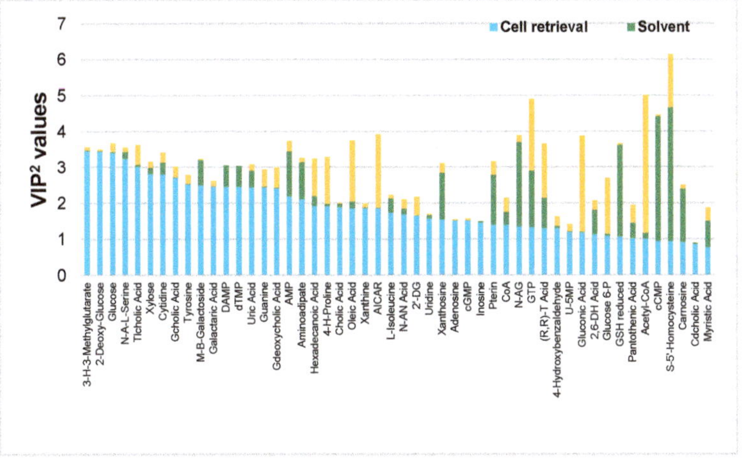

Figure 4. Effect-specific variable importance in the projection (VIP)^2 values for the 50 out of the 106 annotated metabolites ranked according to the impact of the cell retrieval effect.

3.4. Assessing the Recovery and the Variability of the Annotated Metabolites

A larger number of metabolites were found in higher abundance with the following combination of factors: cell filtering, MeOH:H_2O, and F/T cycles (Figure S3). First, nucleosides (guanine, guanosine, cytidine, inosine, xanthosine, and xanthine) were more abundant in sample preparations starting with a filtering step, regardless of what other following procedures were used. One explanation could be that these metabolites leak from cells more easily during the longer centrifugation procedure, as it has already been found for *Escherichia coli* [51]. Second, the MeOH:H_2O condition enables the retrieval of more metabolites in higher proportion. This can be explained by the medium's polarity, since this solvent enables a better extraction of polar compounds as compared to the $CHCl_3$ system. However, it was expected that the addition of $CHCl_3$ and EDTA would prevent metabolite turnover, and oxidation and degradation by metals for nucleotides, CoA, and CoA thioesters derivatives, as described in the literature [37]. This is confirmed for ATP, succinyl-CoA, and CoA which are found in higher proportion in the presence of $CHCl_3$ (Figure S3). Interestingly, ADP was found in higher proportion in MeOH. This result supports that in these conditions ATP gets hydrolyzed into ADP, confirming that EDTA and $CHCl_3$ prevents the degradation of ATP into ADP. For the cell disruption mechanisms, a large majority of metabolites were found in higher abundance in the F/T cycles procedure (Figure S3). Indeed, some nucleotides, CoA and CoA thioester derivatives, and some organic acids (citric acid), which are highly relevant indicators of the metabolic status of the cells, are found more abundantly when F/T was applied. This could be explained by the facilitation of non-enzymatic degradations in the case of bead beating, as previously discussed. Finally, acetyl-CoA, CoA, TMP, and DAMP, were more abundant in either centrifugation-$CHCl_3$-F/T or centrifugation-MeOH:H_2O-F/T. It can be seen that all four metabolites are more abundant after using centrifugation as cell retrieval, MeOH:H_2O (except for CoA, found at the same abundance for both solvent systems used) and F/T cycles (except for TMP and DAMP, found at the same abundance for both cells disrupting mechanisms used). These results support that either centrifugation-$CHCl_3$-F/T or centrifugation-MeOH:H_2O-F/T conditions should be chosen in the case of interest towards these particular metabolites.

The recovery obtained using each sample preparation procedure was assessed on the 106 confirmed metabolites and results are shown in Figure S4. Metabolite integration was manually curated among the features of the preprocessed data and considered as undetected when an intensity value of zero was found (total absence). The recoveries of each sample preparation varied from 88% to 95%. It has been noticed that for each different sample preparation, more metabolites were recovered when MeOH:H_2O solvent was used as compared to MeOH:H_2O:$CHCl_3$ + 1 mM EDTA. The sample preparation Filter-MeOH:H_2O-F/T had the highest recovery percentage (95%) and centrifugation-$CHCl_3$-F/T yielded the lowest recovery percentage (88%). Figure S4 also shows that more metabolites can be recovered when the filtering procedure is used, as opposed to centrifugation.

On the other hand, the RSD is crucial since the most repeatable procedures will guarantee data of higher quality in further metabolomics experiments. It was possible to calculate the RSD on the area of each metabolite obtained in triplicate. These results are presented in Figure 5. The RSD values (X) were classified into five categories: $0 < X \leq 10\%$, $10 < X \leq 20\%$, $20 < X \leq 30\%$, $30 < X \leq 40\%$, or >40%. The sample preparation extracting the largest number of metabolites, i.e., filter-MeOH:H_2O-F/T, presents the worst RSD (>40%) results. On the other hand, we found sample preparation such as centrifugation-$CHCl_3$-F/T and centrifugation-MeOH:H_2O-F/T, encompassing RSDs of $0 < X \leq 10\%$, $10 < X \leq 20\%$ and $20 < X \leq 30\%$ for the highest number of metabolites. Therefore, we would recommend use of the latter two sample preparation procedures when the metabolite extraction rate must be well balanced with repeatability.

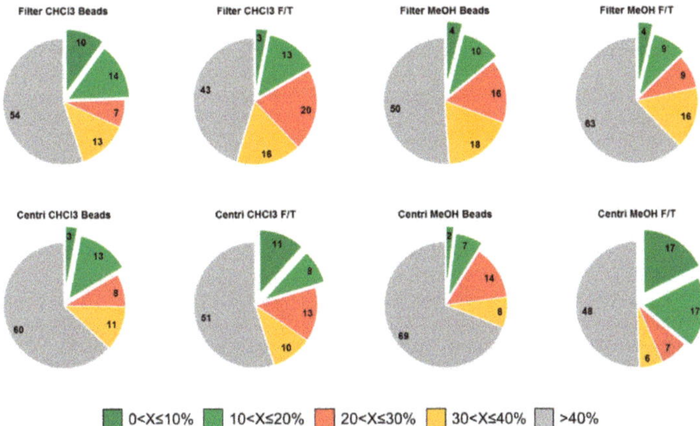

Figure 5. Relative standard deviation (RSD) values calculated on the area reached by each detected metabolite for the eight sample preparations investigated. RSD (X) are classified into 5 categories: $0 < X \leq 10\%$, $10 < X \leq 20\%$, $20 < X \leq 30\%$, $30 < X \leq 40\%$, or $>40\%$.

4. Conclusions

It is currently recognized that sample preparation is a crucial step in untargeted metabolomics workflows as biological outputs will only be drawn from retrieved metabolites. In this work, eight different sample preparation protocols for Gram-negative bacteria *C. crescentus*, combining two cell retrieval systems, two quenching and extraction solvents, and two cell disruption procedures were investigated. The use of a full factorial design to evaluate the contribution of various factors, with the help of an AMOPLS model, provided a potent way to examine their effect on the metabolite recovery rates. Moreover, annotated metabolite variability was assessed with the help of univariate data analysis via the RSD of the area of each metabolite.

The major conclusions about the studied experimental parameters are:

(a) For cell harvesting, the centrifugation procedure could lead to a higher number of metabolites leaking out of the cells as compared to filtering, as the latter is faster, preventing metabolite turnover and degradation and leakage which could take place during centrifugation.

(b) For quenching and extraction solvent, the combination MeOH:H_2O enables a better extraction of polar metabolites.

(c) For cell disruption, bead beating could lead to higher temperature and more degradation because of its rougher nature. The mechanical disruption obtained by this method is often very advantageous when dealing with tissue fractions, but unnecessary in the studied samples, as it leads only to a larger degradation rate.

To summarize, even if filtering enables the retrieval of more metabolites with higher abundances, RSD evaluation demonstrated a detrimental extraction variability for any of the conditions, as compared to the centrifugation procedures. Therefore, centrifugation should be the selected method for cell retrieval. For the quenching and extraction solvent, the choice should be driven by the needs of the study regarding the specific nature of the metabolites to extract. If the target would be acetyl-CoA, an important metabolite for many biological pathways, MeOH should be used. The F/T cycles retrieve more metabolites at higher abundance and present much better RSD values as compared to bead beating for the cell disrupting mechanism. Finally, the combination of centrifugation-MeOH-F/T appears to perform the best among those evaluated in the present study.

Supplementary Materials: The following supplementary materials are available online at http://www.mdpi.com/2218-1989/9/10/193/s1, Table S1: Analytical variability of the 904 retained features after data filtration and processing. More than 90% of them presented an RSD below 30% in the QC samples; Table S2: 106 annotated metabolites' names, adapted name, observed Rt, m/z, CCS, adduct(s), CAS identifier, molecular properties confirmation steps (Rt, MS/MS, CCS, MS/MS, and CCS), and chemical class; Table S3: Ranking of the variables accounting for the cells retrieval effect. The VIP2 values of all metabolites were calculated for each variable (cells retrieval, solvent, and cells disruption); Figure S1: Score plots obtained from the AMOPLS model related to the predictive components bearing the variability coming from respectively (a) cells retrieval (tp1), (b) solvent (tp2), and (c) cells disruption (tp3) effects and the orthogonal component to. Plots were generated according to Table S4; Table S4: Data from the AMOPLS model for the predictive components tp1 (cells retrieval), tp2 (solvent), tp3 (cells disruption), and to (orthogonal component); Figure S2: Effect-specific VIP2 values for the 106 annotated metabolites ranked according to the impact of the cell retrieval effect; Figure S3: Loading plots generated from the AMOPLS predictive components related to the effect of (a) cells retrieval, (b) solvent, and (c) cells disruption on the metabolites extraction. Plots were generated according to Table S5; Table S5: Data from the AMOPLS model for the predictive components pp1 (cells retrieval), pp2 (solvent), pp3 (cells disruption), and po (orthogonal component); Figure S4: Metabolites overall recovery for each sample preparation evaluated. Percentage based on the number of metabolites detected for each condition on the total annotated metabolites (106).

Author Contributions: Conceptualization, J.P., M.B., and V.G.-R.; methodology, J.P., M.B., J.B., S.C., Y.G., and V.G.-R.; software, J.P., J.B., and S.C.; validation, P.H.V. and S.R.; formal analysis, J.P., M.B., J.B., Y.G., and V.G.-R.; investigation, J.P., M.B., S.C., and V.G.-R.; resources, P.H.V. and S.R.; data curation, J.P., J.B., S.C., Y.G., and V.G.-R.; writing—original draft preparation, J.P., M.B., and V.G.-R.; writing—review and editing, J.P., M.B., J.B., S.C., Y.G., P.H.V., V.G.-R., and S.R.; visualization, J.P., J.B., and V.G.-R.; supervision, P.H.V. and S.R.; project administration, P.H.V. and S.R.; funding acquisition, P.H.V. and S.R.

Funding: This research was funded by the Swiss National Science Foundation through grant 31003A_166658 for support through a fellowship to S.C. and by the Swiss Centre for Applied Human Toxicology through grants from the Research Programme 2017–2020 (Core Project 3: Neurotoxicity).

Acknowledgments: Julien Bourquin is warmly thanked for his valuable advice on the Vion TWIMS-QTOF instrument.

Conflicts of Interest: The authors declare no conflict of interest. The funders had no role in the design of the conducted study; in the collection, analyses or interpretation of the data; in the writing of the manuscript or in the decision to publish the results.

References

1. Miggiels, P.; Wouters, B.; Van Westen, G.J.; Dubbelman, A.-C.; Hankemeier, T. Novel technologies for metabolomics: More for less. *TrAC Trends Anal. Chem.* **2018**. [CrossRef]

2. Nemkov, T.; Hansen, K.C.; D'Alessandro, A. A Three-Minute Method for high-throughput quantitative metabolomics and quantitative tracing experiments of central carbon and nitrogen pathways. *Rapid Commun. Mass Spectrom.* **2017**, *31*, 663–673. [CrossRef] [PubMed]

3. Gallart-Ayala, H.; Konz, I.; Mehl, F.; Teav, T.; Oikonomidi, A.; Peyratout, G.; Van Der Velpen, V.; Popp, J.; Ivanisevic, J. A global HILIC-MS approach to measure polar human cerebrospinal fluid metabolome: Exploring gender-associated variation in a cohort of elderly cognitively healthy subjects. *Anal. Chim. Acta* **2018**, *1037*, 327–337. [CrossRef] [PubMed]

4. Boudah, S.; Olivier, M.-F.; Aros-Calt, S.; Oliveira, L.; Fenaille, F.; Tabet, J.-C.; Junot, C. Annotation of the human serum metabolome by coupling three liquid chromatography methods to high-resolution mass spectrometry. *J. Chromatogr. B* **2014**, *966*, 34–47. [CrossRef] [PubMed]

5. Periat, A.; Boccard, J.; Veuthey, J.-L.; Rudaz, S.; Guillarme, D. Systematic comparison of sensitivity between hydrophilic interaction liquid chromatography and reversed phase liquid chromatography coupled with mass spectrometry. *J. Chromatogr. A* **2013**, *1312*, 49–57. [CrossRef]

6. Periat, A.; Krull, I.S.; Guillarme, D. Applications of hydrophilic interaction chromatography to amino acids, peptides, and proteins. *J. Sep. Sci.* **2015**, *38*, 357–367. [CrossRef]

7. Sampsonidis, I.; Witting, M.; Koch, W.; Virgiliou, C.; Gika, H.G.; Schmitt-Kopplin, P.; Theodoridis, G.A. Computational analysis and ratiometric comparison approaches aimed to assist column selection in hydrophilic interaction liquid chromatography–tandem mass spectrometry targeted metabolomics. *J. Chromatogr. A* **2015**, *1406*, 145–155. [CrossRef]

8. Virgiliou, C.; Sampsonidis, I.; Gika, H.G.; Raikos, N.; Theodoridis, G.A. Development and validation of a HILIC- MS/MS multi-targeted method for metabolomics applications. *Electrophoresis* **2015**, *36*, 2215–2225. [CrossRef]

9. Zhang, T.; Creek, D.J.; Barrett, M.P.; Blackburn, G.; Watson, D.G. Evaluation of Coupling Reversed Phase, Aqueous Normal Phase, and Hydrophilic Interaction Liquid Chromatography with Orbitrap Mass Spectrometry for Metabolomic Studies of Human Urine. *Anal. Chem.* **2012**, *84*, 1994–2001. [CrossRef]

10. Zhang, T.; Watson, D.G. High Performance Liquid Chromatographic Approaches to Mass Spectrometry Based Metabolomics. *Curr. Metab.* **2012**, *1*, 58–83.

11. Zhang, R.; Watson, D.G.; Wang, L.; Westrop, G.D.; Coombs, G.H.; Zhang, T. Evaluation of mobile phase characteristics on three zwitterionic columns in hydrophilic interaction liquid chromatography mode for liquid chromatography-high resolution mass spectrometry based untargeted metabolite profiling of Leishmania parasites. *J. Chromatogr. A* **2014**, *1362*, 168–179. [CrossRef]

12. Wernisch, S.; Pennathur, S. Evaluation of coverage, retention patterns and selectivity of seven liquid chromatographic methods for metabolomics. *Anal. Bioanal. Chem.* **2016**, *408*, 6079–6091. [CrossRef]

13. Periat, A.; Guillarme, D.; Veuthey, J.-L.; Boccard, J.; Moco, S.; Barron, D.; Grand-Guillaume-Perrenoud, A. Optimized selection of liquid chromatography conditions for wide range analysis of natural compounds. *J. Chromatogr. A* **2017**, *1504*, 91–104. [CrossRef]

14. Pezzatti, J.; González-Ruiz, V.; Codesido, S.; Gagnebin, Y.; Joshi, A.; Guillarme, D.; Schappler, J.; Picard, D.; Boccard, J.; Rudaz, S.; et al. A scoring approach for multi-platform acquisition in metabolomics. *J. Chromatogr. A* **2019**, *1592*, 47–54. [CrossRef]

15. Schymanski, E.L.; Jeon, J.; Gulde, R.; Fenner, K.; Ruff, M.; Singer, H.P.; Hollender, J. Identifying Small Molecules via High Resolution Mass Spectrometry: Communicating Confidence. *Environ. Sci. Technol.* **2014**, *48*, 2097–2098. [CrossRef]

16. Rochat, B. Proposed Confidence Scale and ID Score in the Identification of Known-Unknown Compounds Using High Resolution MS Data. *J. Am. Soc. Mass Spectrom.* **2017**, *28*, 709–723. [CrossRef]

17. Blaženović, I.; Kind, T.; Ji, J.; Fiehn, O. Software Tools and Approaches for Compound Identification of LC-MS/MS Data in Metabolomics. *Metabolites* **2018**, *8*, 31. [CrossRef]

18. Paglia, G.; Williams, J.P.; Menikarachchi, L.; Thompson, J.W.; Tyldesley-Worster, R.; Halldorsson, S.; Rolfsson, O.; Moseley, A.; Grant, D.; Langridge, J.; et al. Ion Mobility Derived Collision Cross Sections to Support Metabolomics Applications. *Anal. Chem.* **2014**, *86*, 3985–3993. [CrossRef]

19. Ma, X.; Liu, J.; Zhang, Z.; Bo, T.; Bai, Y.; Liu, H. Drift tube ion mobility and four-dimensional molecular feature extraction enable data-independent tandem mass spectrometric 'omics' analysis without quadrupole selection. *Rapid Commun. Mass Spectrom.* **2017**, *31*, 33–38. [CrossRef]

20. Mairinger, T.; Causon, T.J.; Hann, S. The potential of ion mobility–mass spectrometry for non-targeted metabolomics. *Curr. Opin. Chem. Biol.* **2018**, *42*, 9–15. [CrossRef]

21. Hernández-Mesa, M.; Le Bizec, B.; Monteau, F.; García-Campaña, A.M.; Dervilly-Pinel, G. Collision Cross Section (CCS) Database: An Additional Measure to Characterize Steroids. *Anal. Chem.* **2018**, *90*, 4616–4625. [CrossRef]

22. Dwivedi, P.; Schultz, A.J.; Hill, H.H. Metabolic Profiling of Human Blood by High Resolution Ion Mobility Mass Spectrometry (IM-MS). *Int. J. Mass Spectrom.* **2010**, *298*, 78–90. [CrossRef]

23. Want, E.J.; O'Maille, G.; Smith, C.A.; Brandon, T.R.; Uritboonthai, W.; Qin, C.; Trauger, S.A.; Siuzdak, G. Solvent-Dependent Metabolite Distribution, Clustering, and Protein Extraction for Serum Profiling with Mass Spectrometry. *Anal. Chem.* **2006**, *78*, 743–752. [CrossRef]

24. Michopoulos, F.; Lai, L.; Gika, H.; Theodoridis, G.; Wilson, I. UPLC-MS-Based Analysis of Human Plasma for Metabonomics Using Solvent Precipitation or Solid Phase Extraction. *J. Proteome Res.* **2009**, *8*, 2114–2121. [CrossRef]

25. Bruce, S.J.; Tavazzi, I.; Parisod, V.; Rezzi, S.; Kochhar, S.; Guy, P.A. Investigation of Human Blood Plasma Sample Preparation for Performing Metabolomics Using Ultrahigh Performance Liquid Chromatography/Mass Spectrometry. *Anal. Chem.* **2009**, *81*, 3285–3296. [CrossRef]

26. Yang, Y.; Cruickshank, C.; Armstrong, M.; Mahaffey, S.; Reisdorph, R.; Reisdorph, N. New sample preparation approach for mass spectrometry-based profiling of plasma results in improved coverage of metabolome. *J. Chromatogr. A* **2013**, *1300*, 217–226. [CrossRef]

27. Rico, E.; Gonzalez, O.; Blanco, M.E.; Alonso, R.M. Evaluation of human plasma sample preparation protocols for untargeted metabolic profiles analyzed by UHPLC-ESI-TOF-MS. *Anal. Bioanal. Chem.* **2014**, *406*, 7641–7652. [CrossRef]

28. Lorenz, M.A.; Burant, C.F.; Kennedy, R.T. Reducing Time and Increasing Sensitivity in Sample Preparation for Adherent Mammalian Cell Metabolomics. *Anal. Chem.* **2011**, *83*, 3406–3414. [CrossRef]

29. Panopoulos, A.D.; Yanes, O.; Ruiz, S.; Kida, Y.S.; Diep, D.; Tautenhahn, R.; Herrerías, A.; Batchelder, E.M.; Plongthongkum, N.; Lutz, M.; et al. The metabolome of induced pluripotent stem cells reveals metabolic changes occurring in somatic cell reprogramming. *Cell Res.* **2012**, *22*, 168–177. [CrossRef]

30. León, Z.; García-Cañaveras, J.C.; Donato, M.T.; Lahoz, A.; García-Cañaveras, J.C. Mammalian cell metabolomics: Experimental design and sample preparation. *Electrophoresis* **2013**, *34*, 2762–2775. [CrossRef]

31. Kapoore, R.V.; Coyle, R.; Staton, C.A.; Brown, N.J.; Vaidyanathan, S. Cell line dependence of metabolite leakage in metabolome analyses of adherent normal and cancer cell lines. *Metabolomics* **2015**, *11*, 1743–1755. [CrossRef]

32. Rabinowitz, J.D.; Kimball, E. Acidic acetonitrile for cellular metabolome extraction from Escherichia coli. *Anal. Chem.* **2007**, *79*, 6167–6173. [CrossRef]

33. Spura, J.; Reimer, L.C.; Wieloch, P.; Schreiber, K.; Buchinger, S.; Schomburg, D. A method for enzyme quenching in microbial metabolome analysis successfully applied to gram-positive and gram-negative bacteria and yeast. *Anal. Biochem.* **2009**, *394*, 192–201. [CrossRef]

34. Xu, Y.-J.; Wang, C.; Ho, W.E.; Ong, C.N. Recent developments and applications of metabolomics in microbiological investigations. *TrAC Trends Anal. Chem.* **2014**, *56*, 37–48. [CrossRef]

35. Patejko, M.; Jacyna, J.; Markuszewski, M.J. Sample preparation procedures utilized in microbial metabolomics: An overview. *J. Chromatogr. B* **2017**, *1043*, 150–157. [CrossRef]

36. Dettmer, K.; Nurnberger, N.; Kaspar, H.; Gruber, M.A.; Almstetter, M.F.; Oefner, P.J. Metabolite extraction from adherently growing mammalian cells for metabolomics studies: Optimization of harvesting and extraction protocols. *Anal. Bioanal. Chem.* **2011**, *399*, 1127–1139. [CrossRef]

37. Siegel, D.; Permentier, H.; Reijngoud, D.-J.; Bischoff, R. Chemical and technical challenges in the analysis of central carbon metabolites by liquid-chromatography mass spectrometry. *J. Chromatogr. B* **2014**, *966*, 21–33. [CrossRef]

38. Lu, W.; Su, X.; Klein, M.S.; Lewis, I.A.; Fiehn, O.; Rabinowitz, J.D. Metabolite Measurement: Pitfalls to Avoid and Practices to Follow. *Annu. Rev. Biochem.* **2017**, *86*, 277–304. [CrossRef]

39. Vuckovic, D. Current trends and challenges in sample preparation for global metabolomics using liquid chromatography–mass spectrometry. *Anal. Bioanal. Chem.* **2012**, *403*, 1523–1548. [CrossRef]

40. Irnov, I.; Wang, Z.; Jannetty, N.D.; Bustamante, J.A.; Rhee, K.Y.; Jacobs-Wagner, C. Crosstalk between the tricarboxylic acid cycle and peptidoglycan synthesis in Caulobacter crescentus through the homeostatic control of alpha-ketoglutarate. *PLoS Genet* **2017**, *13*, e1006978. [CrossRef]

41. Boccard, J.; Veuthey, J.-L.; Rudaz, S. Knowledge discovery in metabolomics: An overview of MS data handling. *J. Sep. Sci.* **2010**, *33*, 290–304. [CrossRef] [PubMed]

42. Boccard, J.; Rudaz, S. Exploring Omics data from designed experiments using analysis of variance multiblock Orthogonal Partial Least Squares. *Anal. Chim. Acta* **2016**, *920*, 18–28. [CrossRef] [PubMed]

43. Ponzetto, F.; Boccard, J.; Baume, N.; Kuuranne, T.; Rudaz, S.; Saugy, M.; Nicoli, R. High-resolution mass spectrometry as an alternative detection method to tandem mass spectrometry for the analysis of endogenous steroids in serum. *J. Chromatogr. B* **2017**, *1052*, 34–42. [CrossRef] [PubMed]

44. González-Ruiz, V.; Pezzatti, J.; Roux, A.; Stoppini, L.; Boccard, J.; Rudaz, S. Unravelling the effects of multiple experimental factors in metabolomics, analysis of human neural cells with hydrophilic interaction liquid chromatography hyphenated to high resolution mass spectrometry. *J. Chromatogr. A* **2017**, *1527*, 53–60. [CrossRef] [PubMed]

45. Tang, J. Microbial metabolomics. *Curr. Genom.* **2011**, *12*, 391–403. [CrossRef]

46. Shen, Y.; Fatemeh, T.; Tang, L.; Cai, Z. Quantitative metabolic network profiling of Escherichia coli: An overview of analytical methods for measurement of intracellular metabolites. *TrAC Trends Anal. Chem.* **2016**, *75*, 141–150. [CrossRef]

47. van Gulik, W.M.; Canelas, A.B.; Seifar, R.M.; Heijnen, J.J. The Sampling and Sample Preparation Problem in Microbial Metabolomics. In *Metabolomics in Practice: Successful Strategies to Generate and Analyze Metabolic Data*; Lämmerhofer, M., Weckwerth, W., Eds.; Wiley-VCH Verlag Gmbh & Co: Weinheim, Germany, 2013; pp. 1–19.

48. Gagnebin, Y.; Tonoli, D.; Lescuyer, P.; Ponte, B.; De Seigneux, S.; Martin, P.-Y.; Schappler, J.; Boccard, J.; Rudaz, S. Metabolomic analysis of urine samples by UHPLC-QTOF-MS: Impact of normalization strategies. *Anal. Chim. Acta* **2017**, *955*, 27–35. [CrossRef]
49. Cajka, T.; Fiehn, O. Toward Merging Untargeted and Targeted Methods in Mass Spectrometry-Based Metabolomics and Lipidomics. *Anal. Chem.* **2016**, *88*, 524–545. [CrossRef]
50. Wishart, D.S.; Feunang, Y.D.; Marcu, A.; Guo, A.C.; Liang, K.; Vazquez-Fresno, R.; Sajed, T.; Johnson, D.; Li, C.; Karu, N.; et al. HMDB 4.0: The human metabolome database for 2018. *Nucleic Acids Res.* **2018**, *46*, D608–D617. [CrossRef]
51. Nikaido, H.; Bavoil, P.; Hirota, Y. Outer membranes of gram-negative bacteria. XV. Transmembrane diffusion rates in lipoprotein-deficient mutants of Escherichia coli. *J. Bacteriol.* **1977**, *132*, 1045–1047.

Article

Influence of Drying Method on NMR-Based Metabolic Profiling of Human Cell Lines

Irina Petrova, Shenyuan Xu, William C. Joesten, Shuisong Ni and Michael A. Kennedy *

Department of Chemistry and Biochemistry, Miami University, Oxford, OH 45056, USA;
petrovi@miamioh.edu (I.P.); xus2@miamioh.edu (S.X.); joestewc@miamioh.edu (W.C.J.); nis@miamioh.edu (S.N.)
* Correspondence: kennedm4@miamioh.edu; Tel.: +1-513-529-8267

Received: 30 August 2019; Accepted: 28 October 2019; Published: 31 October 2019

Abstract: Metabolic profiling of cell line and tissue extracts involves sample processing that includes a drying step prior to re-dissolving the cell or tissue extracts in a buffer for analysis by GC/LC-MS or NMR. Two of the most commonly used drying techniques are centrifugal evaporation under vacuum (SpeedVac) and lyophilization. Here, NMR spectroscopy was used to determine how the metabolic profiles of hydrophilic extracts of three human pancreatic cancer cell lines, MiaPaCa-2, Panc-1 and AsPC-1, were influenced by the choice of drying technique. In each of the three cell lines, 40–50 metabolites were identified as having statistically significant differences in abundance in redissolved extract samples depending on the drying technique used during sample preparation. In addition to these differences, some metabolites were only present in the lyophilized samples, for example, n-methyl-α-aminoisobutyric acid, n-methylnicotimamide, sarcosine and 3-hydroxyisovaleric acid, whereas some metabolites were only present in SpeedVac dried samples, for example, trimethylamine. This research demonstrates that the choice of drying technique used during the preparation of samples of human cell lines or tissue extracts can significantly influence the observed metabolome, making it important to carefully consider the selection of a drying method prior to preparation of such samples for metabolic profiling.

Keywords: metabonomics; metabolomics; metabolic profiling; NMR; nuclear magnetic resonance spectroscopy; cell line; human cell line; MiaPaCa-2; Panc-1; AsPC-1

---

1. Introduction

Metabonomics is a metabolic profiling technique first described in 1999 [1]. Metabonomics studies, which can be used for biomarker identification for early disease detection and drug development, are accomplished by examining the differences in metabolic profiles of biological fluids between control samples and samples obtained under some form of transformative change or intrinsic difference such as the presence, progression or treatment of disease [2–5]. Metabonomics has been employed in food science to assess food quality and in nutrition studies to identify diet–health relationships by examination of dietary biomarkers [6–10], in fields such as toxicology and immunology to understand the mechanisms of chemical toxicity and viral infections [11,12], to determine how the use of oral antibiotics alters the gut microbiome based on analysis of fecal extracts [13,14], and to assess the consequences of different forms of acute kidney injury [15–17], among many other applications. A recent overview of applications of metabonomic studies was published by Liu and Locasale in 2017 [18]. Metabonomics research has grown exponentially since its inception in 1999, as can be appreciated by considering the number of publications per year determined from a recent literature search of the PubMed citation database (www.ncbi.nlm.nih.gov/pubmed) with the search terms "metabonomics OR metabolomics" specified in the "Title/Abstract" field, which returned over 44,000 publications overall and more than 3000 publications in 2018 alone. Biological fluids, cell line cultures and tissue extracts are common samples used in metabonomics analyses [3,14,16,19–21]. Metabolic profiling studies of human cancer cell lines in particular have played a major role in drug discovery as they allow for comparatively inexpensive high-throughput screening to reveal potential drug targets and to study metabolic responses to various stimuli [21–24]. Metabonomics studies in general can either be targeted, with a focus on specific metabolites, or untargeted where an entire metabolome is analyzed in order to identify and quantify changes in all measurable metabolites [25]. Mass spectrometry (MS) coupled to gas chromatography (GC-MS) or liquid chromatography (LC-MS) and nuclear magnetic resonance spectroscopy (NMR) are the most common analytical techniques used for metabolic profiling [26,27].

Cell lines have been used as models for studying human diseases and metabolic pathways since as early as 1956 [28]. As with metabonomics research itself, the number of metabonomic studies of cell line cultures has also experienced significant growth over the last 15 years, reaching over 400 published studies based on a literature search on the PubMed citation database (www.ncbi.nlm.nih.gov/pubmed) with the search terms "metabonomics OR metabolomics" AND "cell line" OR "cell culture" in the "Title/Abstract" field. Major research areas that employ metabonomics studies of human cell cultures include investigations of many human diseases, pharmacology, systems biology and toxicology [29–32]. Tumor cells have also been valuable for studying metabolic pathways involved in the biology of cancer and modeling of cancerous formations [33–36]. In the area of metabolic profiling-based cancer research, human cancer cell line cultures have been utilized in studies of breast cancer, bladder cancer, prostate cancer, oral cancer, brain cancer and pancreatic cancer, among others [37–42]. Cell cultures are advantageous in that they represent study systems that are capable of self-replication, thereby providing an infinite amount of material for study, and unlike biological samples that come from animal models and human hosts, cell lines can be easily modified, manipulated and studied in a controlled environment, and used in relatively inexpensive high-throughput screening studies compared to animal model studies [22,43]. Because the cells are cultured in isolation from the original tumor environment and pure cell line cultures, there is a lack of communication between tumor cells and other host and tumor stroma components, which allows for simplified characterization of the tumor cells; however, the lack of the presence of a tumor stroma is limiting, since it participates in formation, maintenance of tumors, cancer cell differentiation and cancer development [36]. Consequently, because of the absence of a biologically relevant tumor stroma context, and due to the absence of interaction with an intact immune system, human cancer cells culture studies cannot perfectly replicate cancer cell metabolism expected in intact tumors in vivo [23,36].

Metabolic profiling of human cell cultures typically follows a five-step procedure post cell culture, including sample preparation and processing, data collection, data analysis, identification of metabolites, and determination and analysis of metabolic pathways to which identified metabolites belong [43]. It has been previously reported that the choice of technique used for sample extraction or cell lifting can have an impact on the observed metabolome of the cell culture [44–51]. In light of this fact, it has been suggested that the sample preparation protocols and minimum reporting information should be standardized to provide consistent and accurate results in metabolomic studies [29,48–50,52–54]. One of the steps that has not been given much attention in the literature up to this point is the influence of the method used to dry the extracts prior to preparation of the concentrated cell extract samples for analysis [29,55,56]. The two most widely used methods of drying cell and tissue extracts are speed-vacuum and lyophilization [29,55–57]. In speed-vacuum (SpeedVac) drying, the extract is centrifuged and the solvent evaporated under vacuum at (typically unregulated) room temperature. In lyophilization (freeze-drying), the sample is first frozen with liquid nitrogen, followed by sublimation of the solvent, a process in which the solvent evaporates from its frozen solid state without going through a liquid phase. The temperature inside the trap chamber of a lyophilizer is typically −40 °C to −50 °C and the sample remains frozen throughout the drying process, while temperature inside the speed-vacuum chamber ranges anywhere between 25 °C and 35 °C and, as stated above, is generally unregulated. The vacuum levels in a freeze-dryer typically range between 60 mTorr to 300 mTorr, whereas the vacuum level in a typical SpeedVac is around 10 Torr. Although the relative performance of drying techniques in metabonomics studies has been raised before [57], the issue has not been examined considering the preparation of human cell line extracts with all other procedural steps being the same in the comparison of the two drying techniques.

Here, we investigated how the observed metabolome of three human pancreatic cancer cell lines, MiaPaCa-2, Panc-1 and AsPC-1, was influenced by application of drying either by SpeedVac or lyophilization. Ten replicate extract samples from each cell line were dried using each technique, and the metabolic profiles of the dried sample extracts were measured by NMR spectroscopy. The resulting metabolic profiles were compared to determine the influence of the choice of drying technique on the observed metabolic profiles.

2. Results

2.1. Analysis of MiaPaCa-2 Cells

2.1.1. Representative NMR Data

Representative 1D NMR spectra of samples prepared from redissolved speed vacuum dried and lyophilized cell extracts of MiaPaCa-2 cells used for metabolic profiling analysis are shown in Figure 1 with the spectrum of a SpeedVac-dried sample shown in Figure 1A and the spectrum of a lyopilized sample shown in Figure 1B.

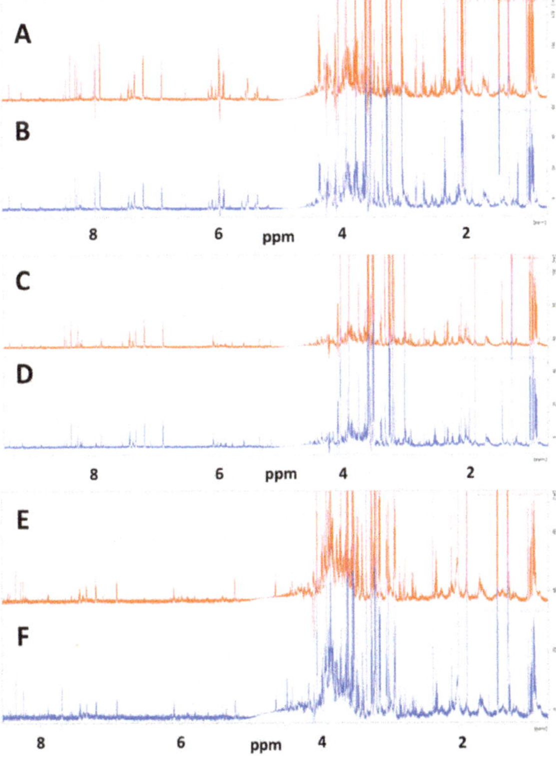

Figure 1. Representative 1D ^{1}H NMR spectra of SpeedVac dried and lyophilized MiaPaCa-2, Panc-1 and AsPC-1 cell extracts. 1D CPMG NMR spectra are shown for representative SpeedVac-dried cell extracts for (**A**) MiaPaCa-2, (**C**) Panc-1 and (**E**) AsPC-1 cells and for representative lyophilized cell extracts for (**B**) MiaPaCa-2, (**D**) Panc-1 and (**F**) AsPC-1 cells.

2.1.2. Unsupervised Principal Component Analysis (PCA) and Partial Least Squares-Discriminant Analysis (PLS-DA) of MiaPaCa-2 Cells

A total of 407 manually defined spectral buckets were included in the analysis of MiaPaCa-2 cells. The PCA scores plot for MiaPaCa-2 samples showed significant cluster separation at 95% confidence interval based on an F-test analysis (Mahalanobis distance = 1.35, F-statistic = 3.38, F-critical = 2.10) (Figure 2A) with 87% of the variance accounted for by the first two principal components (PCs) (PC-1 72%, PC-2 15%). Based on a Welch's t-test, four buckets out of 407 were statistically significant based on the Bonferroni-corrected critical alpha value (0.0001229), and 173 buckets had p-values less than 0.05, which are depicted in the PCA loadings plot in Figure S1A.

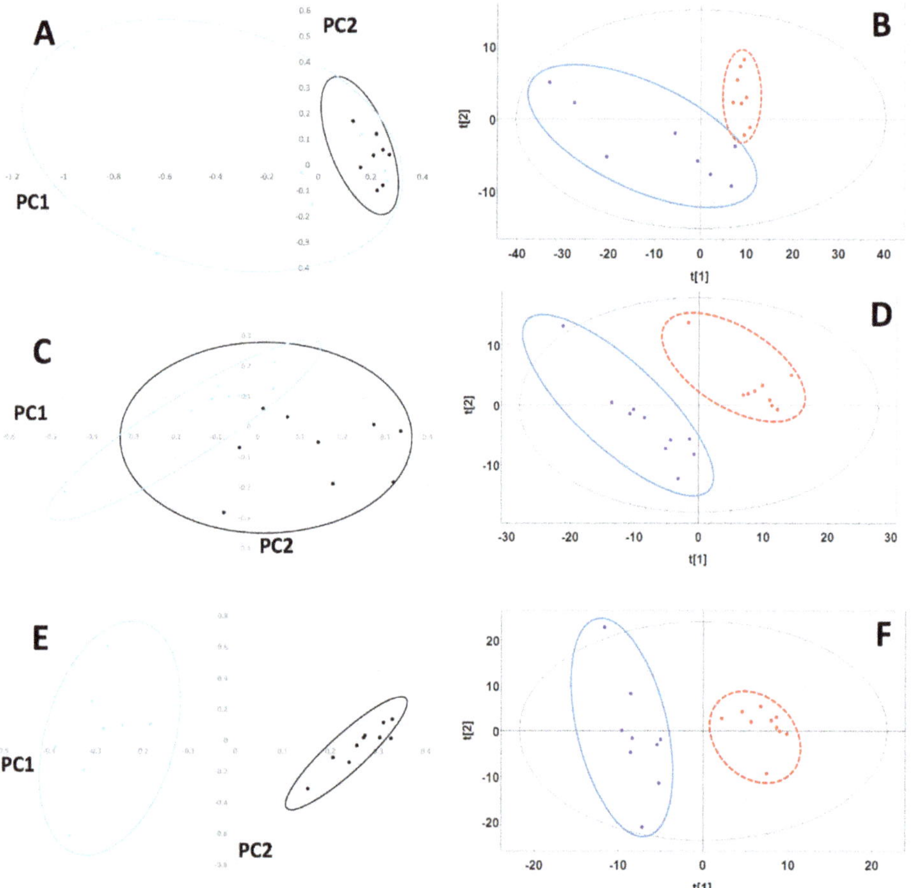

Figure 2. Principal Component Analysis (PCA) and Partial Least Squares-Discriminant Analysis (PLS-DA) scores plots for MiaPaCa-2, Panc-1 and AsPC-1 cell line comparisons. PCA scores plots using the first two principal components for comparisons between lyophilized (blue) and SpeedVac (black) samples from (**A**) MiaPaCa-2, (**C**) Panc-1 and (**E**) AsPC-1 cells. The 95% confidence intervals of the Hotelling's T^2 distributions are indicated by the overlaid ovals. PLS-DA scores plots computed using the first two principal components for (**B**) MiaPaCa-2, (**D**) Panc-1 and (**F**) AsPC-1 cell comparisons. Scores indicating lyophilized samples are colored blue and encircled by a solid blue oval line. Scores indicating SpeedVac samples are colored red and encircled by a dashed red line. The 95% confidence intervals of the Hotelling's T^2 distributions are indicated by the overlaid black solid oval lines.

The PLS-DA scores plot for MiaPaCa-2 also exhibited statistically significant separation between data set clusters based on the F-test (Mahalanobis distance = 3.75, F-statistic = 26.17, F-critical = 2.10) (Figure 2B). PLS-DA cross-validation revealed an excellent fit with the model (R^2 = 0.95) and good predictive power (Q^2 = 0.75) using three PCs. In the PLS-DA, 120 buckets out of 407 had variable importance in projection (VIP) scores greater than 1.0, indicating that these buckets carried the most weight in being responsible for cluster separation.

2.1.3. Prominent Differences in Metabolite Abundances Depending on Drying Method

To be considered significant, a bucket had to have either a *p*-value <0.05 or an AUC >0.70 [58]. In MiaPaCa-2 cells, 53 (Table S1) metabolites out of a total of 74 metabolites identified (Table S2) were present at significantly different abundances depending on drying technique. Out of these 53 metabolites, one metabolite was observed only in the lyophilized samples, zero metabolites were only observed in SpeedVac samples, and 52 metabolites were present in both samples, but at different abundances depending on the drying technique used. Figure 3 shows examples where the differences were apparent from visual comparison of the spectra from each MiaPaCa-2 drying group. Ethanol was detected predominantly in the lyophilized samples (Figure 3A) and made a considerable contribution to the PLS-DA cluster separation between the drying techniques as it had the four highest VIP scored buckets. The absence of metabolites such as ethanol in SpeedVac-dried samples was potentially due to more rapid evaporation at room temperature due to higher vapor pressure at this temperature. On the other hand, formic acid was present in both samples but at a higher abundance in those dried by speed-vacuum (Figure 3B). The absence of metabolites such as formic acid in the lyophilized samples may be due to chemical conversion during speed-vacuum drying at a higher temperature. In addition to the different abundances of the identified metabolites in the MiaPaCa-2 extract samples, an additional 54 buckets were identified that had statistically significant differences in intensities between groups based on at least one of our four criteria depending on the drying technique used to prepare the samples, but could not be assigned to a specific metabolite (Table S3) due to limitations of existing databases. An NMR spectrum of the DMEM media was also collected to determine if carry over from the media influenced the profile of metabolites identified from cell extracts (Figure S2A). Of the 19 metabolites identified in the DMEM media (Table S4), 11 were also detected in cell extracts, but eight were not detected in cell extracts, indicating the 11 metabolites detected in the cell extracts originated from the cells and not from carryover from the media since eight metabolites present in the media were absent from the cell extracts.

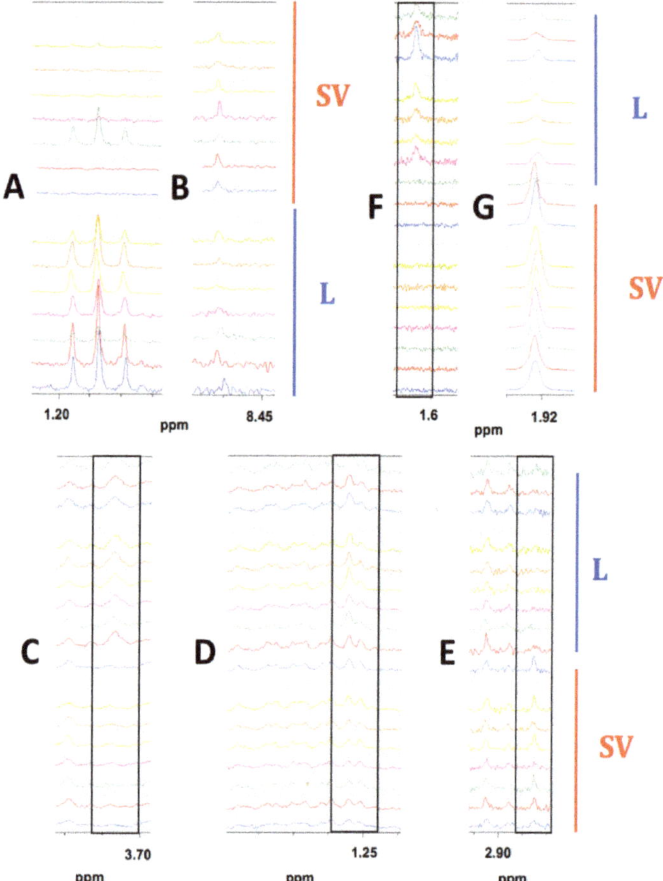

Figure 3. Highlighted spectral differences observed in MiaPaCa-2, Panc-1 and AsPC-1 cell extract comparisons depending on the drying method used for sample preparation. (**A**) Ethanol present in the lyophilized MiaPaCa-2 samples only. (**B**) Formic acid present in both groups of MiaPaCa-2 cell extracts but at higher concentration in SpeedVac samples. The bottom eight spectra were lyophilized samples (L) and the top eight spectra were SpeedVac samples (SV). (**C**) Leucine present only in the lyophilized Panc-1 samples. (**D**) Isoleucine present in both groups of Panc-1 cell extracts but at a higher abundance in the lyophilized samples. (**E**) Trimethylamine present only in SpeedVac dried Panc-1 samples. The bottom nine spectra were SpeedVac samples (SV) and the top 10 spectra were lyophilized samples (L). (**F**) N-methyl-α-aminobutyric acid present only in the lyophilized AsPC-1 samples. (**G**) Acetate was present in both groups but at a higher concentration in SpeedVac samples. The bottom nine spectra were SpeedVac (SV) samples and the top 10 spectra were lyophilized samples (L). Peaks of interest are boxed in each comparison of stacked spectra.

2.2. Analysis of Panc-1 Cells

2.2.1. Representative NMR Data

Representative 1D NMR spectra of samples prepared from redissolved speed vacuum dried and lyophilized Panc-1 cell extracts used for metabolic profiling analysis are shown in Figure 1 with a spectrum of a SpeedVac sample shown in Figure 2C and a spectrum of a lyophilized sample shown in Figure 2D.

2.2.2. Unsupervised Principal Component Analysis (PCA) and Partial Least Squares-Discriminant Analysis (PLS-DA) of Panc-1 Cells

A total of 359 manually defined spectral buckets were included in the analysis of Panc-1 cells. The PCA scores plot for Panc-1 exhibited significant cluster separation (Mahalanobis distance = 3.11, F-statistic = 21.51, F-critical = 1.95) (Figure 2C) with 74% of the variance accounted for by the first two PCs (PC-1 51%, PC-2 23%). Based on a Welch's t-test analysis, 18 buckets out of 359 were statistically significant based on the Bonferroni-corrected critical alpha value (0.0001393), which are depicted in the PCA loadings plot in Figure S2B, and 142 buckets had p-values less than 0.05.

The PLS-DA scores plot for Panc-1 showed statistically significant separation between data sets based on F-test analysis (Mahalanobis distance = 8.47, F-statistic = 159.95, F-critical = 1.95) (Figure 2D). Cross-validation of PLS-DA using three PCs indicated excellent fit of the data to the model ($R^2 = 0.98$) and high predictive capability of the model ($Q^2 = 0.90$). In the PLS-DA, 160 buckets out of 359 had VIP scores greater than 1.0, indicating that these buckets carried the most weight in being responsible for cluster separation.

2.2.3. Prominent Differences in Metabolite Abundances Depending on Drying Method

In Panc-1 cells, 49 metabolites (Table S5) out of a total of 63 identified (Table S6) were present at significantly different abundances depending on drying technique. Out of these 49 metabolites, two metabolites were observed only in the lyophilized samples, two metabolites were only observed in SpeedVac samples, and 47 metabolites were present in both samples, but at different abundances depending on the drying technique used. Figure 3C–E demonstrates differences in metabolite abundances that are visually apparent between drying groups for Panc-1 sample extracts. Leucine was detected only in the lyophilized samples (Figure 3C) and had considerable contribution towards differences between the drying techniques as it had the highest VIP bucket. Isoleucine was present in both samples but at a higher abundance in those dried by lyophilization (Figure 3D). Trimethylamine was detected only in speed-vacuumed samples (Figure 3E). The absence of metabolites such as trimethylamine in the lyophilized samples was potentially due to chemical degradation. Trimethylamine was most likely converted from glycerophosphocholine which was at a higher abundance in the lyophilized samples. In addition to the different abundances of identified metabolites in Panc-1 extract samples prepared by different drying techniques, there were an additional 35 buckets that had statistically significant differences in intensities, based on at least one of our four criteria, depending on the drying technique used to prepare the samples that could not be assigned to a specific metabolite (Table S7). Based on the analysis of the NMR spectrum of the DMEM media (Figure S2A), of the 19 metabolites identified in the DMEM media (Table S3), 11 were also detected in cell extracts, but eight were not detected in cell extracts, indicating the 11 metabolites detected in the cell extracts originated from the cells and not from carryover from the media since eight metabolites present in the media were absent from the cell extracts.

2.3. Analysis of AsPC-1 Cells

2.3.1. Representative NMR Data

Representative 1D NMR spectra of samples prepared from redissolved speed vacuum dried and lyophilized AsPC-1 cell extracts used for metabolic profiling analysis are shown in Figure 1, with a spectrum of a SpeedVac sample shown in Figure 1D and a spectrum of a lyophilized sample shown in Figure 1E.

2.3.2. Unsupervised Principal Component Analysis (PCA) and Partial Least Squares-Discriminant Analysis (PLS-DA) of AsPC-1 Cells

A total of 307 manually defined spectral buckets were included in the analysis of Panc-1 cells. The PCA scores plot for AsPC-1 also displayed clear cluster separation at 95% confidence interval

(Mahalanobis distance = 11.80, F-statistic = 310.40, F-critical = 1.95) (Figure 2E) with 85% of the variance accounted for by the first two PCs (PC-1 51%, PC-2 34%). Based on a Welch's t-test analysis, 21 buckets out of 307 were statistically significant based on the Bonferroni-corrected critical alpha value (0.0001629), which are depicted in the PCA loadings plot in Figure S1C, and 107 buckets had p-values less than 0.05.

The PLS-DA scores plot for AsPC-1 cell line samples showed separation of the two groups that was statistically significant based on F-test analysis (Mahalanobis distance = 7.64, F-statistic = 130.05, F-critical = 1.95) (Figure 3A). PLS-DA cross-validation using two PCs (Figure 2F) generated R^2 = 0.80 indicating excellent data agreement with the model, and Q^2 = 0.94, indicating great model predictive power. In the PLS-DA, 113 buckets out of 307 had VIP scores greater than 1.0, indicating that these buckets carried the most weight in being responsible for cluster separation.

2.3.3. Prominent Differences in Metabolite Abundances Depending on Drying Method

In AsPC-1 cells, 44 metabolites (Table S8) out of a total of 60 identified (Table S9) were present at significantly different abundances depending on drying technique. Out of these 44 metabolites, four metabolites were observed only in the lyophilized samples, zero metabolites were only observed in SpeedVac samples, and 40 metabolites were present in both samples, but at different abundances depending on the drying technique used. Figure 3F–G displays visual differences between each group by comparing spectra from each AsPC-1 sample. N-methyl-α-aminobutyric acid was detected only in the lyophilized samples (Figure 3F). Acetate was found in both samples but at a higher concentration in those dried by speed-vacuum (Figure 3G). The absence of metabolites such as acetate in the lyophilized samples was likely due to chemical conversion at the relatively high temperature compared to lyophilization during speed-vacuum drying. In addition to the different abundances of identified metabolites in AsPC-1 extract samples prepared by different drying techniques, an additional 60 buckets that had statistically significant differences in intensities based on at least one of our four criteria depending on the drying technique used to prepare the samples could not be assigned to a specific metabolite (Table S10). An NMR spectrum of the RPMI media was also collected (Figure S2B) to determine if carry over from the media influenced the profile of metabolites identified from cell extracts. Of the 22 metabolites identified in the RPMI media (Table S11), nine were also detected in cell extracts, but 13 were not detected in cell extracts, indicating the nine metabolites detected in the cell extracts originated from the cells and not from carryover from the media since 13 metabolites present in the media were absent from the cell extracts.

3. Discussion

The data presented here indicate that the choice of drying technique used to prepare extracts from human cell lines for metabolic profiling has a significant impact on the observed metabolic profiles. 70–75% of all the buckets intensities were significantly different depending on the drying technique in the observed metabolome in each of the three cell lines. In the three human pancreatic cancer cell lines investigated in this study, the number of metabolites whose abundances were significantly different depending on drying technique ranged from 44 in AsPC-1 cell extracts to 53 in MiaPaCa-2 cell extracts. A prominent difference in the metabolic profiles depending on drying technique was that some metabolites were present in samples prepared by one drying technique but completely absent in the other (Table 1). For instance, n-methyl-α-aminoisobutyric acid was present in the lyophilized samples but undetected in speed-vacuum dried samples in cell extracts prepared from all three cell lines. Other metabolites detected in lyophilized cell extracts but undetected in SpeedVac samples included 3-hydroyvaleric acid, N-methyl nicotinamide and sarcosine, which was only detected in AsPC1 cells.

On the other hand, one metabolite was present in speed-vacuumed samples but absent in the lyophilized samples (Table 1), i.e., trimethylamine, which was only present in Panc-1 cell extracts prepared by SpeedVac. The absence of trimethylamine in the lyophilized samples could be due to chemical conversion during speed-vacuum drying that takes place at a higher temperature compared to during lyophilization.

Table 1. Summary of metabolite differences in resuspended MiaPaCa-2, Panc-1 and AsPC-1 cell extract samples depending on the drying method used for sample preparation.

Only in Lyophilized	More Abundant in Lyophilized Cell Extracts	More Abundant in Speed-Vacuum Dried Extracts
3-Hydroxyisovaleric acid [3]	1,3,7-trimethyluric acid1 [3]	Acetate [2,3]
N-methyl-α-aminoisobutyric acid [1,2,3]	Creatinine1 [3]	1,9-Dimethyluric acid [1,3]
N-Methylnicotinamide [3]	Methanol1 [3]	Cytidine Triphosphate [3]
Sarcosine [3]	3-Cresotinic acid [3]	Glucose-6-phosphate [3]
Leucine [2]	Threonine [3]	Glucose [3]
Ethanol [1]	1,3-Dihydroxyacetone [3]	1-methyladenosine [1,3]
Only in Speed-Vacuum	L-Aspartic acid [1,3]	Carnosine [3]
Trimethylamine [2]	2-phosphoglycerate [3]	L-Threonine [3]
	L-Methionine [1,3]	Tyramine [3]
	Creatine [2,3]	Mannose-6-phosphate [3]
	Glycerophosphocholine [1,3]	Uracil [3]
	Butanone [3]	D-Alanine1 [3]
	Methylcysteine [3]	Glycine [3]
	Leucine [1,3]	L-Lactic acid [3]
	Capric acid [3]	L-Valine [3]
	Galactaric acid [3]	Isoleucine [3]
	Phosphocreatine [3]	L-Phenylalanine [3]
	Pyruvate [3]	Glycerate [3]
	Uridine diphosphategalactose [1,2]	N-Carbamoyl-β-alanine [3]
	Glucaric acid [1]	Glycerol [3]
	Threonic acid [1]	Trimethylamine [1]
	Uracil [1]	Fumaric acid [1]
	Acetylglycine [1]	Formic acid [1,2]
	L-Phenylalanine [1,2]	1-Butanol [1]
	creatine phosphate [1,2]	2-hydroxybutyrate [1]
	2,3-Diphosphoglyceric acid [1]	Taurine [1]
	L-Tyrosine [1,2]	1-Methyluric acid [1]
	L-Isoleucine [1,2]	NAD+ [1]
	Uridine [1]	L-Malic acid [1]
	Inosine [1,2]	Methanol [2]
	S-Adenosylhomocysteine [1]	L-Aspartic acid [2]
	1-Butanol [2]	Uracil [2]
	S-Adenosylhomocysteine [1,2]	Acetylcholine [2]
	Histidine [2]	L-Methionine [2]
	Glutathione [1,2]	Glycerophosphocholine [2]
	Methionine sulfoxide [1]	Phosphoenolpyruvic acid [2]
	Phosphorylcholine [1]	1-methylguanosine [2]
	Cellobiose [1]	3-Methyladipic acid [2]
	3-Hydroxyisovaleric acid [1]	Oxypurinol [2]
	L-Valine [1,2]	Adenine [2]
	Pyroglutamate [1]	Guanosine [2]
	Acetylphosphate [1]	
	Glucose [1,2]	
	Glucose-6-phosphate [1]	
	Myoinositol [1,2]	
	1-Methylguanine [1]	
	1,1-Dimethylbiguanide [1]	
	Pyruvatoxime [1,2]	
	L-lactic acid [1,2]	
	3,7-Dimethyluric acid [2]	
	Succinic acid [2]	

Table 1. *Cont.*

Only in Lyophilized	More Abundant in Lyophilized Cell Extracts	More Abundant in Speed-Vacuum Dried Extracts
	Adenosine phosphosulfate [2]	
	Mannose 6-phosphate [2]	
	3-Phosphoglyceric acid [2]	
	NAD+[2]	
	N-acetylneuraminic acid [2]	
	Lysine [2]	
	5-aminolevulinate [2]	
	Trimethylamine N-oxide [2]	
	Taurine [2]	

[1] MiPaCa-2 cell extracts, [2] Panc-1 cell extracts, [3] AsPC-1 cell extracts.

The majority of metabolites that were present at different abundances depending on drying technique were present in extracts prepared by both drying techniques (Table 1). Forty-one metabolites were found to be more concentrated in samples dried by speed-vacuum. Fifteen out of 41 metabolites that were present at higher abundances in SpeedVac samples were found only in AsPC-1. The AsPC-1 cell line also had more metabolites present in greater amounts in samples dried by speed-vacuum compared to the MiaPaCa-2 and Panc-1 cells. This could be due to two factors: the differences in cellular composition in cancer cell lines [21] and the different optimal growth media, i.e., RPMI, used to culture the AsPC-1 cells [51]. Sixty-one metabolites were present at higher concentrations in the lyophilized samples. The lower abundances of these metabolites in speed-vacuumed samples was likely caused by more rapid evaporation at room temperature (or higher) during speed vacuum drying compared to freeze-drying, since metabolites generally have higher vapor pressures at higher temperatures, and a higher rate of chemical conversion of heat labile metabolites to different compounds at the higher temperature of the SpeedVac during drying.

About 10–15% of all significant buckets in each cell line could not be identified from the existing databases (i.e., no matching peaks could be identified in from the existing databases) therefore limiting our ability to draw specific conclusions from this missing fraction of data. While some metabolites were more abundant after SpeedVac drying, lyophilization seems to be the generally superior technique compared to centrifugal evaporation under vacuum as it seemed to preserve the original metabolome of the cell lines to a higher degree based on the general observation of including more metabolites, or the same metabolites at higher concentrations compared to SpeedVac prepared samples.

4. Materials and Methods

4.1. Cell Cultures and Preparation of Cell Extracts

MiaPaCa-2, Panc-1, and AsPC-1 were purchased from the American Type Culture Collection (Manassas, VA, USA). MiaPaCa-2 and Panc-1 were grown in high glucose Dulbecco's Modified Eagle Medium, (DMEM) and AsPC-1 cells were grown in Roswell Park Memorial Institute (RPMI) medium. Both media were supplemented with 10% fetal bovine serum and 1% penicillin–streptomycin. Sixty 100 mm × 15 mm petri dishes per cell line were harvested and extracted as previously described by Watanabe et al. [21]. Briefly, the cells were scraped using chilled phosphate-buffered saline (PBS), washed 3× in cold PBS buffer, and then stored in −80 °C prior to cell extraction. Frozen cells thawed on ice, resuspended in 1.5 mL of ice-cold chloroform-methanol-water (1:1:1) solution and vortexed, after which they were chilled on ice for 15 min. The tubes were then centrifuged for 15 min at 15,000× *g*. The top layer (hydrophilic) and the bottom layer (lipophilic) were separated and transferred to new Eppendorf tubes. Hydrophilic extracts were dried and used for NMR spectroscopic analysis.

4.2. Drying Methods

Thirty samples of each cell line were dried in a speed vacuum centrifugal concentrator (UVS800DA THERMO Savant Sunnyvale, CA). Another thirty samples of each cell line were frozen in liquid nitrogen and dried using a lyophilizer (Labconco 7740020, Kansas City, MO, USA). The drying time used for both SpeedVac and lyophilization were kept constant. MiaPaCa-2 and Panc-1 cells were dried for 21 h, and AsPC-1 for 20 h. For lyophilization, it was ensured that the samples remained frozen throughout the sublimation process and did not thaw into a liquid prior to complete lyophilization.

4.3. Preparation of Samples for NMR Data Collection

Following resuspension in 150 µL of buffer (150 mM potassium phosphate at pH 7.4, 1 mM trimethylsilyl propionate (TSP), 0.01% sodium azide in 100% deuterium oxide (the buffer was first prepared in 100% H2O, lyophilized, and then reconstituted in 100% D2O)) [21], cell line extracts were combined in the following manner: ten were combined to generate a representative sample for lyophilized samples for 2D NMR experiments needed for confirmation of metabolite assignments, another ten were combined and used as a representative sample for speed vacuumed samples for 2D NMR experiments needed for confirmation of metabolite assignments, and the last forty were used to create ten replicate samples (two samples combined for each NMR sample to increase the extract concentration for NMR analysis) per drying method for 1D NMR experiments. Overall, a total of 60 samples were prepared for 1D NMR spectroscopy analysis. A total of six representative samples were made for analysis using 2D NMR spectroscopy. Samples of both the DMEM and RPMI media were also prepared for NMR analysis as controls.

4.4. NMR Data Collection

All NMR experiments were recorded at 850.104 MHz at 298K on a Bruker Avance Spectrometer. 150 microliters of redissolved extracts were transferred to 3 mm NMR tubes for 1D experiments. Two hundred and fifty microliters of redissolved extracts were transferred to 5 mm Shigemi tubes for 2D experiments. One-dimensional ^{1}H Carr-Purcell-Meiboom-Gill (CPMG) NMR spectra were collected using a spectral width of 20.0 ppm and 32K points resulting in 1.21 s acquisition time, 512 scans, 4 dummy scans, 1.80 s recycle delay and 1 ms mixing time. Two-dimensional ^{1}H-^{1}H TOCSY NMR experiments were performed using a spectral width of 10.0 ppm and 2K points, 1.5 s recycle time and 60 ms mixing time [14,17,19–21,58–61]. Two-dimensional ^{1}H-^{13}C HSQC NMR experiments were performed using non-uniform sampling in the indirect carbon dimension [62,63].

4.5. Data Analysis

NMR spectra were phase adjusted, baseline corrected, referenced to 0.0 ppm using the internal TSP standard manually using Top-Spin 3.6.1 (Bruker BioSpin, Billerica, MA, USA) and normalized to the total intensity prior to statistical analysis. Manual peak bucketing and principal components analyses (PCA) were performed using AMIX (Bruker BioSpin, Billerica, MA, USA) as we have described previously [21,58,60,61]. PCA scores plot cluster analyses using the Mahalanobis distance and F-test calculations were performed as described previously [61]. Statistical significance analysis of bucket intensity differences were carried out as previously described [60] using a Bonferroni-corrected critical alpha values for p-values calculated using a Welch's t-test and fold change (FC) were calculated using Excel. Partial least squared-discriminant analysis (PLS-DA), and variable importance in projection (VIP) scores were obtained using the SIMCA-P software (Umetrics, Santorious Stedim, Umeå, Sweden). Area under the curve (AUC) of the receiver operating characteristic (AUROC) was conducted using the MetaboAnalyst (www.metaboanalyst.ca) [17,58].

4.6. Metabolite Identification

One-dimensional NMR peaks that were statistically different between the groups were initially and putatively identified using ChenomX Profiler (https://www.chenomx.com/), the Biological Resonance Data Bank (BMRB) [64,65] and the Human Metabolome Database (HMDB) [66–69]. Two-dimensional NMR data were analyzed using the COLMAR software [70,71] (https://ccic.ohio-state.edu/) to confirm metabolite assignments. The confidence in metabolite assignments was ranked using the RANCM scheme developed in our laboratory [72].

5. Conclusions

Sample preparation of cell line cultures for metabolomic analyses generally requires a drying step. In the case of metabolic profiling of human cell line cultures, a chloroform/methanol extraction procedure is normally used to isolate the hydrophobic and hydrophilic fractions that must be dried prior to reconstitution in a buffer appropriate for instrumental analysis. Considering absence of standardized procedures for sample preparation for metabolic profiling studies, we examined whether the choice of drying method had any influence on the observed metabolome of hydrophilic extracts measured by NMR of three different human pancreatic cell line cultures. We determined that the metabolic profiles of each cell line differed significantly depending on the drying technique used during sample preparation. Certain metabolites were only found in samples dried by a specific drying technique, suggesting either that some metabolites evaporated during drying by one method and not the other, or that some metabolites went through chemical transformation in one of the sample preparation methods compared to the other. Metabolites undetected in speed-vacuum-dried samples may have been lost due to more rapid evaporation over time due to their having higher vapor pressure near room temperatures compared to at the relatively low temperature of the sample maintained during lyophilization. Overall, the lyophilized samples contained more metabolites at higher concentrations in two out of the three cell lines tested, i.e., MiaPaCa-2 and Panc-1, appearing to make it the technique of choice for the sample drying step during sample preparation. However, the two cell lines that had more metabolites more abundant in the lyophilized samples, MiaPaCa-2 and Panc-1, were grown on the same DMEM media whereas the one cell line that had more metabolites abundant in the SpeedVac samples were cultured using a different media, i.e., RPMI. Therefore, future experiments should examine whether the choice of media used to culture human cell lines for metabolic profiling can influence the efficiency of metabolite recovery using different drying methods and potentially explain the difference in the abundances of metabolites following one drying method over the other. Another aspect of the research that warrants further investigation is simple replication of the reported experiments. For example, in the case of the MiaPaCa-2 samples, the PCA indicates that the lyophilized samples were widely scattered in the PCA eigenspace of the scores plot in comparison to the SpeedVac samples. In contrast, in the AsPC-1 cells, the lyophilized and SpeedVac samples were tightly clustered and well separated in the PCA scores plot. These disparate observations indicated that the experiments should be repeated to determine if these observations are reproducible. Future work could also include working with the database developers to generate more complete databases since approximately half of the observed NMR resonances could not be assigned to a metabolite. Finally, it would be beneficial to repeat the experiment using an MS-based analytical technique for data collection to achieve more comprehensive metabolic profiles. In conclusion, this research demonstrates that the drying technique used in the preparation of human cell culture samples for metabolic profiling can have a significant effect on the observed metabolomes and the selection of the drying technique used during sample preparation should be given careful consideration when designing the sample preparation strategy.

Supplementary Materials: The following are available online at http://www.mdpi.com/2218-1989/9/11/256/s1, Figure S1. PCA Loadings plots for each cell line comparison; Figure S2. 850 MHz ^{1}H NMR spectrum of DMEM and RPMI media; Table S1. Identified metabolites whose abundance were significantly different depending on drying technique in the MiaPaCa-2 cell line; Table S2. Complete list of identified metabolites in the MiaPaCa-2 cell extracts; Table S3. Full list of unidentified buckets for MiaPaCa-2 that were determined to be significant; Table S4. List of metabolites identified in the NMR spectrum of DMEM media; Table S5. Identified metabolites whose abundance were significantly different depending on drying technique in the Panc-1 cell line; Table S6. Complete list of identified metabolites in the Panc-1 cell extracts; Table S7. Full list of unidentified buckets for Panc-1 that were determined to be significant; Table S8. Identified metabolites that were found significant for AsPC-1 cell line. Only first significant buckets for each metabolite are shown; Table S9. Complete list of identified metabolites in the AsPC-1 cell extracts; Table S10. Full list of unidentified buckets for AsPC-1 that were determined to be significant; Table S11. List of metabolites identified in the NMR spectrum of RPMI media; Raw NMR data has been archived in the public figshare data repository at https://figshare.com/projects/Influence_of_Drying_Method_on_NMR-Based_Metabolic_Profiling_of_Human_Cell_Lines/68144.

Author Contributions: The authors made the following contributions: conceptualization, M.A.K; methodology, I.P., S.N., M.A.K.; software, W.C.J., M.A.K.; validation, I.P., S.X., W.C.J., M.A.K.; formal analysis, I.P., S.X., W.C.J., M.A.K.; investigation, I.P., S.N.; resources, M.A.K.; data curation, I.P., W.C.J.; writing—original draft preparation, I.P., M.A.K.; writing—review and editing, I.P., M.A.K.; visualization, I.P., M.A.K.; supervision, W.C.J., S.N., M.A.K.; project administration, M.A.K.; funding acquisition, M.A.K.

Funding: This research received no external funding.

Acknowledgments: The research was conducted with the support of Miami University. The instrumentation used in this work was obtained with the support of Miami University and the Ohio Board of Regents with funds used to establish the Ohio Eminent Scholar Laboratory where the work was performed.

Conflicts of Interest: The authors declare no conflict of interest.

References

1. Nicholson, J.K.; Lindon, J.C.; Holmes, E. 'Metabonomics': Understanding the metabolic responses of living systems to pathophysiological stimuli via multivariate statistical analysis of biological NMR spectroscopic data. *Xenobiotica* **1999**, *29*, 1181–1189. [CrossRef] [PubMed]

2. Lindon, J.C.; Holmes, E.; Nicholson, J.K. Metabonomics in pharmaceutical R & D. *FEBS J.* **2007**, *274*, 1140–1151. [PubMed]

3. Robertson, D.G. Metabonomics in Toxicology: A Review. *Toxicol. Sci.* **2005**, *85*, 809–822. [CrossRef] [PubMed]

4. Zhang, A.; Sun, H.; Yan, G.; Wang, P.; Wang, X. Metabolomics for Biomarker Discovery: Moving to the Clinic. *BioMed Res. Int.* **2015**, *2015*, 1–6. [CrossRef]

5. Oldiges, M.; Lütz, S.; Pflug, S.; Schroer, K.; Stein, N.; Wiendahl, C. Metabolomics: Current state and evolving methodologies and tools. *Appl. Microbiol. Biotechnol.* **2007**, *76*, 495–511. [CrossRef] [PubMed]

6. Scalbert, A.; Brennan, L.; Manach, C.; Andres-Lacueva, C.; Dragsted, L.O.; Draper, J.; Rappaport, S.M.; Van Der Hooft, J.J.; Wishart, D.S. The food metabolome: A window over dietary exposure. *Am. J. Clin. Nutr.* **2014**, *99*, 1286–1308. [CrossRef]

7. Kim, S.; Kim, J.; Yun, E.J.; Kim, K.H. Food metabolomics: From farm to human. *Curr. Opin. Biotechnol.* **2016**, *37*, 16–23. [CrossRef]

8. Guasch-Ferre, M.; Bhupathiraju, S.N.; Hu, F.B. Use of Metabolomics in Improving Assessment of Dietary Intake. *Clin. Chem.* **2018**, *64*, 82–98. [CrossRef]

9. Brouwer-Brolsma, E.M.; Brennan, L.; Drevon, C.A.; Van Kranen, H.; Manach, C.; Dragsted, L.O.; Roche, H.M.; Andres-Lacueva, C.; Bakker, S.J.L.; Bouwman, J.; et al. Combining traditional dietary assessment methods with novel metabolomics techniques: Present efforts by the Food Biomarker Alliance. *Proc. Nutr. Soc.* **2017**, *76*, 619–627. [CrossRef]

10. Gibbons, H.; Brennan, L. Metabolomics as a tool in the identification of dietary biomarkers. *Proc. Nutr. Soc.* **2017**, *76*, 42–53. [CrossRef]

11. Bouhifd, M.; Hartung, T.; Hogberg, H.T.; Kleensang, A.; Zhao, L. Review: Toxicometabolomics. *J. Appl. Toxicol.* **2013**, *33*, 1365–1383. [CrossRef] [PubMed]

12. Manchester, M.; Anand, A. Metabolomics: Strategies to Define the Role of Metabolism in Virus Infection and Pathogenesis. In *Advances in Virus Research*; Elsevier: Amsterdam, The Netherlands, 2017; Volume 98, pp. 57–81.

13. Romick-Rosendale, L.E.; Goodpaster, A.M.; Hanwright, P.J.; Patel, N.B.; Wheeler, E.T.; Chona, D.L.; Kennedy, M.A. NMR-based metabonomics analysis of mouse urine and fecal extracts following oral treatment with the broad-spectrum antibiotic enrofloxacin (Baytril). *Magn. Reson. Chem.* **2009**, *47*, S36–S46. [CrossRef] [PubMed]

14. Romick-Rosendale, L.E.; Legomarcino, A.; Patel, N.B.; Morrow, A.L.; Kennedy, M.A. Prolonged antibiotic use induces intestinal injury in mice that is repaired after removing antibiotic pressure: Implications for empiric antibiotic therapy. *Metabolomics* **2014**, *10*, 8–20. [CrossRef]

15. Rosendale, L.E.R.; Schibler, K.R.; Kennedy, M.A. A Potential Biomarker for Acute Kidney Injury in Preterm Infants from Metabolic Profiling. *J. Mol. Biomark. Diagn.* **2012**, *3*, 1–7. [CrossRef]

16. Chihanga, T.; Ma, Q.; Nicholson, J.D.; Ruby, H.N.; Edelmann, R.E.; Devarajan, P.; Kennedy, M.A. NMR spectroscopy and electron microscopy identification of metabolic and ultrastructural changes to the kidney following ischemia-reperfusion injury. *Am. J. Physiol.-Ren. Physiol.* **2018**, *314*, F154–F166. [CrossRef]

17. Chihanga, T.; Ruby, H.N.; Ma, Q.; Bashir, S.; Devarajan, P.; Kennedy, M.A. NMR-based urine metabolic profiling and immunohistochemistry analysis of nephron changes in a mouse model of hypoxia-induced acute kidney injury. *Am. J. Physiol. Physiol.* **2018**, *315*, F1159–F1173. [CrossRef]

18. Liu, X.J.; Locasale, J.W. Metabolomics: A Primer. *Trends Biochem. Sci.* **2017**, *42*, 274–284. [CrossRef]

19. Watanabe, M.; Sheriff, S.; Ramelot, T.A.; Kadeer, N.; Cho, J.; Lewis, K.B.; Balasubramaniam, A.; Kennedy, M.A. NMR Based Metabonomics Study of DAG Treatment in a C2C12 Mouse Skeletal Muscle Cell Line Myotube Model of Burn-Injury. *Int. J. Pept. Res. Ther.* **2011**, *17*, 281–299. [CrossRef]

20. Watanabe, M.; Sheriff, S.; Kadeer, N.; Cho, J.; Lewis, K.B.; Balasubramaniam, A.; Kennedy, M.A. NMR based metabonomics study of NPY Y5 receptor activation in BT-549, a human breast carcinoma cell line. *Metabolomics* **2012**, *8*, 854–868. [CrossRef]

21. Watanabe, M.; Sheriff, S.; Lewis, K.B.; Cho, J.; Tinch, S.L.; Balasubramaniam, A.; Kennedy, M.A. Metabolic Profiling Comparison of Human Pancreatic Ductal Epithelial Cells and Three Pancreatic Cancer Cell Lines using NMR Based Metabonomics. *J. Mol. Biomark. Diagn.* **2012**, *3*, 1–17. [CrossRef]

22. Nichols, A.C.; Black, M.; Yoo, J.; Pinto, N.; Fernandes, A.; Haibe-Kains, B.; Boutros, P.C.; Barrett, J.W. Exploiting high-throughput cell line drug screening studies to identify candidate therapeutic agents in head and neck cancer. *BMC Pharmacol. Toxicol.* **2014**, *15*, 66. [CrossRef] [PubMed]

23. Wilding, J.L.; Bodmer, W. Cancer Cell Lines for Drug Discovery and Development. *Cancer Res.* **2014**, *74*, 2377–2384. [CrossRef] [PubMed]

24. Burgess, K.; Rankin, N.; Weidt, S. Chapter 10-Metabolomics. In *Handbook of Pharmacogenomics and Stratified Medicine*; Padmanabhan, S., Ed.; Academic Press: San Diego, CA, USA, 2014; pp. 181–205.

25. Cajka, T.; Fiehn, O. Toward Merging Untargeted and Targeted Methods in Mass Spectrometry-Based Metabolomics and Lipidomics. *Anal. Chem.* **2016**, *88*, 524–545. [CrossRef] [PubMed]

26. Zhang, A.; Sun, H.; Wang, P.; Han, Y.; Wang, X. Modern analytical techniques in metabolomics analysis. *Analyst* **2012**, *137*, 293–300. [CrossRef]

27. Lenz, E.M.; Wilson, I.D. Analytical Strategies in Metabonomics. *J. Proteome Res.* **2007**, *6*, 443–458. [CrossRef]

28. Morgan, H.R. Latent viral infection of cells in tissue culture. I. Studies on latent infection of chick embryo tissues with psittacosis virus. *J. Exp. Med.* **1956**, *103*, 37–47. [CrossRef]

29. Hayton, S.; Maker, G.L.; Mullaney, I.; Trengove, R.D. Experimental design and reporting standards for metabolomics studies of mammalian cell lines. *Cell. Mol. Life Sci.* **2017**, *74*, 4421–4441. [CrossRef]

30. Halama, A. Metabolomics in cell culture—A strategy to study crucial metabolic pathways in cancer development and the response to treatment. *Arch. Biochem. Biophys.* **2014**, *564*, 100–109. [CrossRef]

31. Kumar, A.; Misra, B.B. Challenges and Opportunities in Cancer Metabolomics. *Proteomics* **2019**, e1900042. [CrossRef]

32. Čuperlović-Culf, M.; Barnett, D.A.; Culf, A.S.; Chute, I. Cell culture metabolomics: Applications and future directions. *Drug Discov. Today* **2010**, *15*, 610–621. [CrossRef]

33. Sato, G. Tissue culture: The unrealized potential. *Cytotechnology* **2008**, *57*, 111–114. [CrossRef] [PubMed]

34. Zhao, M.; Sano, D.; Pickering, C.R.; Jasser, S.A.; Henderson, Y.C.; Clayman, G.L.; Sturgis, E.M.; Ow, T.J.; Lotan, R.; Carey, T.E.; et al. Assembly and initial characterization of a panel of 85 genomically validated cell lines from diverse head and neck tumor sites. *Clin. Cancer Res.* **2011**, *17*, 7248–7264. [CrossRef] [PubMed]

35. Lacroix, M.; Leclercq, G. Relevance of Breast Cancer Cell Lines as Models for Breast Tumours: An Update. *Breast Cancer Res. Treat.* **2004**, *83*, 249–289. [CrossRef] [PubMed]

36. Gazdar, A.F.; Girard, L.; Lockwood, W.W.; Lam, W.L.; Minna, J.D. Lung Cancer Cell Lines as Tools for Biomedical Discovery and Research. *J. Natl. Cancer Inst.* **2010**, *102*, 1310–1321. [CrossRef]

37. Dai, X.; Cheng, H.; Bai, Z.; Li, J. Breast Cancer Cell Line Classification and Its Relevance with Breast Tumor Subtyping. *J. Cancer* **2017**, *8*, 3131–3141. [CrossRef]

38. Sun, H.; Zhang, A.-H.; Liu, S.-B.; Qiu, S.; Li, X.-N.; Zhang, T.-L.; Liu, L.; Wang, X.-J. Cell metabolomics identify regulatory pathways and targets of magnoline against prostate cancer. *J. Chromatogr.* **2018**, *1102*, 143–151. [CrossRef]

39. Scheff, N.N.; Ye, Y.; Bhattacharya, A.; Macrae, J.; Hickman, D.N.; Sharma, A.K.; Dolan, J.C.; Schmidt, B.L.; Hickman, D.H. Tumor necrosis factor alpha secreted from oral squamous cell carcinoma contributes to cancer pain and associated inflammation. *Pain* **2017**, *158*, 2396–2409. [CrossRef]

40. Han, S.E.; Park, K.-H.; Lee, G.; Huh, Y.-J.; Min, B.-M. Mutation and aberrant expression of Caveolin-1 in human oral squamous cell carcinomas and oral cancer cell lines. *Int. J. Oncol.* **2004**, *24*, 435–440. [CrossRef]

41. Mohamadi, N.; Kazemi, S.M.; Mohammadian, M.; Toofani Milani, A.; Moradi, Y.; Yasemi, M. Toxicity of Cisplatin-Loaded Poly Butyl Cyanoacrylate Nanoparticles in a Brain Cancer Cell Line: Anionic Polymerization Results. *Asian Pac. J. Cancer Prev.* **2017**, *18*, 629–632.

42. Fujita, M.; Imadome, K.; Imai, T. Metabolic characterization of invaded cells of the pancreatic cancer cell line, PANC-1. *Cancer Sci.* **2017**, *108*, 961–971. [CrossRef]

43. Zhang, A.H.; Sun, H.; Xu, H.Y.; Qiu, S.; Wang, X.J. Cell Metabolomics. *Omics* **2013**, *17*, 495–501. [CrossRef] [PubMed]

44. Kapoore, R.V.; Coyle, R.; Staton, C.A.; Brown, N.J.; Vaidyanathan, S. Cell line dependence of metabolite leakage in metabolome analyses of adherent normal and cancer cell lines. *Metabolomics* **2015**, *11*, 1743–1755. [CrossRef]

45. Sapcariu, S.C.; Kanashova, T.; Weindl, D.; Ghelfi, J.; Dittmar, G.; Hiller, K. Simultaneous extraction of proteins and metabolites from cells in culture. *MethodsX* **2014**, *1*, 74–80. [CrossRef] [PubMed]

46. Sellick, C.A.; Knight, D.; Croxford, A.S.; Maqsood, A.R.; Stephens, G.M.; Goodacre, R.; Dickson, A.J. Evaluation of extraction processes for intracellular metabolite profiling of mammalian cells: Matching extraction approaches to cell type and metabolite targets. *Metabolomics* **2010**, *6*, 427–438. [CrossRef]

47. Sheikh, K.D.; Khanna, S.; Byers, S.W.; Fornace, A.J.; Cheema, A.K. Small molecule metabolite extraction strategy for improving LC/MS detection of cancer cell metabolome. *J. Biomol. Tech. JBT* **2011**, *22*, 1–4.

48. Canavan, H.E.; Cheng, X.; Graham, D.J.; Ratner, B.D.; Castner, D.G. Cell sheet detachment affects the extracellular matrix: A surface science study comparing thermal liftoff, enzymatic, and mechanical methods. *J. Biomed. Mater. Res. Part A* **2005**, *75*, 1–13. [CrossRef]

49. Masson, P.; Alves, A.C.; Ebbels, T.M.D.; Nicholson, J.K.; Want, E.J.; Alves, A. Optimization and Evaluation of Metabolite Extraction Protocols for Untargeted Metabolic Profiling of Liver Samples by UPLC-MS. *Anal. Chem.* **2010**, *82*, 7779–7786. [CrossRef]

50. Tambellini, N.P.; Zaremberg, V.; Turner, R.J.; Weljie, A.M. Evaluation of Extraction Protocols for Simultaneous Polar and Non-Polar Yeast Metabolite Analysis Using Multivariate Projection Methods. *Metabolites* **2013**, *3*, 592–605. [CrossRef]

51. Chihanga, T.; Hausmann, S.M.; Ni, S.; Kennedy, M.A. Influence of media selection on NMR based metabolic profiling of human cell lines. *Metabolomics* **2018**, *14*, 28. [CrossRef]

52. Villas-Bôas, S.G.; Koulman, A.; Lane, G.A. Analytical methods from the perspective of method standardization. In *Advanced Structural Safety Studies*; Springer Science and Business Media LLC: Berlin, Germany, 2007; Volume 18, pp. 11–52.

53. The Standard Metabolic Reporting Structures working group; Lindon, J.C.; Nicholson, J.K.; Holmes, E.; Keun, H.C.; Craig, A.; Pearce, J.T.M.; Bruce, S.J.; Hardy, N.; Sansone, S.-A.; et al. Summary recommendations for standardization and reporting of metabolic analyses. *Nat. Biotechnol.* **2005**, *23*, 833–838.

54. van der Werf, M.J.; Takors, R.; Smedsgaard, J.; Nielsen, J.; Ferenci, T.; Portais, J.C.; Wittmann, C.; Hooks, M.; Tomassini, A.; Oldiges, M.; et al. Standard reporting requirements for biological samples in metabolomics experiments: Microbial and in vitro biology experiments. *Metabolomics* **2007**, *3*, 189–194. [CrossRef] [PubMed]

55. Alvarez-Sanchez, B.; Priego-Capote, F.; de Castro, M.D.L. Metabolomics analysis II. Preparation of biological samples prior to detection. *TrAC Trends Anal. Chem.* **2010**, *29*, 120–127. [CrossRef]

56. León, Z.; García-Cañaveras, J.C.; Donato, M.T.; Lahoz, A. Mammalian cell metabolomics: Experimental design and sample preparation. *Electrophoresis* **2013**, *34*, 2762–2775. [CrossRef]

57. Åkesson, M.; Smedsgaard, J.; Nielsen, J.; Villas-Bôas, S.G.; Højer-Pedersen, J. Global metabolite analysis of yeast: Evaluation of sample preparation methods. *Yeast* **2005**, *22*, 1155–1169.

58. Schmahl, M.J.; Regan, D.P.; Rivers, A.C.; Joesten, W.C.; Kennedy, M.A. NMR-based metabolic profiling of urine, serum, fecal, and pancreatic tissue samples from the Ptf1a-Cre; LSL-KrasG12D transgenic mouse model of pancreatic cancer. *PLoS ONE* **2018**, *13*, e0200658. [CrossRef]

59. Wang, B.; Sheriff, S.; Balasubramaniam, A.; Kennedy, M.A. NMR based metabolomics study of Y2 receptor activation by neuropeptide Y in the SK-N-BE2 human neuroblastoma cell line. *Metabolomics* **2015**, *11*, 1243–1252. [CrossRef]

60. Goodpaster, A.M.; Romick-Rosendale, L.E.; Kennedy, M.A. Statistical significance analysis of nuclear magnetic resonance-based metabonomics data. *Anal. Biochem.* **2010**, *401*, 134–143. [CrossRef]

61. Goodpaster, A.M.; Kennedy, M.A. Quantification and statistical significance analysis of group separation in NMR-based metabonomics studies. *Chemom. Intell. Lab. Syst.* **2011**, *109*, 162–170. [CrossRef]

62. Hyberts, S.G.; Arthanari, H.; Wagner, G. Applications of non-uniform sampling and processing. *Top. Curr. Chem.* **2012**, *316*, 125–148.

63. Delaglio, F.; Grzesiek, S.; Zhu, G.; Pfeifer, J.; Bax, A.; Vuister, G.; Vuister, G.W. NMRPipe: A multidimensional spectral processing system based on UNIX pipes. *J. Biomol. NMR* **1995**, *6*, 277–293. [CrossRef]

64. Ulrich, E.L.; Akutsu, H.; Doreleijers, J.F.; Harano, Y.; Ioannidis, Y.E.; Lin, J.; Livny, M.; Mading, S.; Maziuk, D.; Miller, Z.; et al. BioMagResBank. *Nucleic Acids Res.* **2008**, *36*, D402–D408. [CrossRef] [PubMed]

65. Wishart, D.S.; Watson, M.S.; Boyko, R.F.; Sykes, B.D. Automated H-1 and C-13 chemical shift prediction using the BioMagResBank. *J. Biomol. NMR* **1997**, *10*, 329–336. [CrossRef] [PubMed]

66. Wishart, D.S.; Feunang, Y.D.; Marcu, A.; Guo, A.C.; Liang, K.; Vázquez-Fresno, R.; Sajed, T.; Johnson, D.; Li, C.; Karu, N.; et al. HMDB 4.0: The human metabolome database for 2018. *Nucleic Acids Res.* **2018**, *46*, D608–D617. [CrossRef] [PubMed]

67. Wishart, D.S.; Jewison, T.; Guo, A.C.; Wilson, M.; Knox, C.; Liu, Y.; Djoumbou, Y.; Mandal, R.; Aziat, F.; Dong, E.; et al. HMDB 3.0-The Human Metabolome Database in 2013. *Nucleic Acids Res.* **2013**, *41*, D801–D807. [CrossRef]

68. Wishart, D.S.; Knox, C.; Guo, A.C.; Eisner, R.; Young, N.; Gautam, B.; Hau, D.D.; Psychogios, N.; Dong, E.; Bouatra, S.; et al. HMDB: A knowledgebase for the human metabolome. *Nucleic Acids Res.* **2009**, *37*, D603–D610. [CrossRef]

69. Wishart, D.S.; Tzur, D.; Knox, C.; Eisner, R.; Guo, A.C.; Young, N.; Cheng, D.; Jewell, K.; Arndt, D.; Sawhney, S.; et al. HMDB: The Human Metabolome Database. *Nucleic Acids Res.* **2007**, *35*, D521–D526. [CrossRef]

70. Bingol, K.; Li, D.-W.; Zhang, B.; Brüschweiler, R. Comprehensive Metabolite Identification Strategy Using Multiple Two-Dimensional NMR Spectra of a Complex Mixture Implemented in the COLMARm Web Server. *Anal. Chem.* **2016**, *88*, 12411–12418. [CrossRef]

71. Bingol, K.; Bruschweiler-Li, L.; Li, D.-W.; Brüschweiler, R. Customized Metabolomics Database for the Analysis of NMR 1H–1H TOCSY and 13C–1H HSQC-TOCSY Spectra of Complex Mixtures. *Anal. Chem.* **2014**, *86*, 5494–5501. [CrossRef]

72. Joesten, W.C.; Kennedy, M.A. RANCM: A new ranking scheme for assigning confidence levels to metabolite assignments in NMR-based metabolomics studies. *Metabolomics* **2019**, *15*, 5. [CrossRef]

 *metabolites*

Article

Single Spheroid Metabolomics: Optimizing Sample Preparation of Three-Dimensional Multicellular Tumor Spheroids

Mate Rusz [1], Evelyn Rampler [2,3,4], Bernhard K. Keppler [1], Michael A. Jakupec [1] and Gunda Koellensperger [2,3,4,*]

[1] Institute of Inorganic Chemistry, University of Vienna, Währinger Str. 42, 1090 Vienna, Austria; mate.rusz@univie.ac.at (M.R.); bernhard.keppler@univie.ac.at (B.K.K.); michael.jakupec@univie.ac.at (M.A.J.)

[2] Institute of Analytical Chemistry, University of Vienna, Währinger Str. 38, 1090 Vienna, Austria; evelyn.rampler@univie.ac.at

[3] Vienna Metabolomics Center (VIME), University of Vienna, Althanstraße 14, 1090 Vienna, Austria

[4] Research Network Chemistry Meets Microbiology, Althanstraße 14, 1090 Vienna, Austria

[*] Correspondence: gunda.koellensperger@univie.ac.at; Tel.: +43-1-4277-52303

Received: 9 November 2019; Accepted: 12 December 2019; Published: 14 December 2019

Abstract: Tumor spheroids are important model systems due to the capability of capturing in vivo tumor complexity. In this work, the experimental design of metabolomics workflows using three-dimensional multicellular tumor spheroid (3D MTS) models is addressed. Non-scaffold based cultures of the HCT116 colon carcinoma cell line delivered highly reproducible MTSs with regard to size and other key parameters (such as protein content and fraction of viable cells) as a prerequisite. Carefully optimizing the multiple steps of sample preparation, the developed procedure enabled us to probe the metabolome of single MTSs (diameter range 790 ± 22 μm) in a highly repeatable manner at a considerable throughput. The final protocol consisted of rapid washing of the spheroids on the cultivation plate, followed by cold methanol extraction. ^{13}C enriched internal standards, added upon extraction, were key to obtaining the excellent analytical figures of merit. Targeted metabolomics provided absolute concentrations with average biological repeatabilities of <20% probing MTSs individually. In a proof of principle study, MTSs were exposed to two metal-based anticancer drugs, oxaliplatin and the investigational anticancer drug KP1339 (sodium *trans*-[tetrachloridobis(1*H*-indazole)ruthenate(III)]), which exhibit distinctly different modes of action. This difference could be recapitulated in individual metabolic shifts observed from replicate single MTSs. Therefore, biological variation among single spheroids can be assessed using the presented analytical strategy, applicable for in-depth anticancer drug metabolite profiling.

Keywords: multicellular tumor spheroids; metabolomics; metallodrugs; oxaliplatin; KP1339; method development; LC-MS; IT-139

1. Introduction

Three-dimensional multicellular tumor spheroids (3D MTSs) emerged as essential tools in cancer research with the aim of increasing the efficiency of oncologic drug development. Indeed, MTSs provide a cancer model of intermediate complexity, not replacing animal models entirely, but constituting a substantial improvement compared to two-dimensional (2D) monolayer cell cultures [1,2]. 3D MTSs grown from established cancer cell lines resemble more closely early-stage avascular tumors than 2D monolayer cell cultures in many aspects. For example, nutrient and oxygen concentration gradients as present in tumors are established in 3D MTS models, which in turn results in a concentration of proliferating cancer cells in the outer layers of the MTS, while in the inner zone, deprived of nutrients and oxygen, necrotic cells accumulate (depending on the MTS size). In addition, viable quiescent cells are found in a transition zone between the necrotic core and outer layer of proliferating cells [3]. This inhomogeneity is an important feature of in vivo tumors, which 2D monolayer cell cultures fail to recapitulate as they predominantly contain normoxic, actively proliferating cells [2,4]. Many anticancer drugs exert their cytotoxicity against proliferating cells, with quiescent cells evading the treatment [5]. Thus, a model system such as 3D MTSs, which involve quiescent cells, is of utmost importance.

Another key aspect is the fact that the three-dimensionality intrinsically affects the cell morphology (rather round instead of stretched-out on a plastic surface), which relates to cell-to-cell contacts, stimuli exerted by cell surface receptors, and, ultimately, transcription and protein expression levels [6]. Furthermore, reduced oxygen levels and hypoxia lead to the generation of reactive oxygen species (ROS) and hypoxia-inducible factor-1 (HIF-1) stabilization, which is a major transcription factor responsible for metabolic transformation and tumor progression [7–10].

In the recent past, the combination of cancer models with cutting edge –omics type of analysis showed to be a powerful approach for generating new hypotheses regarding the prediction of drug susceptibility, drug resistance, and mode of action [11,12]. This was accompanied by a reemerging interest in cancer metabolism [13]. Metabolic signatures are accepted as the closest proxy for a phenotype [14,15]. One of the biggest hopes in cancer metabolomics is the discovery of molecular signatures with predictive power in cancer therapy [16]. As a consequence of this renewed interest in metabolism, dedicated workflows were introduced, addressing the needs of preclinical and clinical studies [17–19]. Cancer cell model studies required the development and validation of multi-step sample preparation protocols [20]. While today metabolomics experiments with 2D monolayer cell cultures are routine, sample preparation protocols for 3D MTSs are rarely discussed in detail. Up to date, only a few reports on metabolomics in 3D MTS models exist. The studies involved a range of LC-MS-based metabolomics workflows including lipidomics and fluxomics applications [21–25]. Despite showing the power of combining advanced models such as 3D MTSs and metabolomics, the validation of sample preparation was not addressed comprehensively. Even a detailed description of the experimental design (e.g., whether MTS samples were pooled for the analysis or how gels established in 3D culture were removed upon metabolome extraction) was lacking.

In this work, we focus on the experimental design enabling metabolomics in 3D MTSs, using non-scaffold based cell cultures grown on ultra-low attachment plates from cell suspensions. This approach avoids 3D scaffolds or gels for obtaining a three-dimensional structure, which facilitates metabolomics measurements. More specifically, sample preparation protocols are developed with the goal to provide a validated procedure capable of probing single MTSs; at the same time not compromising on analytical throughput and, thus, the number of biological replicates. The validity of the established preclinical tool-set was shown for the example of metal-based anticancer drug development. Metal-based drugs are a prime example since a clear cut mechanism remains to be elucidated despite extensive clinical use and fundamental research [26,27]. In fact, how the drugs exert their cytotoxicity differs even for the three clinically approved platinum(II) drugs [28–30]. Although massive research efforts resulted in a plethora of promising candidate drugs, the failure rate upon translation into clinics was/is extremely high for metal-based anticancer drugs [31]. Discovering potential metabolic pertubations specific for drug exposure, by measuring molecular signatures in advanced 3D MTS models, bear the potential of accelerating discoveries with regard to the mode of action and susceptibility towards the drug. In this work, a 3D human colon cancer model was studied. Metabolic shifts, as exerted by the clinically established oxaliplatin and KP1339, a promising candidate drug, were investigated.

2. Results

2.1. Establishment and Validation of Sample Preparation Procedures Suitable for Probing the Metabolome of Spheroids

2.1.1. Determining the Minimal Number of MTS Required for Metabolomics Experiment

A uniform size distribution of MTSs was obtained by using the colon cancer model HCT116; seeding of 3×10^3 cells and cultivation for 4 days following the suspension-based procedure described in the experimental section. The average resulting spheroid diameter was 560 ± 30 μm ($n = 56$). Suspension-based 3D cultivation enabled a straightforward establishment of metabolomics workflows otherwise hampered by tedious washing procedures in hydrogel and other scaffold-based techniques. The investigated size range amounted to $9.9 \times 10^3 \pm 3.9 \times 10^3$ cells with $81\% \pm 5\%$ viable cells (Table S1, "cellNumbers" sheet). The estimation was based on disaggregating the MTS with a recombinant enzyme reagent, staining with a dye, and subsequent measurement with a flow cytometer to determine cell counts and viability in the final suspension.

In order to exclude poor extraction efficiencies for MTSs in this work, metabolomics experiments resorted to boiling ethanol extractions, implementing the yeast-derived fully ^{13}C labeled (U^{13}C) internal standard. More specifically, extractions were carried out on single spheroids (three biological replicates, $N = 3$) and pooled spheroid samples, i.e., pools of 5 ($N = 3$), 10 ($N = 3$) and 15 spheroids ($N = 2$), respectively. The applied boiling ethanol protocol is the established gold standard in yeast metabolomics, offering nearly 100% extraction efficiency and recovery for a large panel of primary metabolites [32,33]. In cancer cell monolayer cultures, less tedious cold extraction protocols are established, which demand thorough validation when applied to MTS investigations [20,34].

Targeted metabolomics measurements were performed implementing reversed-phase chromatography coupled to tandem mass spectrometry using a 100%-wettable column providing enhanced separation for branched amino acids and organic acids [35]. Metabolite abundances relative to the isotopically enriched internal standard, i.e., relative response ratios, were addressed in single MTS extractions versus pooled extractions. The validation considered 29 metabolites (amino acids, organic acids, nucleotides, nucleosides). Single spheroid extractions resulted on average in repeatabilities of 28%, while pooled samples showed repeatabilites of 17% and 32% for 5 and 10 MTSs pooled, respectively (Table S1, "metabolitesExtract" sheet). In order to evaluate whether the number of pooled MTSs correlated with cell number and protein content, the cell pellets remaining upon boiling ethanol were submitted to acidic hydrolysis. Absolute amounts of 8 amino acids (alanine, arginine, glycine, histidine, lysine, phenylalanine, proline, tyrosine) were determined. The selected amino acids were quantitatively recovered [36] by the applied sample preparation protocol and were used for traceable protein quantification. The obtained absolute amino acid amounts displayed a strong linear correlation with the number of MTSs (Table S1, "aminoAcids_pellet", and Figure A3). This linear correlation (coefficient of determination above 0.99) was a prerequisite for further evaluation of metabolome abundances in single versus pooled MTS samples, depending on the assumption of linear correlation between cell number, protein concentration, and the number of spheroids (for a uniformly sized MTS sample set). Upon transforming the linear regressions of metabolite abundances from intracellular cell extracts versus number of spheroids by normalizing the metabolite abundances to average abundance found in single MTS samples, the parameters of the linear regression became comparable: assuming an ideal scenario (as represented by 100% extraction efficiency and recovery regardless of whether pooled or single samples are investigated) for these plots, a slope of 1 is expected and can be observed for the investigated metabolites, the ideal case of slope of 1 is nearly met (Table S1, "regressionNormalized", Figures A2 and A3). On average, the slope of the normalized regression is 1.11 and the average coefficient of determination is 0.94. Overall, the strong linear correlation of the relative responses with the number of uniform-size spheroids indicates that metabolomics experiments, even from single spheroids, are highly repeatable.

These findings are captured in a correlation matrix (Figure 1), which reveals that most of the investigated metabolites have a strong positive correlation with the amino acids measured from the corresponding pellet as well as from the number of MTSs they were extracted from.

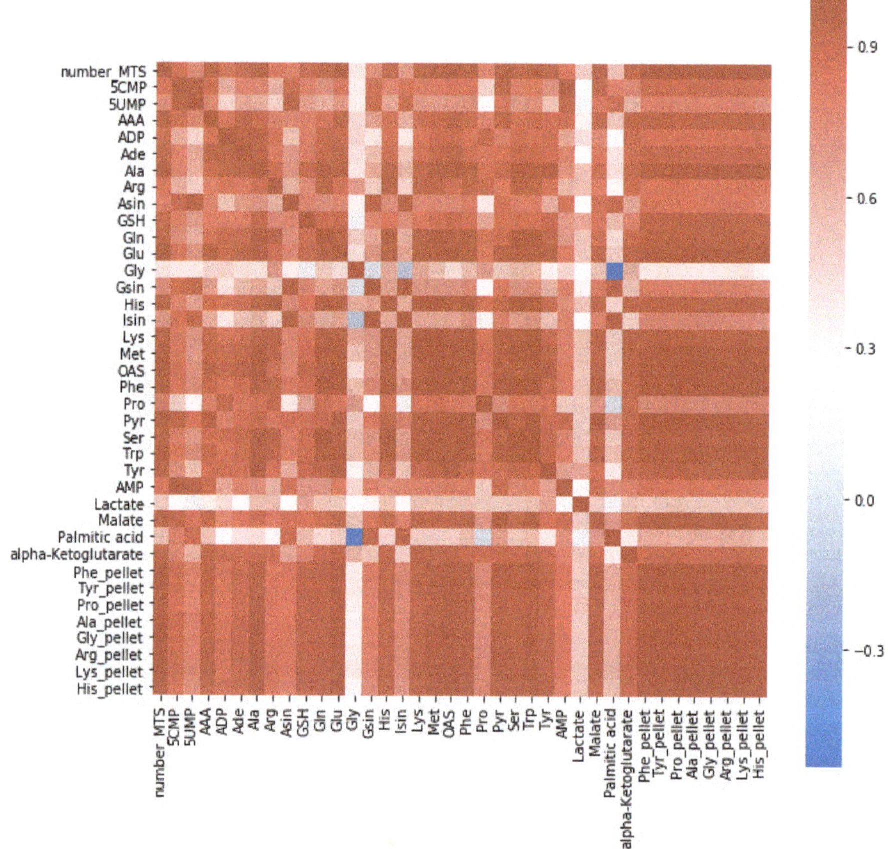

Figure 1. Correlation matrix of number of multicellular tumor spheroids (MTS), relative responses ($^{12}C/^{13}C$) of 29 metabolites from boiling ethanolic extracts with ^{13}C internal standardization, absolute amounts of 8 amino acids [nmol] measured from hydrolyzed pellet (indicated with the "pellet" suffix) from samples containing 1 ($N = 3$), 5 ($N = 3$), 10 ($N = 3$), and 15 MTS ($N = 2$) from the same population. Most metabolites show a strong correlation with all the amino acids hydrolyzed from the pellet and the number of MTSs.

2.1.2. Speeding Up the Sample Preparation

In the next step, alternative extraction protocols were addressed with the aim of increasing analytical throughput without compromising overall accuracy. The workflow based on boiling ethanol, involved the transfer of MTSs into tubes, three washing steps, and quenching by flash freezing with liquid nitrogen, which was followed by hot extraction. As a drawback, the procedure is rather time-consuming and on top of that, only a very limited number of MTS samples can be prepared in parallel. In fact, during collection and washing, only a limited number of spheroids can be handled at a given time. Only after the handled samples are quenched can be the next MTS collected. This often implies different durations of incubation at uncontrolled temperatures and CO_2 partial pressure, which might lead to systematic metabolic biases. Thus, it is key to decrease the time until quenching, as only a high degree of synchronization enables reasonable study sizes (number of replicates) and investigations of cells upon multiple metabolic perturbations [20].

In this work, the reduction of cell manipulations was addressed as follows: First, comparative metabolomics experiments compared different strategies regarding the first washing steps. Instead of transferring MTSs to tubes for repeated washing, single-step washing was performed on the cultivation plate directly. (For the sake of clarity the first approach was denoted as "OFF", while the single washing step on the well-plate was denoted as "ON".) Second, the extraction procedure was optimized. More specifically, the hot extraction (boiling ethanol, "BE") was compared to a cold methanol-based extraction ("CM"). In the past, it was shown that the harsh conditions of hot ethanol were not a stringent requirement for monolayer mammalian cell cultures [34], where mild cold extractions proved to be valid strategies.

The already-mentioned studies [18–22] with 3D cultures involving LC-MS-based metabolomics workflows, including lipidomics and fluxomics applications utilized washing of the samples and organic solvents (methanol, acetonitrile) with an aqueous proportion for extraction, mostly in cold state. To the best of our knowledge, a thorough evaluation of 3D cancer cell models is still lacking.

Figure 2 depicts the two sample preparation strategies involving washing, quenching, internal standardization using ^{13}C internal standards and extraction. The validation experiments were carried out using replicates of single spheroids. The cultivation was designed to produce large spheroids (diameter = 744 ± 15 µm, 4-day long cultivation, and 10^4 cells seeded) representing a "worst-case scenario" for efficient extraction. Four different sample preparation strategies were compared, namely OFF and ON plate washing, both followed by either BE or CM, respectively, all involving U^{13}C internal standardization upon extraction. The extracts were measured as described in [37]. In brief, a hydrophilic interaction liquid chromatography (HILIC) separation at pH 4 was used combined with high-resolution MS. Targeted absolute quantification of 26 metabolites based on external calibration with internal standardization served as validation of the sample preparation.

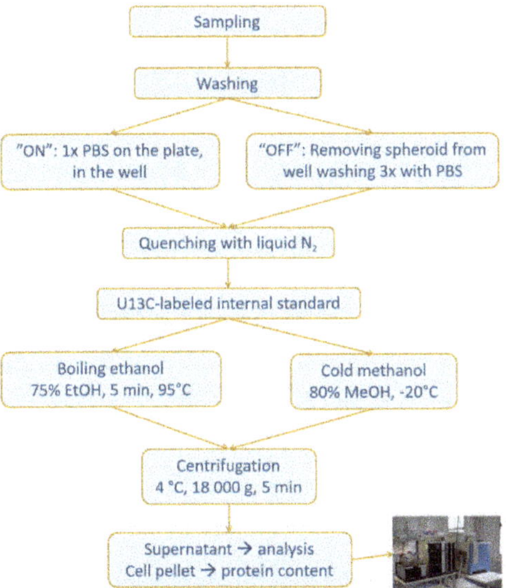

Figure 2. The study design for speeding up the sample preparation. Two different washing procedures ("ON": washing the multicellular tumor spheroid once with PBS on the 96-well plate vs. "OFF" washing three-times in an Eppendorf-tube after transfer from cultivation well) and two alternative extractions (boiling ethanol vs. cold methanol extraction) were tested on large multicellular tumor spheroids (744 ± 15 µm), which in combination resulted in four sample groups.

In addition to the biological repeatability, the technical repeatability was assessed by injections of a pooled sample over the measurement series (denoted as QC). Figure 3 summarizes the quantitative metabolome data. Overall, it was found that the accelerated workflow with the reduced washing steps was key to improve repeatability, as this resulted in the lowest average relative standard deviations (18% and 19% for ON/BE and ON/CM, respectively. See Table 1.). Moreover, it could be shown that the cold methanol (CM) extraction was comparable to boiling ethanol (BE) in terms of extraction efficiency, as the quantitative values were in good agreement (within their uncertainty).

Table 1. Average relative standard deviations based on absolute concentrations of 26 analytes in four investigated sample preparation strategies for single multicellular tumor spheroids. The four different sample preparation methods involved two washing procedures ("ON": washing the multicellular tumor spheroid once with PBS on the 96-well plate vs. "OFF" washing three times in an Eppendorf tube after transfer from cultivation well) and two alternative extractions are combined (boiling ethanol "BE" vs. cold methanol extraction "CM"). OFF/BE ($N = 6$), ON/BE ($N = 4$), OFF/CM ($N = 5$), ON/CM ($N = 7$) One pooled sample from independent OFF/CM and ON/CM samples, QC was a mixture of one OFF/CM and ON/CM sample and was measured as quality control throughout the measurement ($N = 4$). Measurement was performed with liquid chromatography (HILIC separation) high-resolution Orbitrap mass spectrometry. All extractions utilized U^{13}C internal standardization.

Sample Preparation	Average RSD [%]
OFF/BE	34%
ON/BE	19%
OFF/CM	24%
ON/CM	18%
QC	7%

Thus it can be concluded that the protocol ON/CM offers a fast and convenient fit-for-purpose method to generate 60 biological replicates in parallel. The 60 replicates were treated and washed within a few minutes (depending on the operator) and then were quenched at the same time by flash freezing them with liquid nitrogen. Overall, a 96-well plate accommodates up to 60 spheroid cultivations since wells at the edges are not used to avoid the "edge effect" (artifacts due to inhomogeneity in evaporation rates).

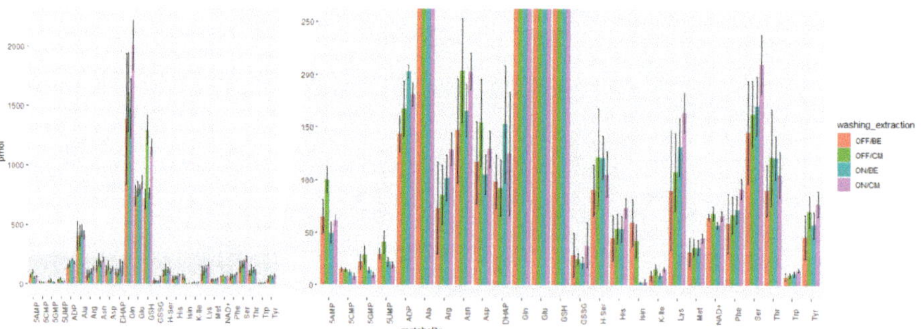

Figure 3. Barplots of the absolute amounts (pmol) of 26 selected metabolites extracted from single multicellular tumor spheroids with four different sample preparation method as well as two washing procedures ("ON": washing the multicellular tumor spheroid once with PBS on the 96-well plate vs. "OFF" washing three times in an Eppendorf-tube after transfer from the culture plate) and two alternative extractions (boiling ethanol, "BE" vs. cold methanol extraction "CM"). OFF/BE ($N = 6$), ON/BE ($N = 4$), OFF/CM ($N = 5$), ON/CM ($N = 7$) Measurement with liquid chromatography (HILIC separation) high-resolution Orbitrap mass spectrometry. All extractions utilized U^{13}C internal standardization. For the abbreviations of compounds, see Appendix A.

2.2. Assessing Metabolic Shifts in Single MTS Exposed to Metal-Based Anticancer Drugs

Finally, the optimized ON/CM workflow was applied in a proof of principle study addressing metabolic perturbations in single MTSs due to exposure of metal-based anticancer drugs. Again human colon cancer cells (HCT116) were selected. The spheroids were grown from 10^4 cells for 8 days (790 ± 22 μm). Two drugs of distinctly different proposed modes of action were investigated, namely the clinically established oxaliplatin and the investigational anticancer drug sodium *trans*-[tetrachloridobis(1H-indazole)ruthenate(III)] (denoted as KP1339). While oxaliplatin exerts its cytotoxic effects through DNA damage and ribosome biogenesis stress [38], there is growing evidence that KP1339 is not a DNA-damaging agent but its primary mode of action is through endoplasmic reticulum stress [39]. Furthermore, KP1339 shows a prodrug nature, as it is thought to be preferentially reduced in the more reductive milieu of solid tumors to the active Ru(II) form. Finally, while oxaliplatin is considered a bona fide immunogenic cell death inducer [40], KP1339 also exhibits the hallmarks of immunogenic cell death [41,42].

The choice of incubation time and drug concentration was based on previous studies using monolayer cultures, which showed prominent metabolomic shifts only after 24 h exposure to sub-cytotoxic drug concentrations [37]. Specifically, the applied concentrations were 20 and 200 μM for oxaliplatin and KP1339, respectively. As the dissolution and thus the application of KP1339 involved dimethyl sulfoxide (DMSO), an additional control group resembling the DMSO background in the medium was included. Figure A4 and Table S3 (protein_μg sheet) show the protein concentrations obtained in replicate single MTSs for the different conditions under investigation. The plotted protein content was assessed by measuring the concentration in the remaining cell pellets of the individual MTS samples. As can be readily observed, oxaliplatin treatment resulted in a minor reduction of the overall protein content; otherwise the average mean protein concentration was comparable (within its uncertainty) for all conditions, pointing towards comparable growth rates. On top of that, the data clearly showed that using the average protein content of parallel MTS cultivations would compromise the quality of comparative metabolomics experiments. When aiming at the investigation at single MTS level, it is a requirement to normalize metabolic abundances to the corresponding protein content obtained from the same sample. Finally, the ON/CM sample preparation procedure optimized for single MTS analysis comprised the addition of yeast ^{13}C standards. The measurements relied on hydrophilic interaction chromatography—Orbitrap MS and included external calibrants with internal standards for a large panel of metabolites [37]. Implementing the internal standardization approach together with the use of high-resolution Orbitrap MS enabled us to perform targeted and non-targeted metabolomics in a single analytical run.

As suggested by the standardization initiative of the Metabolomics Society, pooled MTS extracts were used as quality control samples [43,44]. Combining positive and negative data, absolute concentration values were obtained. Only analytes with technical repeatability obtained from repeated injections of the QC sample below 30% were considered, which resulted in a total number of 58 remaining analytes. The average relative standard deviation (RSD) for the 58 metabolites was 6.5% for the technical replicates and 13% for biological replicates considering each four group (Table S3, pmolMetabolite sheet). The average RSD after the protein content normalization was 15.2% (Table S3, pmolMetabolitePerμgProtein sheet).

Unsupervised statistical analysis of the targeted, protein content normalized data revealed that there is group clustering according to the type of drug treatment in the case of KP1339 treatment (PCA plot in Figure A5 and heat map in Figure 4). The KP1339-treated samples showed significant regulation of 19 metabolites (proline, propionyl-L-carnitine, ribulose-5-phosphate/ribose-5-phosphate, adenosine monophosphate, lactate, aspartate, reduced glutathione, guanosine monophosphate, glutamine, inosine, glutamate, N-acetylserine, adenosine, dihydroxyacetone phosphate, asparagine, mevalonic acid, alanine, and sarcosine; Table S3, ("significant_cmpds_KP1339") compared to six metabolites (adenosine, guanosine monophosphate, cytidine monophosphate, nicotinamide adenine dinucleotide phosphate (oxidized), uridine monophosphate, and uracil; from Table S3 "significant_cmpds_oxaliPt") in oxaliplatin-treated samples. The stronger metabolic change caused by the ruthenium drug (KP1339) treatment compared to oxaliplatin treatment was further confirmed by hierarchical cluster analysis, where KP1339-treated samples were separated from both controls, which was not the case for oxaliplatin-treated samples (Figure 4). Overall, the fact that we see a stronger effect in the metabolome with KP1339 treatment than with oxaliplatin treatment is not surprising, since in a study with monolayer cell cultures of the same cell line, we observed considerably milder effects with oxaliplatin, as well [37].

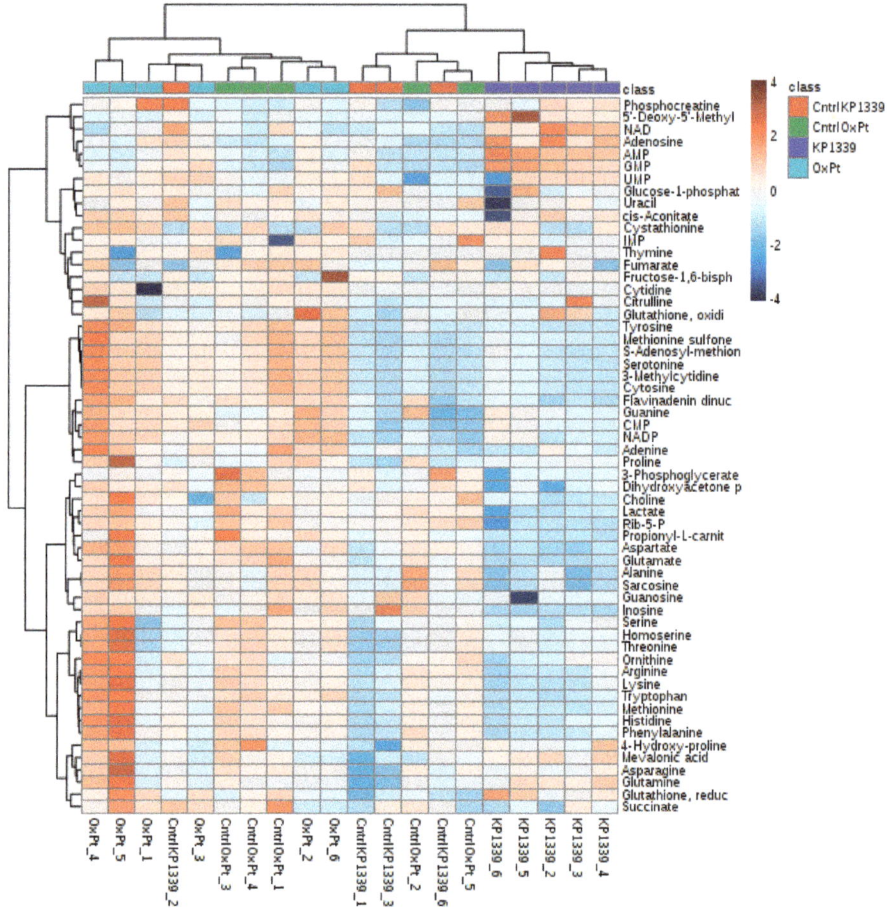

Figure 4. Heatmap displaying absolute amounts of metabolites normalized to total protein content (pmol/µg) of samples from single 3D MTS. Spheroids were treated for 24 h with either 200 µM KP1339, 20 µM oxaliplatin (OxPt), medium (CntrlOxPt) or medium with 0.5% DMSO (CntrlKP1339). Numbers in sample names refer to independent biological replicates. Extraction with cold methanol and internal standardization, measurement by HILIC high-resolution Orbitrap MS.

Pathway enrichment analysis using targeted data revealed that oxaliplatin exposure affected purine metabolism (GMP and adenosine being most significantly down-regulated; glutamine, IMP, inosine, guanosine, guanine, adenine, and AMP were also affected) and pyrimidine synthesis (UMP and CMP being most significantly down-regulated; but cytidine, uracil, glutamine and thymine were also affected), which is in agreement with the accepted mechanism of action of DNA targeting [45] (Figure 5a, Table S3, "pathways_pathways_oxaliplatin"). These findings were also supported by other significant metabolic shifts involving DNA building blocks. Other retrieved pathways could be related to redox stress such as the pentose phosphate pathway, glutathione metabolism, and nicotinamide metabolism, but also purine metabolism, which is in accordance with the ribosome biogenesis stress only recently proposed as the primary reason for the cytotoxic effect [38]—a hypothesis generated based on transcriptomic analysis. In our work, the measurement of the metabolome provided additional evidence supporting this hypothesis, as the levels of many RNA monomers were significantly altered upon drug exposure (Table S3, "pathways_pathways_oxaliplatin"), and RNA being one of the major component of ribosomes. This was further supported by the fact that the "aminoacyl-tRNA biosynthesis" pathway was also among the affected pathways (see Table S3 for the complete list). Finally, the MetaboAnalyst 4.0 Pathway Analysis module [46] revealed another interesting pathway to be further investigated, namely, the "pantothenate and CoA biosynthesis" pathway (through uracil downregulation). A recent study [29] addressed signature genes for patients responding to oxaliplatin therapy by machine learning. Among the most accurate signature genes for oxaliplatin treatment was PANK3 which encodes for pantothenate kinase, a key regulatory enzyme in the biosynthesis of coenzyme A (CoA). A seminal study correlating transcriptomics and metabolomics for the NCI60 cell line panel showed the involvement of the TCA cycle, pyruvate metabolism, lipoprotein uptake and nucleotide synthesis in platinum sensitivity [47,48]. Again, the metabolomics data of this study support the generated hypothesis as the TCA cycle, purine and pyrimidine metabolism, and pyruvate metabolism were indicated by pathway enrichment analysis.

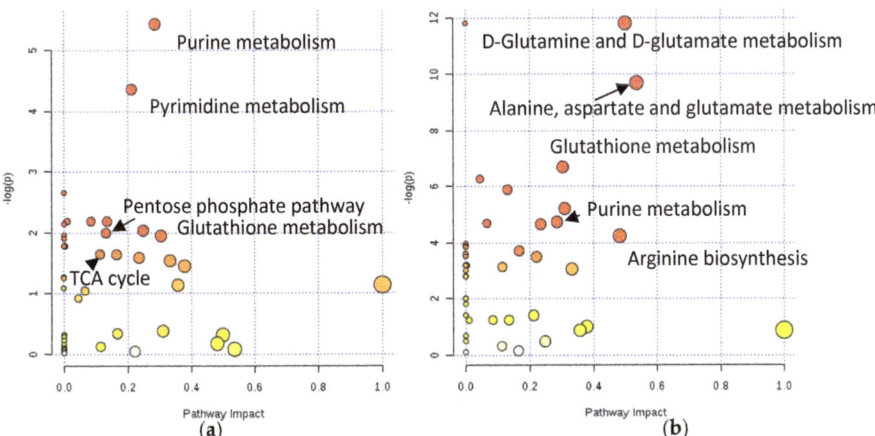

Figure 5. KEGG pathways with MetaboAnalyst affected by oxaliplatin treatment (**a**) and KP1339 treatment (**b**) using pathway enrichment and topology analysis with the MetaboAnalyst pathway analysis module.

The hypothesized difference in the mode of action of the two investigated drugs was reflected by the distinct metabolic shifts, which were observed upon exposing MTSs to the candidate ruthenium drug KP1339. There are indications that this drug is a GRP78 inhibitor, an ER stress sensing chaperone [39,49]. It not only induces ER stress and unfolded protein response, but it has also been shown that it exhibits the hallmarks of immunogenic cell death, calreticulin exposure, and ATP secretion among others [42]. Less is known about the metabolic effects of this compound. In this work, we could individuate the impact on biochemical pathways related to redox stress: glutathione metabolism (glutathione, oxidized glutathione and NADP+ are up-, glutamate down-regulated), purine metabolism, pentose phosphate pathway (ribulose 5-phosphate/ribose 5-phosphate down-regulated), which could be explained by the hypothesized reduction of the drug. Redox stress and elevation of ROS protective proteins [42,50] were confirmed by previous proteomics studies. Finally, pathways such as glycerophospholipid metabolism (strong choline down-regulation) and various amino acid metabolism-related pathways were pinpointed (glutamine and glutamate metabolism, alanine, aspartate and glutamate metabolism, arginine biosynthesis). Altered amino acid synthesis is related to the fact that the unfolded protein response induces changes in the expression of genes of amino acid transport and synthesis [51].

3. Discussion

In this work, we present a carefully optimized sample preparation workflow for metabolomics experiments in 3D MTS samples. To the best of our knowledge, this is the first study performing extraction optimization ($n \geq 4$) for single tumor spheroids. The comparison of different extraction strategies, namely ON/BE, OFF/BE, OFF/CM, ON/CM, revealed that cold 80% methanol extraction with a single washing step on the plate was most promising for single tumor spheroid analysis. Using the presented workflow increased throughput and convenience as well as resulted in superior analytical figures of merit including the highest analyte concentration and lowest RSDs observed for 26 metabolites in the pmol to nmol range. The sample preparation strategy is limited to spheroids with diameters >400–500 μm and with maximum diameters of 900–1000 μm due to potential error-prone handling and growth limitation, respectively. Other studies in tumor spheroid analysis were performed using various culturing and extraction protocols such as (1) gelatinous cultivation of spheroids and methanol-water extraction [21], (2) magnetized cells to form 3D cultures and cold acetonitrile, 70% methanol or 80% acetone extraction [22], (3) gel-based ultra-low attachment plates of 20–25 spheroids per sample followed by derivatization and chloroform/methanol extraction [23]. Overall, most –omics mass spectrometry-based studies on tumor spheroid analysis rely on extraction strategies with cold organic solvents such as methanol, ethanol or isopropanol [21–24]. This is consistent with our tumor spheroid workflow based on non-scaffold based cultivation on ultra-low attachment plates and the optimized cold 80% methanol extraction with a single washing step. The obtained strong linear correlation with the number of (uniform-size) spheroids and relative responses indicates highly repeatable metabolomics readout from single spheroids. The use of ^{13}C labeled standards for quantification represents a significant advance compared to label-free quantification approaches as the internal standardization allowed to account for sample losses during sample preparation, storage and measurement. Additionally, ^{13}C metabolites represent an interesting normalization strategy for comprehensive non-targeted metabolomics experiments on tumor spheroids. The novel procedure enabled us to probe individual spheroids with excellent biological repeatability (average RSD of 18%) in a considerable throughput. An experiment with a single 96-well plate allows the investigation of up to 60 single spheroids (since wells at the edges are not used). However, high throughput analysis is feasible as there is no theoretical limit in the number of biological replicates or investigated conditions when multiple plates are combined.

The proposed methodology enabled us to investigate different conditions of 3D human cancer models in a parallelized manner comparable to the degree obtained in 2D monolayer cell cultures. The applicability of the method was shown on the example of the metal-based anticancer drugs KP1339 and oxaliplatin. Comparison of the two drugs ($n \geq 4$) revealed stronger metabolic changes caused by the ruthenium drug (KP1339) treatment compared to oxaliplatin treatment, which is consistent with the literature on monolayer cell cultures [37]. We observed significant metabolic changes with purine and pyrimidine pathways after oxaliplatin treatment, which is in agreement with the accepted mechanism of action of DNA targeting of oxaliplatin [45] and consistent with the proposed induction of ribosome biogenesis stress [38]. Further biological interpretation of the data is beyond the scope of this work.

The presented optimized sample preparation workflow is the ideal starting point for single spheroid metabolomics experiments. A relatively large proportion of clinically tested drugs fail in phase 3 of the trials due to insufficient efficacy or unacceptable toxicity, which means a substantial financial loss [6,52]. Since tumor spheroids are three-dimensional models derived from established human cell lines, it is possible to capture more of the complexity of a tumor compared to standard 2D monolayer cultures [1]. Ultimately, metabolomics studies on tumor spheroids could be integrated into anticancer drug screening, helping to prevent late clinical failure of drug candidates. Future studies will focus on in-depth metabolomics analysis of different anticancer drugs effects using tumor spheroids. Overall, we strongly believe that drug development will benefit significantly from the new discovery tools provided by the unique combination of 3D MTSs and metabolomics.

4. Materials and Methods

4.1. Cell Culture

4.1.1. Cultivation of Spheroids

The human colon carcinoma cell line HCT116 was kindly provided by Brigitte Marian, Institute of Cancer Research, Department of Medicine I, Medical University of Vienna. HCT116 cells were cultured as adherent monolayers in 75 cm^2 flasks (StarLab Germany) in McCoy's 5a medium (Sigma-Aldrich) supplemented with 10% fetal calf serum (FCS) (BioWest) and 4 mM L-glutamine without antibiotics at 37 °C (StarLab) under a humidified atmosphere containing 5% CO_2. All cell culture media and reagents were obtained from Sigma-Aldrich (Vienna, Austria), and all plastic dishes, plates and flasks were from StarLab (Germany) unless stated otherwise. For spheroid generation, HCT116 cells were harvested from culture flasks by trypsinization, resuspended in their respective supplemented medium, and seeded in 200 µL on ultra-low attachment round-bottom 96-well plates (Nunclon Sphera^{TM}, Thermo Fisher Scientific). To avoid effects caused by evaporation, the outermost wells were not used for spheroid production and filled with 200 mL of PBS instead. In the experiment where 1, 5, 10, 15 spheroids were investigated, the spheroids were seeded at a density of 3 × 10^3 viable cells per well in 200 µL. Plates were incubated at 37 °C with 5% CO_2 for 96 h to allow spheroid formation. In the experiment where washing procedures and extraction methods were compared, 10 × 10^3 viable cells were seeded in 200 µL medium and cultivated at 37 °C with 5% CO_2 for 96 h to allow spheroid formation. In the proof of concept experiment with metallodrugs, 10 × 10^3 viable cells were seeded in 200 µL medium and cultivated at 37 °C with 5% CO_2 for 96 h to allow spheroid formation, and then 100 µL medium was aspirated and exchanged for fresh medium. After 192 h (8 days) of total incubation, the diameter of the spheroids was measured, 100 µL of medium was exchanged for the respective treatment solution and 24 h of incubation followed either with the drug or with control medium. KP1339 was first dissolved in DMSO, and stock solutions were prepared in the respective medium with FCS and glutamine supplement. Oxaliplatin was dissolved freshly before the experiments in the supplemented medium only.

4.1.2. Microscopy

An Olympus CKX41 (Olympus, Vienna, Austria) microscope was used to measure the diameter of the spheroids in horizontal and vertical directions with cellF 2.7 imaging software and the average diameter was calculated.

4.1.3. Cell Number Estimation

Spheroids grown from 3 × 10^3 viable cells for 96 h were transferred from the plate to Eppendorf-tubes as single spheroids. After washing once with PBS, dissociation in 100 µL TrypLE Express (Gibco, Vienna, Austria) followed for 15 min at 37 °C. Samples were thoroughly pipetted to disperse any remaining cell aggregates, and 100 µL colorless McCoy's 5a medium (Sigma-Aldrich) with 2% FCS were added followed by 0.8 mL Guava ViaCount reagent (Merck/Millipore, Germany). Samples were measured immediately by using a Guava Soft flow cytometer (Merck/Millipore, Darmstadt, Germany).

4.2. Methods for Determining the Minimal Number of MTS Required for a Metabolomics Experiment

4.2.1. Sampling 1, 5, 10, 15 Multicellular Tumor Spheroids

3D MTS were seeded with 3×10^3 cells in 200 µL McCoys medium and cultivated for 4 days, reaching a diameter of 560 ± 30 µm. Upon extraction, spheroids were transferred by pipetting to a Petri-dish, where little droplets of PBS had been pre-pipetted. Each spheroid was transferred sequentially into three fresh PBS droplets corresponding to three washing steps. After that, each spheroid was transferred into a conical screw cap vial (Bioquote Limited, York, United Kingdom). The vial was put into liquid nitrogen in order to quench the enzymatic activity and put on wet ice, while more spheroids were collected. As soon as the last spheroid for a given sample was washed and sampled to the vial, it was put on −20 °C. When all the samples were collected, they were stored at −80 °C until boiling ethanol extraction.

4.2.2. Internal Standardization

A fully ^{13}C labeled yeast extract of *Pichia pastoris* (2 billion cells) from ISOtopic Solutions e.U., (Vienna, Austria) was reconstituted in 2 mL of water and added in equal amounts to the samples. The final dilution of the internal standard for the measurement was 1:10.

4.2.3. Boiling Ethanol Extraction for 1×-5×-10×-15× MTSs

The protocol was implemented from the works [32,33]. Shortly, 75% ethanol was prepared from ethanol, abs. 100%, HPLC grade (Chem-Lab, Vienna, Austria), H_2O (Sigma, Vienna, Austria, LC-MS-grade), and pre-heated in a clean glass beaker in a 95 °C water bath. The 50 µL of U^{13}C internal standard was added to samples. Once the ethanol–water was boiling in the glass beaker, the hot extraction solvent was added to the conical screw cap vials (Bioquote Limited, York, United Kingdom) to dilute the internal standard amount 1:10 and, subsequently, a 5-min incubation in the 95 °C water bath followed. The vials were closed, vortexed, and centrifuged (14,000 rcf, 4 °C, 5 min). The supernatant was carefully transferred to an MS-vial (Macherey-Nagel, Vienna, Austria), without disturbing the pellet. Extracts were evaporated until dryness. Extracts and resulting pellets were stored at −80 °C until measurement.

4.2.4. Metabolomics Measurement of Extracts 1×-5×-10×-15× MTSs

Samples were reconstituted in 100 µL LC-MS-grade H_2O and analyzed with reversed phase liquid chromatography coupled to high-resolution Orbitrap mass spectrometry. Waters Acquity HSS T3 (2.1 × 150 mm, 1.8 µm) column was used, mobile phase A was H_2O (0.1% formic acid), and mobile phase B was 100% methanol. The following gradient was used at a flow rate of 0.3 mL min^{-1} and 40 °C: 0.0–2.0 min 0% B, 2–6 min 0–40% B, 6–8 min 40–100% B, 8–11 min 100% B, and at 11.1 min switch to 0% B, 11.1–15 min 0% B. The injection volume was 10 µL. The Vanquish Duo UHPLC-system (Thermo Fisher Scientific) was used.

High-resolution mass spectrometry was done with a high field Thermo Scientific™ Q Exactive HF™ quadrupole-Orbitrap mass spectrometer equipped with an electrospray source. The ESI source parameters were the following: sheath gas 40, auxiliary gas 3, spray voltage 2.8 kV in negative and 3.5 kV in the positive mode, capillary temperature 280 °C, S-Lens RF level 30 and auxiliary gas heater 320 °C. Spectral data were acquired in profile mode. Resolution = 60.000, mass range = 60–900 m/z, AGC target 10^6, both in positive and negative polarities. Quantification was done through the areas of extracted ion chromatograms of [M+H]$^+$ and [M-H]$^-$ with 5 ppm mass tolerance on the U^{12}C to U^{13}C ratio.

4.2.5. Acidic Hydrolysis of the Pellet and Amino Acid Analysis

Acidic hydrolysis of the pellet resulting from the boiling ethanol extraction of samples of 1, 5, 10, and 15 spheroids pooled was based on a procedure described in detail elsewhere [53,54]. In short,

1 mL 6 M hydrochloric acid (Fluka, TraceSELECT from ≥37%) was added to the pellets, already in the conical screw cap vials (Bioquote Limited, York, United Kingdom); screw caps were tightly closed and the samples were hydrolyzed at 100 °C for 24 h. The hydrolyzed samples were evaporated to dryness and stored at −80 °C until analysis. Procedural blanks were also prepared to investigate possible background. Upon measurement, samples were reconstituted in 1 mL H_2O. A 1-to-10 dilution was done by taking 50 µL from the sample, adding 50 µL ISTD and 400 µL acetonitrile. NIST SRM 2389a amino acid standard (Sigma-Aldrich) was used to prepare an external calibration with internal standard and achieve absolute concentrations. The eight amino acids were subject to quantitative extraction and measurement according to [36] (alanine, arginine, glycine, histidine, lysine, phenylalanine, proline, tyrosine) were measured with a method described in detail elsewhere [35]. Shortly, the Agilent Infinity LC-System coupled to an Agilent 6490 triple quadrupole mass spectrometer was used with a HILIC separation, Waters Acquity BEH Amide (1.7 µm, 100 × 0.786 mm) column with 10 mM ammonium formate (pH 3.25) as eluent A and 80% acetonitrile 20% 10 mM ammonium formate (pH 3.25) as eluent B. Multiple reaction monitoring transitions were acquired in positive polarity.

4.3. Methods for Comparison Boiling Ethanol (BE) and Cold Methanol (CM) Extraction and Washing Procedures (OFF vs. ON) of 3D MTS

For this experiment, 3D MTS were grown as described above from 10×10^3 cells for 96 h.

4.3.1. Transfer of Spheroids

Sampling and transfer of spheroids were carried out with a 200 µL (Eppendorf, Vienna, Austria) pipette. The tip of the pipette tip was cut with a scissor (washed before in methanol:H_2O 50:50) so that the opening was slightly enlarged. The spheroid was taken up with the surrounding solution by the suction generated by the pipette and transferred from the plate into an Eppendorf tube.

4.3.2. "BE-OFF" Sample Preparation

The spheroids were transferred from the plate into conical screw cap vials (Bioquote Limited, York, United Kingdom), where they were washed three times with phosphate-buffered saline (PBS, Sigma-Aldrich, Vienna, Austria), quenched immediately with liquid nitrogen and stored at −80 °C until extraction. Upon extraction, 20 µL $U^{13}C$ internal standard was added as well as 180 µL 75% ethanol and put on 95 °C in the water bath for 5 min. After 5 min of sonication, samples were vortexed and centrifuged (14,000 rcf, 5 min, 4 °C) and supernatants transferred into MS-vials. Samples were measured directly without evaporation.

4.3.3. "CM-OFF" Sample Preparation

The spheroids were transferred from the plate into an Eppendorf tube, where they were washed three times with PBS and quenched immediately in liquid nitrogen and stored at −80 °C until extraction. Upon extraction, 20 µL $U^{13}C$ internal standard was added as well as 180 µL cold 80% methanol (−20 °C). After 5 min of sonication, samples were vortexed and centrifuged (14,000 rcf, 5 min, 4 °C) and supernatants transferred into MS-vials. Samples were measured directly without evaporation.

4.3.4. "BE-ON" Sample Preparation

The medium of spheroids was carefully aspirated by a multichannel pipette, while spheroids remained in the wells. Subsequently, the spheroids were washed with 200 µL PBS and quenched by liquid nitrogen. The spheroids still in the wells were stored at −80 °C until extraction. Plates were kept on ice during extraction. Upon extraction, 20 µL $U^{13}C$ internal standard was added to the well, then 80 µL 75% ethanol (room temperature, not hot yet) was added; in this amount, the spheroid was transferred to conical screw cap vials (Bioquote Limited, York, United Kingdom). One hundred µL more 75% ethanol was added to wash the well and transferred to the conical screw cap vials (Bioquote Limited, York, United Kingdom). Samples were kept on ice until the extraction of all samples was completed. Samples were vortexed and put on 95 °C in the water bath for 5 min, then 5 min of sonication followed. Finally, samples were centrifuged (14,000 rcf, 5 min, 4 °C), and supernatants were transferred into MS-vials. Samples were measured directly without evaporation.

4.3.5. "CM-ON" Sample Preparation

The medium of spheroids was carefully aspirated by a multichannel pipette, while spheroids remained in the wells. Subsequently, the spheroids were washed with 200 µL PBS and quenched by liquid nitrogen. The spheroids still in the wells were stored at −80 °C until extraction. Plates were kept on ice during extraction. Upon extraction, 20 µL of the U^{13}C internal standard was added, then 80 µL cold 80% methanol was added (−20 °C); in this amount, the spheroid was transferred to an Eppendorf-tube and 100 µL more cold 80% methanol was added to wash the well and transferred to the Eppendorf tube. Samples were kept on ice until the extraction of all samples was completed. After 5 min of sonication, samples were vortexed and centrifuged (14,000 rcf, 5 min, 4 °C), and the supernatants were transferred into MS-vials. Samples were measured directly without evaporation.

4.3.6. LC-MS Method Applied for Proof of Principle Experiment with Metallodrugs

The metabolite standards were purchased from Sigma-Aldrich or Fluka (Vienna, Austria) except for malic acid which was purchased from Merck (Vienna, Austria). The metabolomics experiment also included an external calibration with a calibration mix of 133 substances and internal standardization with U^{13}C-labeled yeast extract. Within every 10 injections, a blank was injected, as well as a pooled QC from extracts (all four groups represented in each: 200 µM KP1339-treated, 20 µM oxaliplatin-treated, control, control with 0.5% dimethylsulfoxid (DMSO)). We used high resolution Orbitrap MS coupled to Vanquish UPLC and with a HILIC separation, as described elsewhere [37]. MS-data were acquired with positive–negative switching and the extracted ion chromatograms were evaluated with Thermo Trace Finder, with the help of external calibration and the internal standard, absolute amounts were calculated in pmol. The normalization with total protein content resulted in pmol metabolite per µg protein.

4.3.7. Total Protein Content Determination

The protein concentration was assessed from the precipitate dissolved in 200 µL 0.05 M NaOH. For that purpose, the commercially available micro BCA protein assay kit (Thermo Fisher Scientific, Pierce Biotechnology, Rockford, IL, USA) was employed, according to the manufacturer's instructions.

4.4. Data Analysis

4.4.1. Targeted Metabolomics Data Treatment and Normalization

Targeted data evaluation of high-resolution mass spectrometry data of metabolites was performed in Thermo Trace Finder 4.1, while for the triple quadrupole data of amino acids hydrolyzed from pellet Agilent MassHunter was used. ^{13}C internal standardization using external calibration was employed for metabolites. All calibration curves were linear.

The dataset with absolute metabolite amounts and total protein contents was exported to Python, where the normalization with protein content was carried out, as well as filtering the dataset based on the performance of pooled quality control samples. (Missing values > 50%, relative standard deviation above 30%).

4.4.2. Exploratory Data Analysis

The MetaboAnalyst 4.0 Statistical Analysis module was used for further exploratory data analysis of dataset with absolute metabolite amounts normalized to total protein content. The default missing values imputation was used with a small number; autoscaling was applied before multivariate analysis.

4.4.3. Pathway Analysis

The MetaboAnalyst 4.0 [46] Pathway Analysis module was used, which combines pathway enrichment analysis and pathway topology analysis to identify the most relevant pathways involved in the conditions under study. As data input, the whole dataset with protein content normalized concentrations was used. Normalized concentrations were autoscaled; missing values imputed with a small number. The following parameters were used: pathway enrichment with "global test", pathway topology analysis with "relative-betweenness centrality", pathway library was homo sapiens KEGG, KEGG version Oct 2019.

Supplementary Materials: The following are available online at http://www.mdpi.com/2218-1989/9/12/304/s1, Table S1: 1x_5x_10x_15x_MTS. Table S2: washing extractions comparison, Table S3: proof Experiment Metallodrugs.

Author Contributions: Conceptualization, M.R. and G.K.; methodology M.R., M.A.J. and G.K.; formal analysis, M.R.; software, M.R.; validation, M.R.; investigation, M.R.; resources, M.A.J. and G.K.; data curation, M.R. and G.K.; writing—original draft preparation, M.R., E.R.; writing—review and editing, M.R., M.A.J., G.K.; visualization, M.R.; supervision, G.K.; M.R.; project administration, M.R., M.A.J. and G.K.; funding acquisition, B.K.K. and G.K.

Funding: This research received funding from Vienna Fund for Innovative Interdisciplinary Cancer Research. Open Access Funding by the University of Vienna.

Acknowledgments: This work is supported by the University of Vienna, the Faculty of Chemistry, the Vienna, Metabolomics Center (VIME; http://metabolomics.univie.ac.at/), the research platform Chemistry Meets Microbiology, the Mass Spectrometry Center (MSC) and Translational Cancer Therapy Research Cluster of the University of Vienna. We would like to thank for Yasin El Abiead, Michaela Schwaiger-Haber and Debora Wernitznig for helpful discussions and for supporting this work.

Conflicts of Interest: The authors declare no conflict of interest.

Abbreviations

Abbreviations of metabolites and their explanation used in this work.

Abbreviation	Explanation
5AMP, AMP	adenosine monophosphate
5CMP, CMP	cytidine monophosphate
5GMP, UMP	guanosine monophosphate
5UMP	uridine monophosphate
AAA	alpha aminoadipic acid
Ade	adenine
ADP	adenosine diphosphate
Ala	alanine
Arg	arginine
Asin	adenosine
Asn	aspragine
Asp	aspartic acid
DHAP	dihydroxyacetone phosphate
Gln	glutamine
Glu	glutamate
Gly	glycine
GMP	guanosine monophosphate
GSH	glutathione (reduced)
Gsin	guanosine
GSSG	Glutathione (oxidized)
His	histidine
H-Ser	homoserine
IMP	inosine monophosphate
Isin	inosine
K-Ile	ketoisoleucine
Lys	lysine

Met	methionine
NAD+	nicotinamide adenine dinucleotide (oxidized)
NADP+	nicotinamide adenine dinucleotide phosphate (oxidized)
OAS	O-acetylserine
Phe	phenylalanine
Pro	proline
Pyr	pyruvate
Rib-5-P	ribose 5-phosphate/ribulose 5-phosphate
Ser	serine
Thr	threonine
Trp	tryptophan
Tyr	tyrosine

Appendix A

Table S1. Summary of the experiment which had the aim of determining the minimal number of MTS required for a metabolomics experiment. Relative responses of 1, 5, 10 and 15 MTS from the boiling ethanol extracts. Absolute amounts of amino acids [nmol] from hydrolysed pellets. Cell numbers and viability. Parameters of the normalized linear regressions.

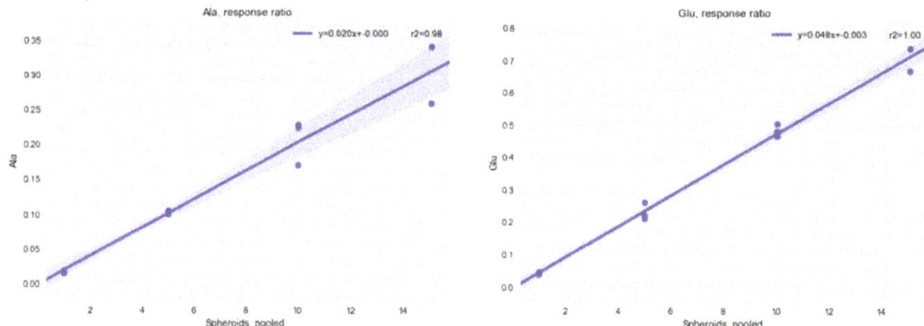

Figure A1. Relative responses (^{12}C/^{13}C) of alanine and glutamate as selected examples from a single ($N = 3$), 5 ($N = 3$), 10 ($N = 3$), 15 ($N = 2$) multicellular tumor spheroids. Sample preparation applied boiling ethanol extraction with U^{13}C internal standardization and LC high-resolution mass spectrometry. Investigated metabolites showed high linear correlation.

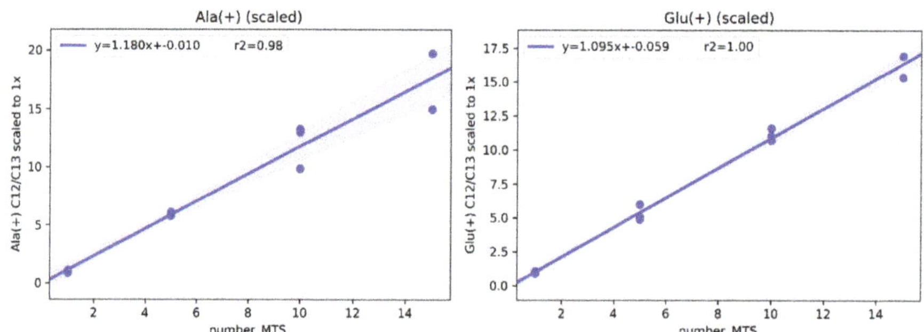

Figure A2. Scaled relative responses (^{12}C/^{13}C) of alanine and glutamate as selected examples from a single ($N = 3$), 5 ($N = 3$), 10 ($N = 3$), 15 ($N = 2$) multicellular tumor spheroids: relative responses were scaled to the mean relative response in single multicellular tumor spheroids ($N = 3$). Sample preparation applied boiling ethanol extraction with U^{13}C internal standardization and LC high-resolution mass spectrometry. Alanine and glutamate are shown from the investigated compounds.

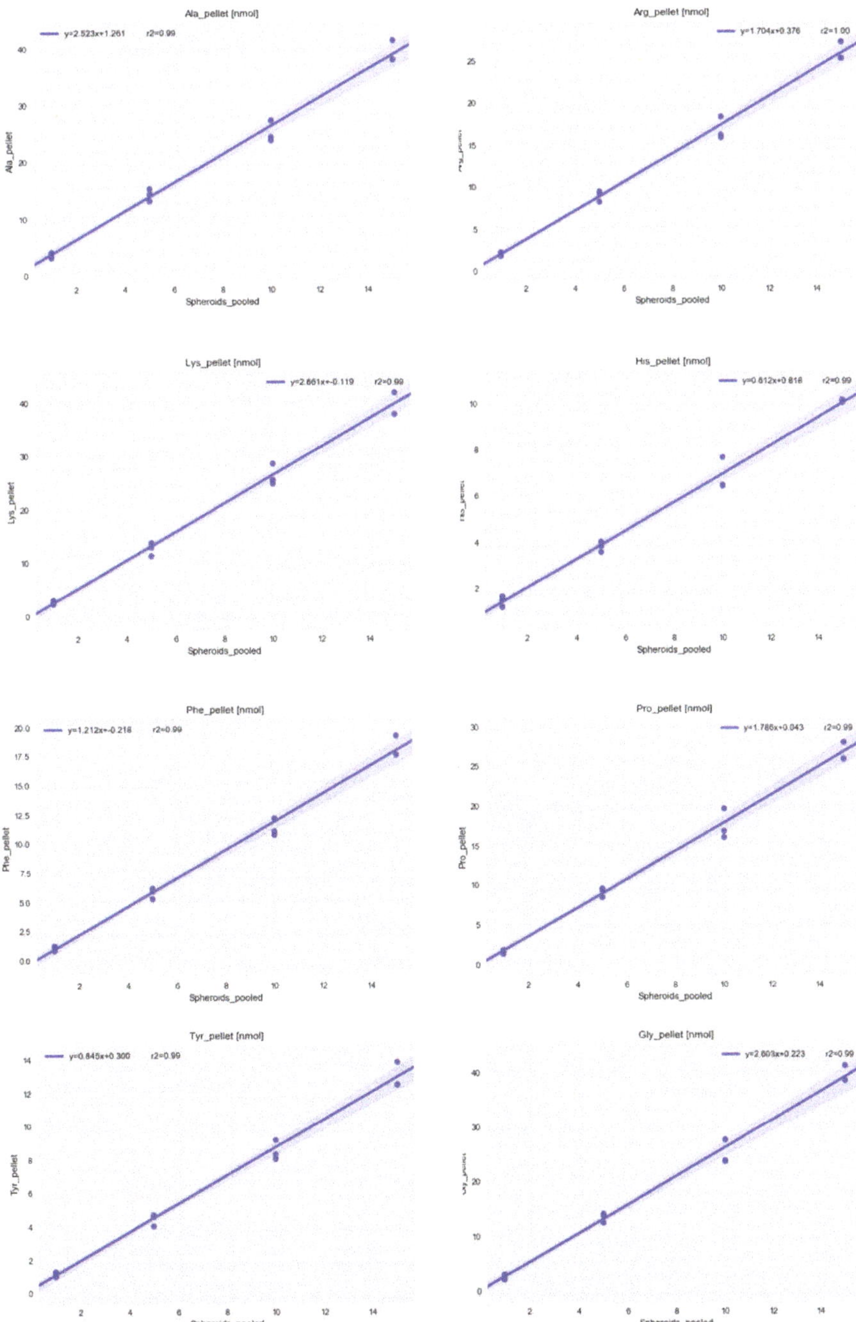

Figure A3. Absolute amounts in nmol of amino acids after acidic hydrolysis of the pellet containing 1 ($N = 3$), 5 ($N = 3$), 10 ($N = 3$) and 15 ($N = 2$) multicellular tumor spheroids. The pellets are resulting from the high molecular fraction of the boiling ethanol extraction and centrifugation. Amino acid amounts from the proteins show a strong linear correlation with the number of spheroids in the sample.

Table S2. Summary of the experiment comparing the different sample preparation strategies with three washing steps vs. single washing on the plate and boiling ethanol vs. cold methanol extraction. Summary of the absolute amounts of selected metabolites. Individual absolute amounts in samples.

Table S3. Summary of the proof of principle experiment with metallodrugs. Absolute amounts of metabolites in samples and RSDs. Absolute amounts of metabolites normalized to total protein content and RSDs. Absolute amounts of metabolites normalized to total protein content with missing value imputation. Absolute amounts of metabolites normalized to total protein content with missing value imputation and autoscaling. Total protein content in the samples. KEGG pathways affected by Oxaliplatin treatment. KEGG pathways affected by KP1339 treatment. Significant compounds ($p < 0.05$) based on oxaliplatin treatment. Significant compounds ($p < 0.05$) based on KP1339 treatment.

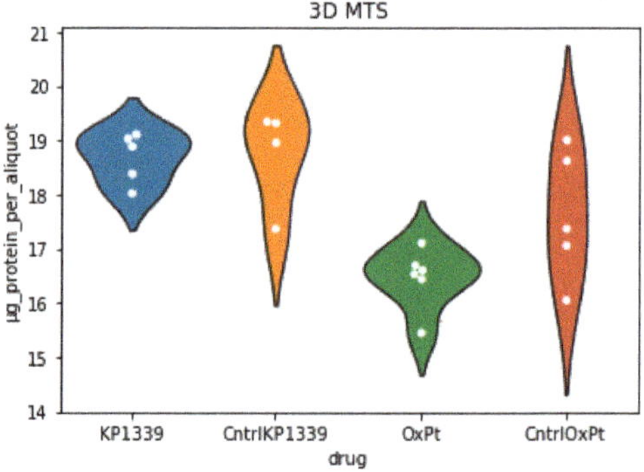

Figure A4. Swarm and violinplot of total protein content [µg] in single 3D MTS cultures (10×10^3 cells seeded, 8 days cultivation diameter 790 ± 22 µm at time of treatment) after 24 h of incubation with 200 µM KP1339, 20 µM Oxaliplatin, growth medium (CntrlOxPt), growth medium with 0.5% DMSO (CntrlKP1339). Protein content is determined by the Thermo micro BCA kit from the high molecular fraction in the pellet.

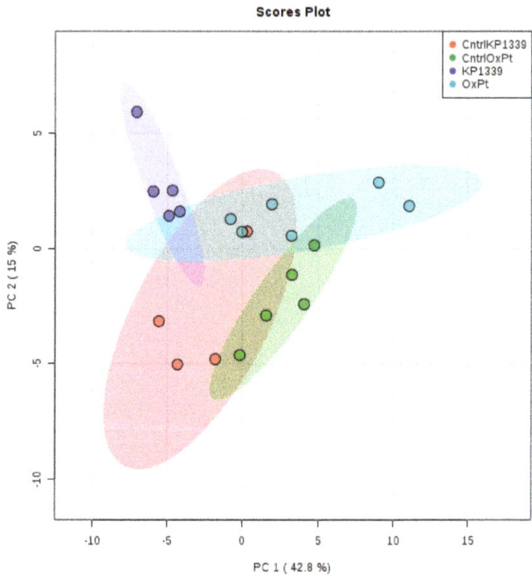

Figure A5. PCA scores plot of absolute amounts of metabolites normalized to protein content of samples from single 3D MTS. Spheroids were treated for 24 h with either 200 µM KP1339, 20 µM Oxaliplatin (OxPt), medium (CntrlOxPt) or medium with 0.5% DMSO (CntrlKP1339). Extraction with cold methanol and internal standardization, measurement by HILIC high-resolution Orbitrap MS.

References

1. Stock, K.; Estrada, M.F.; Vidic, S.; Gjerde, K.; Rudisch, A.; Santo, V.E.; Barbier, M.; Blom, S.; Arundkar, S.C.; Selvam, I.; et al. Capturing tumor complexity in vitro: Comparative analysis of 2D and 3D tumor models for drug discovery. *Sci. Rep.* **2016**, *6*, 28951. [CrossRef] [PubMed]

2. Hirschhaeuser, F.; Menne, H.; Dittfeld, C.; West, J.; Mueller-Klieser, W.; Kunz-Schughart, L.A. Multicellular tumor spheroids: An underestimated tool is catching up again. *J. Biotechnol.* **2010**, *148*, 3–15. [CrossRef] [PubMed]

3. Mehta, G.; Hsiao, A.Y.; Ingram, M.; Luker, G.D.; Takayama, S. Opportunities and Challenges for use of Tumor Spheroids as Models to Test Drug Delivery and Efficacy. *J. Control. Release* **2012**, *164*, 192–204. [CrossRef] [PubMed]

4. Costa, E.C.; Moreira, A.F.; de Melo-Diogo, D.; Gaspar, V.M.; Carvalho, M.P.; Correia, I.J. 3D tumor spheroids: An overview on the tools and techniques used for their analysis. *Biotechnol. Adv.* **2016**, *34*, 1427–1441. [CrossRef] [PubMed]

5. Stewart, D.J. Mechanisms of resistance to cisplatin and carboplatin. *Crit. Rev. Oncol. Hematol.* **2007**, *63*, 12–31. [CrossRef]

6. Edmondson, R.; Broglie, J.J.; Adcock, A.F.; Yang, L. Three-Dimensional Cell Culture Systems and Their Applications in Drug Discovery and Cell-Based Biosensors. *Assay Drug Dev. Technol.* **2014**, *12*, 207–218. [CrossRef]

7. Tafani, M.; Sansone, L.; Limana, F.; Arcangeli, T.; De Santis, E.; Polese, M.; Fini, M.; Russo, M.A. The Interplay of Reactive Oxygen Species, Hypoxia, Inflammation, and Sirtuins in Cancer Initiation and Progression. *Oxid. Med. Cell. Longev.* **2016**, *2016*, 18. [CrossRef]

8. Kondoh, M.; Ohga, N.; Akiyama, K.; Hida, Y.; Maishi, N.; Towfik, A.M.; Inoue, N.; Shindoh, M.; Hida, K. Hypoxia-Induced Reactive Oxygen Species Cause Chromosomal Abnormalities in Endothelial Cells in the Tumor Microenvironment. *PLoS ONE* **2013**, *8*, e80349. [CrossRef]

9. Wang, Q.-S.; Zheng, Y.-M.; Dong, L.; Ho, Y.-S.; Guo, Z.; Wang, Y.-X. Role of mitochondrial reactive oxygen species in hypoxia-dependent increase in intracellular calcium in pulmonary artery myocytes. *Free Radic. Biol. Med.* **2007**, *42*, 642–653. [CrossRef]

10. Hanahan, D.; Weinberg, R.A. Hallmarks of Cancer: The Next Generation. *Cell* **2011**, *144*, 646–674. [CrossRef]

11. Armitage, E.G.; Southam, A.D. Monitoring cancer prognosis, diagnosis and treatment efficacy using metabolomics and lipidomics. *Metabolomics* **2016**, *12*, 146. [CrossRef] [PubMed]

12. do Valle, Í.F.; Menichetti, G.; Simonetti, G.; Bruno, S.; Zironi, I.; Durso, D.F.; Mombach, J.C.M.; Martinelli, G.; Castellani, G.; Remondini, D. Network integration of multi-tumour omics data suggests novel targeting strategies. *Nat. Commun.* **2018**, *9*, 4514. [CrossRef] [PubMed]

13. Beger, R.D. A Review of Applications of Metabolomics in Cancer. *Metabolites* **2013**, *3*, 552–574. [CrossRef] [PubMed]

14. Fiehn, O. Metabolomics—The link between genotypes and phenotypes. *Plant Mol. Biol.* **2002**, *48*, 155–171. [CrossRef] [PubMed]

15. Patti, G.J.; Yanes, O.; Siuzdak, G. Metabolomics: The apogee of the omic triology. *Nat. Rev. Mol. Cell Biol.* **2012**, *13*, 263–269. [CrossRef] [PubMed]

16. Armitage, E.G.; Barbas, C. Metabolomics in cancer biomarker discovery: Current trends and future perspectives. *J. Pharm. Biomed. Anal.* **2014**, *87*, 1–11. [CrossRef]

17. Beger, R.D. Interest is high in improving quality control for clinical metabolomics: Setting the path forward for community harmonization of quality control standards. *Metabolomics* **2018**, *15*, 1. [CrossRef]

18. Cui, L.; Lu, H.; Lee, Y.H. Challenges and emergent solutions for LC-MS/MS based untargeted metabolomics in diseases. *Mass Spectrom. Rev.* **2018**, *37*, 772–792. [CrossRef]

19. Schwaiger, M.; Schoeny, H.; Abiead, Y.E.; Hermann, G.; Rampler, E.; Koellensperger, G. Merging metabolomics and lipidomics into one analytical run. *Analyst* **2018**, *144*, 220–229. [CrossRef]

20. León, Z.; García-Cañaveras, J.C.; Donato, M.T.; Lahoz, A. Mammalian cell metabolomics: Experimental design and sample preparation. *Electrophoresis* **2013**, *34*, 2762–2775. [CrossRef]

21. Coloff, J.L.; Murphy, J.P.; Braun, C.R.; Harris, I.S.; Shelton, L.M.; Kami, K.; Gygi, S.P.; Selfors, L.M.; Brugge, J.S. Differential Glutamate Metabolism in Proliferating and Quiescent Mammary Epithelial Cells. *Cell Metab.* **2016**, *23*, 867–880. [CrossRef] [PubMed]

22. Fan, T.W.-M.; El-Amouri, S.S.; Macedo, J.K.A.; Wang, Q.J.; Song, H.; Cassel, T.; Lane, A.N. Stable Isotope-Resolved Metabolomics Shows Metabolic Resistance to Anti-Cancer Selenite in 3D Spheroids versus 2D Cell Cultures. *Metabolites* **2018**, *8*, 40. [CrossRef] [PubMed]

23. Jones, D.T.; Valli, A.; Haider, S.; Zhang, Q.; Smethurst, E.A.; Schug, Z.T.; Peck, B.; Aboagye, E.O.; Critchlow, S.E.; Schulze, A.; et al. 3D Growth of Cancer Cells Elicits Sensitivity to Kinase Inhibitors but Not Lipid Metabolism Modifiers. *Mol. Cancer Ther.* **2019**, *18*, 376–388. [CrossRef] [PubMed]

24. Vorrink, S.U.; Ullah, S.; Schmidt, S.; Nandania, J.; Velagapudi, V.; Beck, O.; Ingelman-Sundberg, M.; Lauschke, V.M. Endogenous and xenobiotic metabolic stability of primary human hepatocytes in long-term 3D spheroid cultures revealed by a combination of targeted and untargeted metabolomics. *FASEB J.* **2017**, *31*, 2696–2708. [CrossRef]

25. Sato, M.; Kawana, K.; Adachi, K.; Fujimoto, A.; Yoshida, M.; Nakamura, H.; Nishida, H.; Inoue, T.; Taguchi, A.; Takahashi, J.; et al. Spheroid cancer stem cells display reprogrammed metabolism and obtain energy by actively running the tricarboxylic acid (TCA) cycle. *Oncotarget* **2016**, *7*, 33297–33305. [CrossRef]

26. Eljack, N.D.; Ma, H.-Y.M.; Drucker, J.; Shen, C.; Hambley, T.W.; New, E.J.; Friedrich, T.; Clarke, R.J. Mechanisms of cell uptake and toxicity of the anticancer drug cisplatin. *Metallomics* **2014**, *6*, 2126–2133. [CrossRef]

27. Gibson, D. The mechanism of action of platinum anticancer agents—What do we really know about it? *Dalton Trans.* **2009**, *48*, 10681–10689. [CrossRef]

28. Johnstone, T.C.; Suntharalingam, K.; Lippard, S.J. The Next Generation of Platinum Drugs: Targeted Pt (II) Agents, Nanoparticle Delivery and Pt (IV) Prodrugs. *Chem. Rev.* **2016**, *116*, 3436–3486. [CrossRef]

29. Mucaki, E.J.; Zhao, J.Z.L.; Lizotte, D.J.; Rogan, P.K. Predicting responses to platin chemotherapy agents with biochemically-inspired machine learning. *Signal Transduct. Target. Ther.* **2019**, *4*, 1. [CrossRef]

30. Wang, D.; Lippard, S.J. Cellular processing of platinum anticancer drugs. *Nat. Rev. Drug Discov.* **2005**, *4*, 307–320. [CrossRef]

31. Mjos, K.D.; Orvig, C. Metallodrugs in Medicinal Inorganic Chemistry. *Chem. Rev.* **2014**, *114*, 4540–4563. [CrossRef] [PubMed]

32. Canelas, A.B.; ten Pierick, A.; Ras, C.; Seifar, R.M.; van Dam, J.C.; van Gulik, W.M.; Heijnen, J.J. Quantitative evaluation of intracellular metabolite extraction techniques for yeast metabolomics. *Anal. Chem.* **2009**, *81*, 7379–7389. [CrossRef] [PubMed]

33. Gonzalez, B.; François, J.; Renaud, M. A rapid and reliable method for metabolite extraction in yeast using boiling buffered ethanol. *Yeast* **1997**, *13*, 1347–1355. [CrossRef]

34. Dettmer, K.; Nürnberger, N.; Kaspar, H.; Gruber, M.A.; Almstetter, M.F.; Oefner, P.J. Metabolite extraction from adherently growing mammalian cells for metabolomics studies: Optimization of harvesting and extraction protocols. *Anal. Bioanal. Chem.* **2011**, *399*, 1127–1139. [CrossRef] [PubMed]

35. Hermann, G.; Schwaiger, M.; Volejnik, P.; Koellensperger, G. 13C-labelled yeast as internal standard for LC–MS/MS and LC high resolution MS based amino acid quantification in human plasma. *J. Pharm. Biomed. Anal.* **2018**, *155*, 329–334. [CrossRef] [PubMed]

36. Hoofnagle, A.N.; Whiteaker, J.R.; Carr, S.A.; Kuhn, E.; Liu, T.; Massoni, S.A.; Thomas, S.N.; Townsend, R.R.; Zimmerman, L.J.; Boja, E.; et al. Recommendations for the generation, quantification, storage and handling of peptides used for mass spectrometry-based assays. *Clin. Chem.* **2016**, *62*, 48–69. [CrossRef]

37. Galvez, L.; Rusz, M.; Schwaiger, M.; Abiead, Y.E.; Hermann, G.; Jungwirth, U.; Berger, W.; Keppler, B.K.; Jakupec, M.; Koellensperger, G. Preclinical studies on metal based anticancer drugs as enabled by integrated metallomics and metabolomics. *Metallomics* **2019**, *11*, 1716–1728. [CrossRef]

38. Bruno, P.M.; Liu, Y.; Park, G.Y.; Murai, J.; Koch, C.E.; Eisen, T.J.; Pritchard, J.R.; Pommier, Y.; Lippard, S.J.; Hemann, M.T. A subset of platinum-containing chemotherapeutic agents kill cells by inducing ribosome biogenesis stress rather than by engaging a DNA damage response. *Nat. Med.* **2017**, *23*, 461–471. [CrossRef]

39. Schoenhacker-Alte, B.; Mohr, T.; Pirker, C.; Kryeziu, K.; Kuhn, P.-S.; Buck, A.; Hofmann, T.; Gerner, C.; Hermann, G.; Koellensperger, G.; et al. Sensitivity towards the GRP78 inhibitor KP1339/IT-139 is characterized by apoptosis induction via caspase 8 upon disruption of ER homeostasis. *Cancer Lett.* **2017**, *404*, 79–88. [CrossRef]

40. Terenzi, A.; Pirker, C.; Keppler, B.K.; Berger, W. Anticancer metal drugs and immunogenic cell death. *J. Inorg. Biochem.* **2016**, *165*, 71–79. [CrossRef]

41. Kepp, O.; Senovilla, L.; Vitale, I.; Vacchelli, E.; Adjemian, S.; Agostinis, P.; Apetoh, L.; Aranda, F.; Barnaba, V.; Bloy, N.; et al. Consensus guidelines for the detection of immunogenic cell death. *Oncoimmunology* **2014**, *3*, e955691. [CrossRef] [PubMed]

42. Wernitznig, D.; Kiakos, K.; Del Favero, G.; Harrer, N.; Machat, H.; Osswald, A.; Jakupec, M.A.; Wernitznig, A.; Sommergruber, W.; Keppler, B.K. First-in-class ruthenium anticancer drug (KP1339/IT-139) induces an immunogenic cell death signature in colorectal spheroids in vitro. *Met. Integr. Biomet. Sci.* **2019**, *11*, 1044–1048. [CrossRef] [PubMed]

43. Broadhurst, D.; Goodacre, R.; Reinke, S.N.; Kuligowski, J.; Wilson, I.D.; Lewis, M.R.; Dunn, W.B. Guidelines and considerations for the use of system suitability and quality control samples in mass spectrometry assays applied in untargeted clinical metabolomic studies. *Metabolomics* **2018**, *14*, 72. [CrossRef] [PubMed]

44. Dunn, W.B.; Broadhurst, D.I.; Edison, A.; Guillou, C.; Viant, M.R.; Bearden, D.W.; Beger, R.D. Quality assurance and quality control processes: Summary of a metabolomics community questionnaire. *Metabolomics* **2017**, *13*, 50. [CrossRef]

45. Alcindor, T.; Beauger, N. Oxaliplatin: A review in the era of molecularly targeted therapy. *Curr. Oncol.* **2011**, *18*, 18–25. [CrossRef]

46. Chong, J.; Xia, J. MetaboAnalystR: An R package for flexible and reproducible analysis of metabolomics data. *Bioinformatics* **2018**, *34*, 4313–4314. [CrossRef]

47. Cavill, R.; Kamburov, A.; Ellis, J.K.; Athersuch, T.J.; Blagrove, M.S.C.; Herwig, R.; Ebbels, T.M.D.; Keun, H.C. Consensus-Phenotype Integration of Transcriptomic and Metabolomic Data Implies a Role for Metabolism in the Chemosensitivity of Tumour Cells. *PLoS Comput. Biol.* **2011**, *7*, e1001113. [CrossRef]

48. Von Stechow, L.; Ruiz-Aracama, A.; van de Water, B.; Peijnenburg, A.; Danen, E.; Lommen, A. Identification of Cisplatin-Regulated Metabolic Pathways in Pluripotent Stem Cells. *PLoS ONE* **2013**, *8*, e76476. [CrossRef]

49. Gifford, J.B.; Huang, W.; Zeleniak, A.E.; Hindoyan, A.; Wu, H.; Donahue, T.R.; Hill, R. Expression of GRP78, Master Regulator of the Unfolded Protein Response, increases chemoresistance in pancreatic ductal adenocarcinoma. *Mol. Cancer Ther.* **2016**, *15*, 1043–1052. [CrossRef]

50. Kreutz, D.; Bileck, A.; Plessl, K.; Wolrab, D.; Groessl, M.; Keppler, B.K.; Meier, S.M.; Gerner, C. Response Profiling Using Shotgun Proteomics Enables Global Metallodrug Mechanisms of Action To Be Established. *Chemistry* **2017**, *23*, 1881–1890. [CrossRef]

51. Gonen, N.; Meller, A.; Sabath, N.; Shalgi, R. Amino Acid Biosynthesis Regulation during Endoplasmic Reticulum Stress Is Coupled to Protein Expression Demands. *IScience* **2019**, *19*, 204–213. [CrossRef] [PubMed]

52. Scannell, J.W.; Blanckley, A.; Boldon, H.; Warrington, B. Diagnosing the decline in pharmaceutical R & D efficiency. *Nat. Rev. Drug Discov.* **2012**, *11*, 191–200. [PubMed]

53. Hermann, G.; Møller, L.H.; Gammelgaard, B.; Hohlweg, J.; Mattanovich, D.; Hann, S.; Koellensperger, G. In vivo synthesized 34 S enriched amino acid standards for species specific isotope dilution of proteins. *J. Anal. At. Spectrom.* **2016**, *31*, 1830–1835. [CrossRef]

54. Rampler, E.; Dalik, T.; Stingeder, G.; Hann, S.; Koellensperger, G. Sulfur containing amino acids—Challenge of accurate quantification. *J. Anal. At. Spectrom.* **2012**, *27*, 1018–1023. [CrossRef]

 *metabolites*

Article

Modified Protocol of Harvesting, Extraction, and Normalization Approaches for Gas Chromatography Mass Spectrometry-Based Metabolomics Analysis of Adherent Cells Grown Under High Fetal Calf Serum Conditions

Raphaela Fritsche-Guenther [1,2], Anna Bauer [1,2], Yoann Gloaguen [1,2,3], Mario Lorenz [4,5] and Jennifer A. Kirwan [1,2,*]

[1]   Berlin Institute of Health Metabolomics Platform, Berlin Institute of Health (BIH), 10178 Berlin, Germany; raphaela.fritsche@mdc-berlin.de (R.F.-G.); anna.bauer@mdc-berlin.de (A.B.); yoann.gloaguen@mdc-berlin.de (Y.G.)

[2]   Max-Delbrück-Center for Molecular Medicine (MDC) in the Helmholtz Association, 13125 Berlin, Germany

[3]   Core Unit Bioinformatics, Berlin Institute of Health (BIH), 10178 Berlin, Germany

[4]   German Centre for Cardiovascular Research (DZHK), partner site Berlin, 13353 Berlin, Germany; mario.lorenz@charite.de

[5]   Charité-Universitätsmedizin Berlin, corporate member of Freie Universität Berlin, Humboldt-Universität zu Berlin, and Berlin Institute of Health, Medizinische Klinik für Kardiologie und Angiologie, Campus Mitte, 10117 Berlin, Germany

*   Correspondence: Jennifer.kirwan@mdc-berlin.de; Tel.: +49- (0)30-9460-3088

Received: 13 November 2019; Accepted: 11 December 2019; Published: 18 December 2019

Abstract: A gas chromatography mass spectrometry (GC-MS) metabolomics protocol was modified for quenching, harvesting, and extraction of metabolites from adherent cells grown under high (20%) fetal calf serum conditions. The reproducibility of using either 50% or 80% methanol for quenching of cells was compared for sample harvest. To investigate the efficiency and reproducibility of intracellular metabolite extraction, different volumes and ratios of chloroform were tested. Additionally, we compared the use of total protein amount versus cell mass as normalization parameters. We demonstrate that the method involving 50% methanol as quenching buffer followed by an extraction step using an equal ratio of methanol:chloroform:water (1:1:1, *v/v/v*) followed by the collection of 6 mL polar phase for GC-MS measurement was superior to the other methods tested. Especially for large sample sets, its comparative ease of measurement leads us to recommend normalization to protein amount for the investigation of intracellular metabolites of adherent human cells grown under high (or standard) fetal calf serum conditions. To avoid bias, care should be taken beforehand to ensure that the ratio of total protein to cell number are consistent among the groups tested. For this reason, it may not be suitable where culture conditions or cell types have very different protein outputs (e.g., hypoxia vs. normoxia). The full modified protocol is available in the Supplementary Materials.

Keywords: GC-MS; 20% FCS; harvesting; extraction; metabolites; normalization

---

1. Introduction

Reproducibility and robustness when performing metabolomics experiments are essential to guarantee that resulting biological information is both accurate and meaningful, and to reduce the likelihood of false discovery due to technical variation [1]. Cell cultures are often used as models for testing biological concepts. Therefore, an optimized and standardized protocol for efficient quenching, harvesting, and extraction of cells is required. There are many studies focusing on sample preparation methods, and many cell-dependent methods have been optimized for the application of cell metabolomics for analyzing metabolic fingerprints [2–10]. However, no preparation method is broadly applicable for all cell types or cultivation conditions. We had previously discovered that, in our hands, certain cell extraction methods were inappropriate for the cell types we were using, typically because they did not adequately separate the protein layer from the solvents. This was assumed to be a result of inappropriate volumes and ratios of the solvents to adequately extract the cell lipid content.

To prepare cells for metabolomics analyses, quenching at the point of harvesting aims to inactivate intracellular enzymes and stop metabolism to avoid degradation and alterations of sample composition. Sample preparation should be highly reproducible, robust, and fast to allow high-throughput studies [5]. As is known, metabolic reactions occur in milliseconds [11]. Therefore, it is necessary to quench metabolism quickly, most often by using ice-cold organic solvents. After the quenching and harvesting step, metabolite extraction is crucial and often the rate-limiting step. Extraction is the process by which specific compounds, or whole classes of compounds, e.g., polar metabolites are selectively separated from others (e.g., lipids, proteins) and is normally conducted to yield cleaner sample preparations for analysis. There has been much discussions over which extraction solvents are the best for quenching and measurement of metabolites [12]. Biphasic, liquid-liquid extraction is often used to extract metabolites, typically based on Folch or Bligh–Dyer methods [13–15]. Polar solutions like methanol (MeOH) in water (H_2O) are often paired with nonpolar organic solvents, such as chloroform ($CHCl_3$), to form a two-phase system. The chemical properties, volumes, and solvent ratios of the organic and aqueous solvents must be considered carefully. These parameters can significantly affect the extraction efficiency of metabolites and the experimental reproducibility.

Fetal calf serum (FCS) contains a large number of nutritional and macromolecular factors such as amino acids, sugars, lipids, and hormones essential for cell culture growth. For most cell lines, 10% FCS is used; however, some primary cells, such as human umbilical vein endothelial cells (HUVECs), need to be cultured in higher FCS conditions (e.g., medium with 20% FCS) [16–22]. To our knowledge, there have been no reports to date documenting the robustness and efficiency of harvesting and extraction protocols for metabolic profiling of adherent cells grown under high FCS conditions. Ideally, this would be undertaken using a full design-of-experiments approach investigating and fully testing multiple parameters to optimize the method. However, due to limitations in both resources and numbers of primary cell cultures, an incremental optimization approach can be taken to reach a predefined reproducibility and metabolite detection threshold. This approach enables an early exclusion of conditions that do not fit one's criteria, therefore making the most out of the material available.

Therefore, the aim of this study was to compare and enhance our existing protocol for a reliable and reproducible harvesting and extraction of adherent cells grown in media supplemented with 20% FCS. As a comparison, standard cell culture conditions (10% FCS) were also investigated. All analyses were conducted by gas chromatography-mass spectrometry (GC-MS). The use of either 50% or 80% MeOH for quenching of cells were compared for sample harvest. In addition, two different ratios and three different final volumes of MeOH to $CHCl_3$ to H_2O were compared to investigate the efficiency and reproducibility of intracellular metabolite extraction. To achieve robust normalization of the samples, total protein amount versus absolute cell mass was tested.

2. Results

2.1. Preliminary Experiments on Phase Separation

A previous protocol used in our lab for adherent cells growing under 10% FCS was a modified Bligh–Dyer extraction to separate lipid and polar metabolites [23,24]. The protocol includes the use of 50% MeOH for cell quenching followed by an extraction using a 1:0.4:1, *v/v/v* MeOH:$CHCl_3$:H_2O ratio. Routinely, 3 mL of the polar phase is collected and used for further GC-MS measurement. For HUVECs cultured under conditions of 20% FCS, we discovered that this protocol was not sufficient to properly extract the cells. We discovered a fluffy protein pellet rather than the compacted, well demarcated protein pellet we normally achieve. This had consequences for the reproducibility of the measured total protein amount resulting in a relative standard deviation (RSD) for HUVECs of over 25% (European Medicine Agency (EMA) recommendation is RSD < 15%, http://www.ema.europa.eu/). To check this was not a cell-specific effect, the HCT116 colorectal carcinoma cell line was cultured and extracted in an identical way. Results were similar for the HCT116 cells with total protein RSDs also over 25% for high and standard FCS conditions, as seen in Table 1 and Figure 1 50_LOW. This suggests that extraction may not be complete and thus the technical reproducibility was larger than necessary. This prompted us to look at the entire process of harvesting and extraction.

In our study, four different extraction conditions were tested. The four conditions can be summarized here as: (1) 50_LOW: 50% MeOH quenching solution with 1 mL of $CHCl_3$ for extraction with a final ratio of 1:0.4:1 ($MeOH:CHCl_3:H_2O$) and collection of 3 mL polar phase for analysis; (2) 50_MEDIUM: 50% MeOH quenching solution with 2.5 mL of $CHCl_3$ for extraction with a final ratio of 1:1:1 ($MeOH:CHCl_3:H_2O$) and collection of 3 mL polar phase for analysis; (3) 80_HIGH: 80% MeOH quenching solution with 4 mL of $CHCl_3$ for extraction with a final ratio of 1:1:1 ($MeOH:CHCl_3:H_2O$) and collection of 6 mL polar phase for analysis; and (4) 50_HIGH: 50% MeOH quenching solution with 4 mL of $CHCl_3$ for extraction with a final ratio of 1:1:1 ($MeOH:CHCl_3:H_2O$) and collection of 6 mL polar phase for analysis.

HCT116

10% FCS — 50_LOW   50_MEDIUM   50_HIGH   80_HIGH

20% FCS — 50_LOW   50_MEDIUM   50_HIGH   80_HIGH

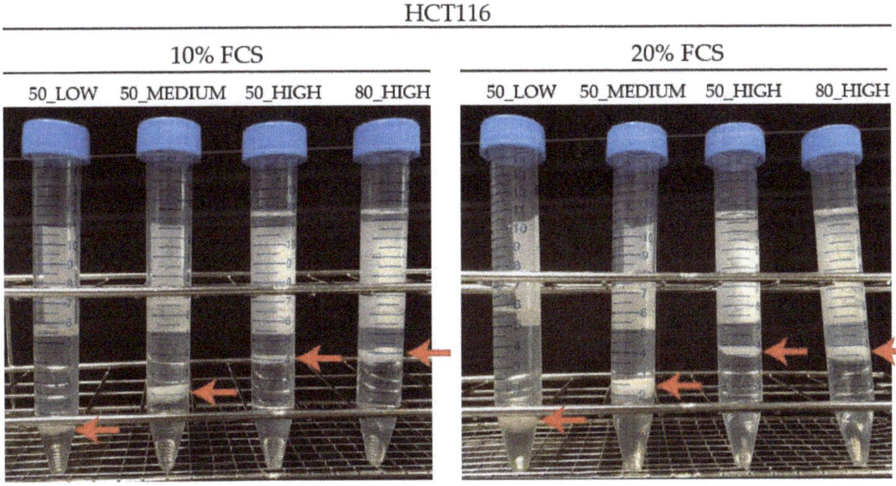

Figure 1. Extraction of HCT116 cells cultured in 10% and 20% fetal calf serum (FCS) conditions using different quenching solvents, extraction ratios, and polar volumes (protein layer shown by extra arrow on picture).

Table 1. Relative standard deviation (RSD) of measured protein amount of HCT116 cells cultured in 10% and 20% FCS conditions using different quenching solvents, extraction ratios, and polar volumes.

Condition Name	50_LOW	50_MEDIUM	50_HIGH	80_HIGH
Biological replicates 10% FCS	4	3	4	5
Biological replicates 20% FCS	5	3	4	4
HCT116 (10% FCS)	51	20	14	13
HCT116 (20% FCS)	30	42	16	8

2.2. Experiment 1: Determining the Optimum Concentration of MeOH as a Quenching and Extraction Solution for Cell Harvesting

The quenching solution used is important as it has a dual function of being the first extraction step and being required to quench metabolism quickly. Many reports suggest 80% MeOH for metabolite quenching and extraction [7,25,26]; our current protocol uses 50% MeOH (50_LOW). Therefore, both 50% and 80% MeOH solutions were tested as quenching and extraction buffers for harvesting of adherent cells cultured in 10% and 20% FCS. The modified Bligh–Dyer extraction step which follows the quenching step will influence the results. In a first test, we used the same volume of 1 mL $CHCl_3$ for the extraction step (as used in our current protocol) and observed that for 80% MeOH, no phase separation occurred, as seen in Figure S1. This is assumed to be due to the relative ratios of $CHCl_3$ and to polar solvents. Increasing the ratio of $MeOH:CHCl_3:H_2O$ to 1:1:1 *v/v/v* resulted in a phase separation. In order to keep the ratio of the final solvent composition constant, the extracted volumes of H_2O and MeOH needed to be adjusted accordingly to account for whether 50% or 80% MeOH was used in the first extraction step.

The performance of individual quenching solvents was determined by a range of measures including (i) number of detected metabolites, (ii) median RSD of all detected metabolites (a measure of reproducibility), (iii) the percentage of metabolites with an individual RSD < 30% (a generally accepted tolerance limit for GC-MS metabolomics for any individual metabolite [27]), and (iv) the sum of normalized peak area for all annotated metabolites (henceforth referred to as sum of area), a proxy measure of the overall sensitivity of the method.

In the HCT116 cells, the conditions were equivalent in terms of number of metabolites detected, as seen in Figure S2, except for the detection of fructose-6-phosphate, when 80% MeOH was used with 10% culture conditions, as seen in Table S1, comparing 50_HIGH and 80_HIGH. For the HUVECs, a lower number of metabolites was found, as seen in Table S2 and Figure S2; 34 metabolites for 50_HIGH and 36 for 80_HIGH. We assume this lower number of metabolites is due to the generally lower average cell count achieved by the HUVECs (mean cell counts: male HUVECs: 1.5×10^6 cells; female HUVECs: 3×10^6 cells; HCT116: 6×10^6 cells). The metabolites detected also had some very minor differences (e.g., proline was only detected in 50_HIGH and tyrosine, uracil, and ribose-5-phosphate were only detected in 80_HIGH).

The reproducibility of both quenching solvents was then assessed using RSDs as an indicator. For HCT116 cells and HUVECs in both culture conditions, 50% MeOH performed better overall as a quenching solvent compared to 80% MeOH, comparing 50_HIGH and 80_HIGH in Table 2, Table 3, Figure 2, Table S1, Table S2, Table S3, Table S4, and Figure S3. In HCT116 cells 50_HIGH has a lower median RSD compared to 80_HIGH (14% compared to 23%, and 26% compared to 30%, for 10% FCS and 20% FCS, respectively). In addition, there was a higher percentage of individual metabolites with RSD < 30% in the 50_HIGH (79% compared to 67%, and 73% compared to 53%, for 10% FCS and 20% FCS, respectively). In line, the sum of the area was higher in 50_HIGH compared to 80_HIGH for both FCS culture conditions (Figure 3; 74 compared to 56, and 35 compared to 26, for 10% FCS and 20% FCS, respectively). In HUVECs, the median RSD was lower in 50_HIGH (13%) compared to 80_HIGH conditions (36%), with a higher percentage of individual metabolite with RSDs < 30% (74% 50_HIGH compared to 44% 80_HIGH). The measured sum of area for HUVECs was also better for 50_HIGH (3.8 compared to 3.2) According to these results, 50% MeOH was deemed to be a better quenching solvent for both types of cells. However, for particular metabolites, 80% MeOH may be more appropriate.

Table 2. Median relative standard deviation (RSD) per metabolite of HCT116 cells cultured in 10% and 20% FCS conditions using different quenching solvents, extraction ratios, and polar volumes.

Condition Name	50_LOW	50_MEDIUM	50_HIGH	80_HIGH
Biological replicates 10% FCS	4	3	4	5
Biological replicates 20% FCS	5	3	4	4
HCT116 (10% FCS)	68%	22%	14%	23%
HCT116 (20% FCS)	33%	55%	26%	30%

Table 3. Percentage of metabolites with a relative standard deviation (RSD) < 30% of HCT116 cells cultured in 10% and 20% FCS conditions using different quenching solvents, extraction ratios, and polar volumes.

Condition Name	50_LOW	50_MEDIUM	50_HIGH	80_HIGH
Biological replicates 10% FCS	4	3	4	5
Biological replicates 20% FCS	5	3	4	4
HCT116 (10% FCS)	0%	73%	79%	67%
HCT116 (20% FCS)	15%	3%	73%	53%

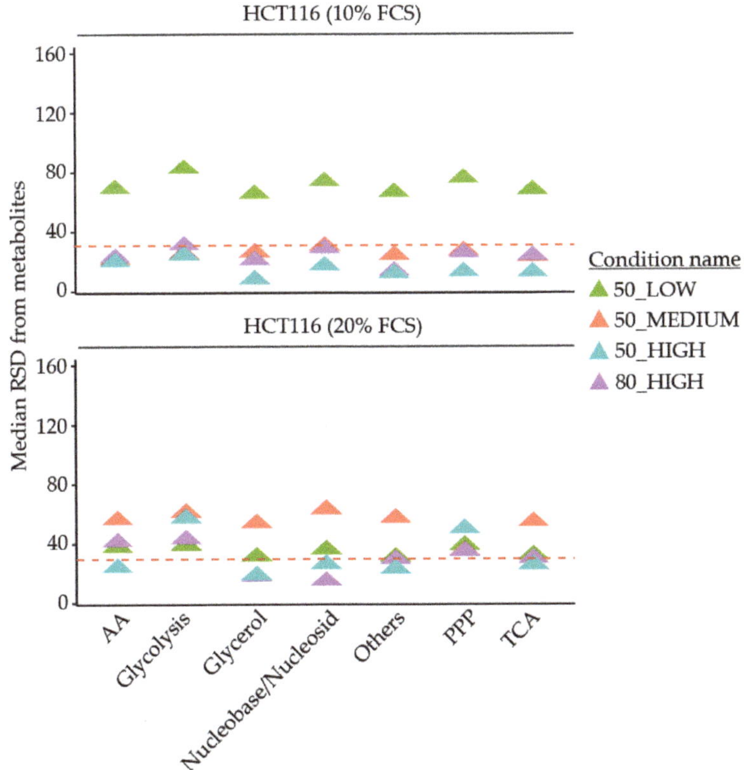

Figure 2. Median relative standard deviation (RSD) of individual metabolites separated by biological metabolite pathways of HCT116 cells cultured in 10% and 20% FCS conditions using different quenching solvents, extraction ratios, and polar volumes. The dashed line represents the maximum 30% RSD threshold advised by the Federal Drug Administration (FDA). AA: Amino acids. PPP: Pentose phosphate pathway. TCA: Tricarboxylic acid cycle.

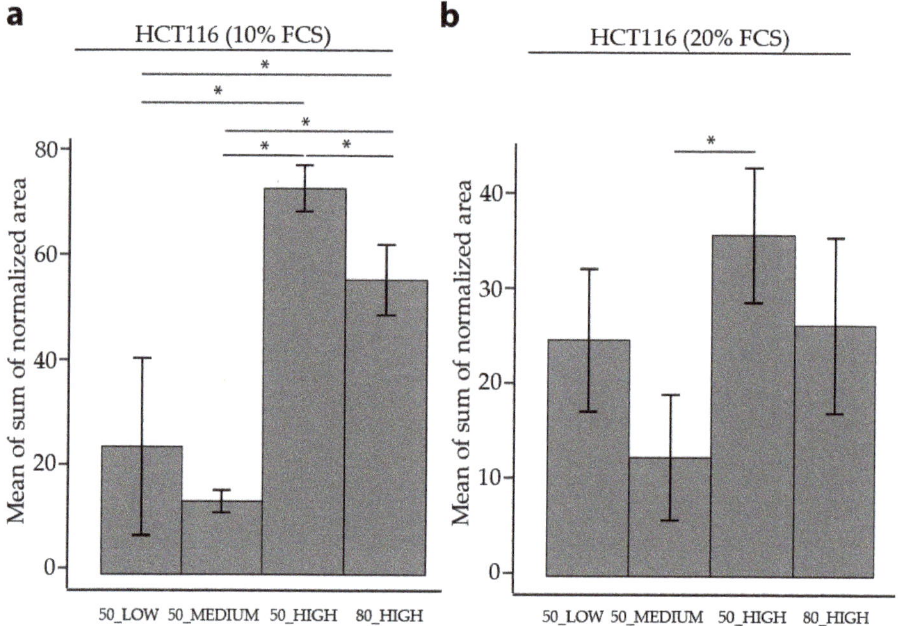

Figure 3. Comparison of the mean of sum of normalized peak area for different quenching and extraction methods. Data from a minimum three out of five HCT116 cells cultured in (**a**) 10% and (**b**) 20% FCS. The error bars represent the standard deviation of the biological replicates. Samples were analyzed using an unpaired Student's *t*-test with $p < 0.05$ deemed as statistically significant.

2.3. Experiment 2: Determination of the Optimal Volume and Ratio of MeOH:CHCl$_3$:H$_2$O for a Metabolite Extraction Solution

The ratio of MeOH:CHCl$_3$:H$_2$O is important for: (i) the efficiency and reproducibility of extraction; (ii) a clear phase separation; and (iii) a robust method for protein collection for normalization or for concurrent proteomics (see Section 2.1). Different ratios, final volumes, and polar phase volume collection for extraction were evaluated, as seen in Table 4. This step was carried out before 50_HIGH was deemed to be optimum for quenching, but, for convenience, is presented here in the same order in which it appears in the standard operating protocol. As the different extraction methods led to different final volumes, leading to different theoretical concentrations, more volume was analyzed in some samples to normalize and compensate for this concentration difference.

Table 4. Experimental set up for the conditions analyzed in the study.

Condition Name	50_LOW (▲)	50_MEDIUM (▲)	50_HIGH (▲)	80_HIGH (▲)
5 mL of MeOH harvest	50% (2.5 + 2.5) *	50% (2.5 + 2.5) *	50% (2.5 + 2.5) *	80% (4.0 + 1.0) *
MeOH [mL] for extraction	-	-	1.5	-
H$_2$O [mL] for extraction	-	-	1.5	3.0
CHCl$_3$ [mL] for extraction	1.0	2.5	4.0	4.0
Final volume [mL]	6.0	7.5	12	12
Final ratio *v/v/v*	1:0.4:1	1:1:1	1:1:1	1:1:1
Final polar volume [mL]	5.0	5.0	8.0	8.0
Used polar volume [mL]	3.0	3.0	6.0	6.0

*: mL of MeOH + H$_2$O.

The evaluation of the different ratios and volumes revealed no significant changes in either the number of detected metabolites, as seen in Table 5, or particular metabolite classes independent of the analyzed cells and growth conditions, as seen in Figure S2.

Table 5. Number of annotated central carbon metabolites of HCT116 cells cultured in 10% and 20% FCS conditions using different quenching solvents, extraction ratios, and polar volumes.

Condition Name	50_LOW	50_MEDIUM	50_HIGH	80_HIGH
Biological replicates 10% FCS	4	3	4	5
Biological replicates 20% FCS	5	3	4	4
HCT116 (10% FCS)	38/45	39/45	38/45	39/45
HCT116 (20% FCS)	41/45	41/45	40/45	40/45

In HCT116 cells, the use of 50_MEDIUM compared to 50_LOW led to lower RSDs (20% compared to 51%) for total protein measurement in cells grown under 10% FCS. HCT116 cells cultured with 20% FCS showed increased RSDs (42% compared to 30%) in the 50_MEDIUM condition, as seen in Table 1. When using 80_HIGH conditions, the pellet was more compact, as seen in Figure 1, and easier to extract, which is reflected in a low RSD (13% for 10% FCS and 8% for 20% FCS) of the measured protein amount in HCT116 cells. When using 50_HIGH conditions, a compact protein pellet could be observed, leading to a more efficient protein extraction, and resulting in RSDs (14% for 10% FCS and 16% for 20% FCS) similar to 80_HIGH conditions. For HUVECs, the RSD of the measured total protein was reduced when using 80_HIGH compared to 50_MEDIUM and 50_HIGH conditions, as seen in Table S5. 50_MEDIUM compared to 50_LOW conditions showed a lower RSD (15% compared to 49%, respectively).

The reproducibility of GC-MS analysis as measured by the median RSD of the individual metabolites or combined as biological pathways, and the percentage of metabolites with RSD < 30%, revealed the best results when using 50_HIGH conditions for HCT116 cells, as seen in Table 2, Table 3 and Figure 2. Likewise, in HUVECs, the best results were shown using 50_HIGH conditions, as seen in Table S3, Table S4 and Figure S3.

Independent of the FCS setting in HCT116 cells, the highest sum of area was found using 50_HIGH conditions, as seen in Figure 3. For HUVECs, the highest sum of area was found in cells in the 50_MEDIUM condition; however, the variability was high, as seen in Figure S4. Comparing 80_HIGH and 50_HIGH conditions measured as a separate batch, a higher sum of area was found using 50% MeOH as quenching buffer.

In summary, our results indicate that using an extraction ratio of 1:1:1, *v/v/v* MeOH:CHCl$_3$:H$_2$O and a polar phase volume of 6 mL achieved good results for samples used for both GC-MS measurement and protein extraction for cells grown under 10% and 20% FCS. Lower individual and combined RSDs and higher peak areas could be detected. The direct comparison of 50% or 80% MeOH as quenching buffer showed a higher sum of area and lower median RSDs when using 50_HIGH conditions for the analyzed cells and FCS conditions. The number of annotated metabolites was equal across all methods. The full modified protocol is available in the Supplementary Materials.

2.4. Experiment 3: Total Protein versus Cell Mass as a Normalization Strategy

Adherent cells need to be detached from the bottom of the cell culture flask for further metabolomics analyses. Trypsinization is known to lead to alteration of metabolism, so we employed direct quenching of cells with physical scraping as an alternative [5,28]. However, this complicates the utilization of the cell count as a normalization parameter. Therefore, we compared the use of total protein amount versus the cell mass as a normalization strategy.

After normalizing to total protein, HCT116 cells grown under 10% FCS showed a slightly lower median RSD of 24% for individual metabolite detection compared to normalization with cell mass (median RSD 28%), as seen in Table 6. For HCT116 cells cultivated in 20% FCS, a median RSD of 35% and 25%, respectively, was found if normalized to total protein or cell mass. For HUVECs, the median RSD was lower when using protein (52%) for normalization compared to cell mass (61%), as seen in Table S6. The percentage of individual metabolite RSDs < 30% threshold was higher when using total protein (26%) compared to cell mass (18%) for normalization. Judging from Table 6 and Table S6, neither of the normalization methods was superior. However, total protein was considerably easier to measure.

Table 6. Median relative standard deviation (RSD) and percentage of metabolites with a RSD < 30% from four replicates of HCT116 cells cultured in 10% and 20% FCS conditions.

	HCT116 (10% FCS)		HCT116 (20% FCS)	
	Cell Mass	Protein	Cell Mass	Protein
Median RSD [%]	28	24	25	35
RSD < 30% [%]	55	60	60	33

3. Discussion

Metabolomics has long been used for the analysis of human body fluids for clinical indications [29,30]. More recently, it is also being applied to cells or tissue [30]. Metabolite profiling of adherent growing mammalian cells is challenging, particularly due to the special requirements with respect to the sampling procedure. While harvesting of suspension cultures can generally be achieved through rapid centrifugation, the adherent cells first need to be detached from the bottom of the cell culture flask [30]. With a large number of metabolites to be analyzed simultaneously, it becomes difficult to find a single optimal extraction method. In this work, we set out to solve a particular challenge, which was the visible suboptimal extraction of cells grown under high (20%) FCS conditions. We focused on optimizing a liquid-liquid extraction protocol for the harvest and extraction of cells grown both under high (20%) or normal (10%) FCS conditions to enable reproducible and robust analysis of the detected metabolites. We compared our existing protocol using 50% MeOH to 80% MeOH as a quenching buffer, followed by an extraction using different ratios and volumes of $CHCl_3$ and polar phases. Environmental and personal safety should also be considered and has resulted in a move away from $CHCl_3$ towards other nonorganic solvents such as methyl-tert-butylether (MTBE) in recent years [31]. MTBE is both highly volatile and floats on the surface of the resulting biphasic mixture. As we were interested in analyzing only the polar metabolites in this study, it was excluded from the test solvents. We also looked at whether cell mass or total protein represented a better normalization strategy, since cell count cannot easily be used for adherent cells harvested for a metabolomics approach [32].

Direct quenching of the metabolism in adherent cells needs extraction with organic solvents such as MeOH. Various reports on the best solvent composition for metabolite extraction suggests 80% MeOH in H_2O [7,26]; however, 50% MeOH is also used for metabolite extraction of adherent cells grown under 10% FCS [23]. We tried using both 50% MeOH and 80% MeOH as quenching buffers, followed by an extraction using different ratios and volumes of $CHCl_3$ and polar phases in cells cultured with 20% FCS compared to standard cell culture conditions. Using the 50_HIGH method (50% MeOH followed by extraction using a 1:1:1, *v/v/v* MeOH:$CHCl_3$:H_2O ratio and collection of 6 mL polar phase) was considered to be the best of the extraction methods we tested. Using a higher ratio of $CHCl_3$ and higher extraction solvent volumes gives a better phase separation and results in a dense protein pellet allowing a robust normalization. The percentage of highly polar solutions in the final ratio is also likely to be important for efficient extraction of polar metabolites and is reflected in the quenching results. The results may be different if nonpolar metabolites are also considered.

Normalization methods for cell culture are often problematic. Normalization is used to eliminate inter-run variability and biological variation [25]. Cell counts are inaccurate and are often performed on separate culture plates to the ones that are analyzed [33]. This is labor-intensive, impractical, and inefficient, especially with regard to large studies. Using cell count for normalization requires cell detachment using trypsin solutions, which can lead to metabolite leakage and changes in the metabolic pattern [5]. Total protein or cell mass has the advantage over cell counts that they can be performed using the same plate as that for the metabolomics analysis. One difficulty in using cell mass for normalization is the removal of the quenching buffer after centrifugation. Remnants of the quenching solution left in the vials will affect the final recorded cell mass, leading to more technical error. Further, the process of weighing the samples is technically laborious. Metabolites are contained in the buffer, meaning that it needs to be first removed for the weighing process and then re-added. This all needs to take place as quickly as possible, while keeping the sample as cold as possible so that it has minimal effects on reproducibility. As total protein can be measured following extraction, it can be done with more ease, although it has the disadvantage that samples cannot then be extracted according to their mass. They can, however, be reconstituted to equivalent concentrations before being analyzed if a drying down step is employed immediately after extraction. We have shown in the results that the extraction of the protein was reproducible enough to be considered an effective normalization method when the 50_HIGH method was used. For large numbers of samples, protein extraction is easier since the critical step of weighing and cooling is missing. Therefore, we recommend using total protein as a potential normalization strategy. We are aware of certain cell culture conditions that reduce the protein amount per cell. For this reason, full consideration should be given to whether the total protein is likely to be a good reflection of cell count or cell volume before it is used. In our hands, we have seen certain culture conditions, e.g., hypoxia, where cell protein production is not equivalent in the same cell type as in normoxia. In these circumstances, cell protein would probably not be recommended as a suitable normalization strategy.

In this study, we have tested and improved our existing method to enhance our reproducibility when analyzing adherent mammalian cells cultured under 20% FCS. For a truly optimized method, a design of experiments (DoE) approach is recommended. However, due to the difficulties of working with large numbers of cell cultures per batch, the cost of repeated experiments and the statistical understanding required to pursue a DoE approach, we instead chose to trial and optimize our existing method. We are aware that this incremental optimization approach may not result in the truly optimal method; however, it enables a faster development that limits material waste. The method is deemed to be optimal once certain reproducibility and metabolite detection criteria are reached. This represents a more realistic scenario for the thousands of labs required to run samples each year without the time and resources to fully optimize methods for each sample type.

In summary, the final modified 50_HIGH quenching, extraction, and normalization method (50% MeOH for harvest followed by extraction using a 1:1:1, *v/v/v* MeOH:CHCl$_3$:H$_2$O ratio and collection of 6 mL polar phase) can be recommended for the investigation of intracellular metabolites from adherent human cells grown under standard (10%) or high (20%) FCS conditions using a GC-MS platform.

4. Materials and Methods

4.1. Cell Culture

Two different human cell types were used in this study. We were interested mainly in a robust analysis of HUVEC male and female twin pair cells, but since these are human derived and therefore in limited supply, HCT116 cells were used to enable more replicates to be tested. For the same reason, a mix of male and female HUVECs were used.

The colorectal carcinoma cell line HCT116 was obtained from ATCC (American Type Culture Collection, Teddington, UK). The HCT116 cell line was maintained in DMEM (Dulbecco's Modified Eagle Medium, Thermo Fischer Scientific, Waltham, MA, USA) supplemented with 10% or 20% FCS (Thermo Fischer Scientific, Waltham, MA, USA), 1% penicillin/streptomycin (Thermo Fischer Scientific, Waltham, MA, USA), 2 mM glutamine (Thermo Fischer Scientific, Waltham, MA, USA) and 1 g/L glucose (Sigma-Aldrich, St. Louis, MO, USA). HUVECs were isolated as previously described [17]. All cells were incubated in a humidified atmosphere of 5% CO_2 in air at 37 °C.

4.2. Experiment 1: Determination of the Optimum Concentration of Methanol as a Quenching and Extraction Solution for Cell Harvesting

Cells were rapidly washed (20 s) with washing buffer (140 mM NaCl, 5 mM HEPES, pH 7.4, 37 °C) before they were quenched by ice-cold MeOH solution. Two solutions were tested: either 5 mL of 50% or 80% MeOH in H_2O with a final concentration of 2 µg/mL cinnamic acid (for use as an internal standard). Immediately as the MeOH solution was added to the culture plate, cells were scraped into the methanol solution and the methanolic lysates were collected. The extracts were agitated to complete cell lysis and centrifuged to separate the layers (see Section 4.3).

4.3. Experiment 2: Determination of the Optimal Volume and Ratio of MeOH:CHCl₃:H₂O for a Metabolite Extraction Solution

After cell harvest, $CHCl_3$ was added to the methanolic cell extracts, shaken for 60 min at 4 °C, and centrifuged at $4149 \times g$ for 10 min at 4 °C to separate the phases. Both the final ratio of MeOH:CHCl₃:H₂O and the final total volume was tested. Test conditions are summarized in Table 4.

The polar phase was collected and dried overnight at 30 °C at a speed of $1550 \times g$ at 0.1 mbar using a rotational vacuum concentrator (RVC 2-33 CDplus, Christ, Osterode am Harz, Germany). Samples were pooled after extraction and used as a quality control (QC) sample to test the technical variability of the instrument. QC samples were prepared alongside the samples in the same way. To generate backup samples, the dried polar phases were resuspended and split into two aliquots.

To further test whether the results of changing the ratios of MeOH:CHCl₃:H₂O are an effect of the ratio or of the total volume, we investigated changing the ratios while maintaining the volume. For the harvesting condition of 50% MeOH, a final volume of 6 mL was tested, resulting in 3 mL of polar phase that could be collected for analysis. For the harvesting condition using 80% MeOH, a final extraction volume of 12 mL was tested with 6 mL polar phase collected and dried for analysis. To perform a direct comparison of the same volume we used 5 mL of 50% MeOH for cell quenching followed by increased extraction volumes (4 mL CHCl₃, additional 1.5 mL MeOH and 1.5 mL H_2O) leading to a 1:1:1, *v/v/v* ratio of MeOH:CHCl₃:H₂O. With the increased volumes, 6 mL of polar phase could be collected and used for GC-MS measurement. The 80_LOW condition (5 mL 80% MeOH for quenching and 1 mL of CHCl₃ for extraction) was discarded from the optimization tests since no phase separation could be achieved, as seen in Figure S1.

4.4. Experiment 3: Total Protein versus Cell Mass as a Normalization Strategy

To measure cell mass, at the point immediately following quenching and harvest of the cells, the quenched cells (50% MeOH) were centrifuged for 10 min at $10,000 \times g$ at 4 °C and the supernatant was carefully removed and collected into a new 15 mL Falcon tube stored on ice. The cell pellet, also kept on ice, was immediately weighed as fast as possible and the collected supernatant was added back to it, ready for metabolite extraction. The metabolites were extracted using a high volume of CHCl₃ (4 mL) as determined in the previous section. After collection of the polar phases (6 mL), proteins were precipitated for each sample by addition of 8 mL 100% MeOH to avoid phase separation followed by centrifugation at $16,000 \times g$ speed for 10 min. The supernatant was carefully discarded. The pellet was air dried at room temperature and used for protein lysis and protein determination as described in Section 4.5.

4.5. Protein Extraction and Determination

For measurement of total protein amount, the pellet was resuspended in 8 M urea buffer (in 50 mM HEPES, pH 8.5). The protein concentration was determined using a bicinchoninic acid (BCA) assay (Thermo Fischer Scientific, Waltham, MA, USA) following the manufacturer's instructions. In brief,

2 µL from each protein lysate was added to 2 µL reagent A, followed by the addition of 100 µL reagent B. After 30 min, absorbance was read at 562 nm using an Infinite M200Pro plate reader (TECAN, Männedorf, Switzerland). In order to calculate a calibration curve, a dilution series of bovine serum albumin (BSA, 1 mg/mL to 0.05 mg/mL) was included in the measurement.

4.6. GC-MS Metabolomics Measurement of Key Central Carbon Pathway Metabolites

All polar cell extracts were stored dry at −80 °C until analysis. Extracts were removed from the freezer and further dried in a rotational vacuum concentrator for 60 min before further processing to ensure there was no residual water which may influence derivatization efficiency. Dried cell extracts were dissolved in 15 µL of methoxyamine hydrochloride solution (40 mg/mL in pyridine) and incubated for 90 min at 30 °C with constant shaking, followed by the addition of 50 µL of N-methyl-N-[trimethylsilyl]trifluoroacetamide (MSTFA) and incubation at 37 °C for 60 min. The extracts were centrifuged for 10 min at 10,000× g, and aliquots of 25 µL were transferred into glass vials for GC-MS measurement. An identification mixture for reliable compound identification was prepared and derivatized in the same way and an alkane mixture for reliable retention index calculation was included [23]. Metabolite analysis was performed on a Pegasus 4D GC×GC TOF-MS-System (LECO Corporation, St. Joseph, MN, USA) complemented with an auto-sampler (Gerstel MPS DualHead with CAS4 injector, Mühlheim an der Ruhr, Germany). The samples were injected in split mode (split 1:5, injection volume 1 µL) in a temperature-controlled injector with a baffled glass liner (Gerstel, Mühlheim an der Ruhr, Germany). The following temperature program was applied during sample injection: for 2 min the column was allowed to equilibrate at 68 °C, a first temperature gradient was started with a rate of increase of 5 °C/min until a maximum of 120 °C was reached. Subsequently, the temperature gradient was changed such that the rate of increase was 7 °C/min up to a maximum temperature of 200 °C. This was increased to a 12 °C/min gradient up to a maximum temperature of 320 °C which was then held for 7.5 min. Gas chromatographic separation was performed on an Agilent 7890 (Agilent Technologies, Santa Clara, CA, USA), equipped with a VF-5ms column (Agilent Technologies, Santa Clara, CA, USA) of 30 m length, 250 µm inner diameter, and 0.25 µm film thickness. Helium was used as the carrier gas with a flow rate of 1.2 mL/min. The spectra were recorded in a mass range of 60 to 600 m/z with 10 spectra/s. The GC-MS chromatograms were processed with ChromaTOF software (LECO Corporation, St. Joseph, MN, USA) including baseline assessment, peak picking, and computation of the area.

4.7. Data Analysis

An in-house-created library and reference search including 45 most relevant metabolites from the central carbon metabolism (CCM) were used, as seen in Table S7. The data were exported and merged by an in-house R script. The metabolites were considered valid when they appeared in a minimum of three out of five biological replicates (BR) in HCT116, two or three in female HUVECs, and two BR for male HUVECs. Male and female HUVECs were analyzed separately because of the availability of the material. The peak area of each metabolite was calculated by normalization to the internal standard, cinnamic acid, and additionally to the protein content (or cell mass). The performance of the different MeOH percentages were evaluated according to the following criteria: (1) number of detected metabolites, (2) relative peak area, and (3) reproducibility. The technical variation of the GC-MS run is shown in Table S8.

Supplementary Materials: The following are available online at http://www.mdpi.com/2218-1989/10/1/2/s1, Table S1. Individual metabolite relative standard deviation (RSD) of HCT116 cells cultured in 10% and 20% FCS conditions using different quenching solvents, extraction ratios, and polar volumes. *: phosphate. Table S2. Individual metabolite relative standard deviation (RSD) of HUVECs cultured in 20% FCS conditions using different quenching solvents, extraction ratios and polar volumes for the different protocols. Cells were measured in different batches (dashed line). *: 50_Medium. Table S3. Median of the relative standard deviation (RSD) per metabolite of HUVECs cultured in 20% FCS conditions using different quenching solvents, extraction ratios, and polar volumes for the different protocols. Cells were measured in different batches. Table S4. Percentage of metabolites with a relative standard deviation (RSD) < 30% of HUVECs cultured in 20% FCS conditions using different quenching solvents, extraction ratios, and polar volumes for the different protocols. Cells were measured in different batches. Table S5. Relative standard deviation (RSD) of measured protein amount of HUVECs cultured in 20% FCS conditions using different quenching solvents, extraction ratios, and polar volumes for the different protocols. Cells were measured in different batches. Table S6. Median of the relative standard deviation (RSD) per metabolite and percentage of metabolites with a RSD < 30% of HUVECs (three biological replicates) cultured in 20% FCS conditions. Table S7. List of metabolite derivatives and their biological group used for reference search. AA: Amino acids. PPP: Pentose phosphate pathway. TCA: Tricarboxylic acid cycle. TMS: Trimethylsilyl derivatives. MeOX: Methoxyamine hydrochloride. Table S8. Technical variation during gas chromatography mass spectrometry (GC-MS) run of four pooled samples. RSD: Relative standard deviation. Figure S1. 50% and 80% of MeOH quenching buffer and different ratios of $MeOH:CHCL_3:H_2O$ were mixed for extraction steps. Sudan I was added to aid visualization of phase separation. No phase separation could be observed using 80_LOW condition. Therefore, the condition was not used for further testing. Figure S2. Recovery of annotated metabolites per biological group for different quenching and extraction methods. (a) HCT116 cells were cultivated in 10% FCS (in minimum three out of five biological replicates). (b) HCT116 cells were cultivated in 20% FCS (in minimum four out of five biological replicates). (c) HUVECs were cultured in 20% FCS (in minimum two out of three biological replicates). HCT116 cells were measured in one batch while HUVECs were measured in different batches (dashed line). The column data represents the number of annotated metabolites for each protocol. AA: Amino acids. PPP: Pentose phosphate pathway. TCA: Tricarboxylic acid cycle. Figure S3. Median relative standard deviation (RSD) of individual metabolites separated by metabolite classes from in minimum two out of three replicates of HUVECs cultured in 20% FCS conditions using different quenching solvents, extraction ratios, and polar volumes. The dashed line represents the maximum 30% RSD threshold advised by the *Food and Drug Administration* (FDA). Condition details see Table 3. Cells were measured in different batches reflected by the three graphs. AA: Amino acids. PPP: Pentose phosphate pathway. TCA: Tricarboxylic acid cycle. Figure S4. Comparison of the mean of sum of normalized peak area for different quenching and extraction methods. HUVECs were cultured in 20% FCS. The cells were measured in 2 different batches (dashed line). Data from in minimum two out of three biological replicates. The peak area was normalized to cinnamic acid and protein amount. Due to the low number of biological replicates no significances were measured. File S1. Standard operating protocol for harvest and extraction of adherent cells grown under 10% or 20% fetal calf serum (FCS) conditions version 1.

Author Contributions: Conceptualization, R.F.-G. and J.A.K.; methodology, R.F.-G.; software, Y.G., R.F.-G.; formal analysis, R.F.-G.; investigation, R.F.-G.; resources, M.L., J.A.K., R.F.-G.; data curation, R.F.-G.; writing—original draft preparation, R.F.-G.; writing—review and editing, J.A.K., A.B., Y.G., M.L.; visualization, R.F.-G.; supervision, J.A.K.; project administration, J.A.K.; funding acquisition, J.A.K.

Funding: This research received no external funding.

Acknowledgments: We acknowledge Alina Eisenberger (Berlin Institute of Health Metabolomics Platform, Berlin Institute of Health (BIH), Berlin, Germany and Max-Delbrück-Center for Molecular Medicine (MDC) in the Helmholtz Association, Berlin, Germany), Cornelia Bartsch as well as Angelika Vietzke (Charité–Universitätsmedizin Berlin, corporate member of Freie Universität Berlin, Humboldt-Universität zu Berlin, and Berlin Institute of Health, Medizinische Klinik für Kardiologie und Angiologie, Campus Mitte, Berlin, Germany) for technical support. We thank the Clinic for Gynecology and Obstetrics, Charité–University Medicine Berlin for support in obtaining the umbilical cords.

Conflicts of Interest: The authors declare no conflict of interest.

References

1. Kapoore, R.V.; Vaidyanathan, S. Towards quantitative mass spectrometry-based metabolomics in microbial and mammalian systems. *Philos. Trans. A Math. Phys. Eng. Sci.* **2016**, *374*. [CrossRef] [PubMed]

2. Sellick, C.A.; Hansen, R.; Maqsood, A.R.; Dunn, W.B.; Stephens, G.M.; Goodacre, R.; Dickson, A.J. Effective quenching processes for physiologically valid metabolite profiling of suspension cultured Mammalian cells. *Anal. Chem.* **2009**, *81*, 174–183. [CrossRef] [PubMed]

3. Sellick, C.A.; Hansen, R.; Stephens, G.M.; Goodacre, R.; Dickson, A.J. Metabolite extraction from suspension-cultured mammalian cells for global metabolite profiling. *Nat. Protoc.* **2011**, *6*, 1241–1249. [CrossRef]

4. Bolten, C.J.; Kiefer, P.; Letisse, F.; Portais, J.C.; Wittmann, C. Sampling for metabolome analysis of microorganisms. *Anal. Chem.* **2007**, *79*, 3843–3849. [CrossRef]

5. Dettmer, K.; Nurnberger, N.; Kaspar, H.; Gruber, M.A.; Almstetter, M.F.; Oefner, P.J. Metabolite extraction from adherently growing mammalian cells for metabolomics studies: Optimization of harvesting and extraction protocols. *Anal. Bioanal. Chem.* **2011**, *399*, 1127–1139. [CrossRef] [PubMed]

6. Danielsson, A.P.; Moritz, T.; Mulder, H.; Spegel, P. Development and optimization of a metabolomic method for analysis of adherent cell cultures. *Anal. Biochem.* **2010**, *404*, 30–39. [CrossRef]

7. Dietmair, S.; Timmins, N.E.; Gray, P.P.; Nielsen, L.K.; Kromer, J.O. Towards quantitative metabolomics of mammalian cells: Development of a metabolite extraction protocol. *Anal. Biochem.* **2010**, *404*, 155–164. [CrossRef]

8. Lorenz, M.A.; Burant, C.F.; Kennedy, R.T. Reducing time and increasing sensitivity in sample preparation for adherent mammalian cell metabolomics. *Anal. Chem.* **2011**, *83*, 3406–3414. [CrossRef] [PubMed]

9. Kuehnbaum, N.L.; Britz-McKibbin, P. New advances in separation science for metabolomics: Resolving chemical diversity in a post-genomic era. *Chem. Rev.* **2013**, *113*, 2437–2468. [CrossRef] [PubMed]

10. Chrysanthopoulos, P.K.; Goudar, C.T.; Klapa, M.I. Metabolomics for high-resolution monitoring of the cellular physiological state in cell culture engineering. *Metab. Eng.* **2010**, *12*, 212–222. [CrossRef] [PubMed]

11. Winder, C.L.; Dunn, W.B.; Schuler, S.; Broadhurst, D.; Jarvis, R.; Stephens, G.M.; Goodacre, R. Global metabolic profiling of Escherichia coli cultures: An evaluation of methods for quenching and extraction of intracellular metabolites. *Anal. Chem.* **2008**, *80*, 2939–2948. [CrossRef] [PubMed]

12. Sapcariu, S.C.; Kanashova, T.; Weindl, D.; Ghelfi, J.; Dittmar, G.; Hiller, K. Simultaneous extraction of proteins and metabolites from cells in culture. *MethodsX* **2014**, *1*, 74–80. [CrossRef] [PubMed]

13. Bligh, E.G.; Dyer, W.J. A rapid method of total lipid extraction and purification. *Can. J. Biochem. Physiol.* **1959**, *37*, 911–917. [CrossRef] [PubMed]

14. Folch, J.; Ascoli, I.; Lees, M.; Meath, J.A.; Le, B.N. Preparation of lipide extracts from brain tissue. *J. Biol. Chem.* **1951**, *191*, 833–841. [PubMed]

15. Folch, J.; Lees, M.; Stanley, G.H.S. A simple method for the isolation and purification of total lipides from animal tissues. *J. Biol. Chem.* **1957**, *226*, 497–509.

16. Chaipinyo, K.; Oakes, B.W.; van Damme, M.P.I. Effects of growth factors on cell proliferation and matrix synthesis of low-density, primary bovine chondrocytes cultured in collagen I gels. *J. Orthop. Res.* **2002**, *20*, 1070–1078. [CrossRef]

17. Lorenz, M.; Blaschke, B.; Benn, A.; Hammer, E.; Witt, E.; Kirwan, J.; Fritsche-Guenther, R.; Gloaguen, Y.; Bartsch, C.; Vietzke, A.; et al. Sex-specific metabolic and functional differences in human umbilical vein endothelial cells from twin pairs. *Atherosclerosis* **2019**, *291*, 99–106. [CrossRef]

18. Gupta, R.K.; Irie, R.F.; Morton, D.L. Antigens on human tumor cells assayed by complement fixation with allogeneic sera. *Cancer Res.* **1978**, *38*, 2573–2580.

19. Marin, V.; Kaplanski, G.; Gres, S.; Farnarier, C.; Bongrand, P. Endothelial cell culture: Protocol to obtain and cultivate human umbilical endothelial cells. *J. Immunol. Methods* **2001**, *254*, 183–190. [CrossRef]

20. Rosner, D.; McCarthy, N.; Bennett, M. Rapamycin inhibits human in stent restenosis vascular smooth muscle cells independently of pRB phosphorylation and p53. *Cardiovasc. Res.* **2005**, *66*, 601–610. [CrossRef]

21. Hinrichs, S.; Scherschel, K.; Kruger, S.; Neumann, J.T.; Schwarzl, M.; Yan, I.; Warnke, S.; Ojeda, F.M.; Zeller, T.; Karakas, M.; et al. Precursor proadrenomedullin influences cardiomyocyte survival and local inflammation related to myocardial infarction. *Proc. Natl. Acad. Sci. USA* **2018**, *115*, E8727–E8736. [CrossRef] [PubMed]

22. Lang, S.; Herrmann, M.; Pfeifer, C.; Brockhoff, G.; Zellner, J.; Nerlich, M.; Angele, P.; Prantl, L.; Gehmert, S.; Loibl, M. Leukocyte-reduced platelet-rich plasma stimulates the in vitro proliferation of adipose-tissue derived mesenchymal stem cells depending on PDGF signaling. *Clin. Hemorheol. Microcirc.* **2017**, *67*, 183–196. [CrossRef] [PubMed]

23. Pietzke, M.; Zasada, C.; Mudrich, S.; Kempa, S. Decoding the dynamics of cellular metabolism and the action of 3-bromopyruvate and 2-deoxyglucose using pulsed stable isotope-resolved metabolomics. *Cancer Metab.* **2014**, *2*, 9. [CrossRef] [PubMed]

24. Fritsche-Guenther, R.; Zasada, C.; Mastrobuoni, G.; Royla, N.; Rainer, R.; Rossner, F.; Pietzke, M.; Klipp, E.; Sers, C.; Kempa, S. Alterations of mTOR signaling impact metabolic stress resistance in colorectal carcinomas with BRAF and KRAS mutations. *Sci. Rep.* **2018**, *8*, 9204. [CrossRef] [PubMed]

25. Cao, B.; Aa, J.; Wang, G.; Wu, X.; Liu, L.; Li, M.; Shi, J.; Wang, X.; Zhao, C.; Zheng, T.; et al. GC-TOFMS analysis of metabolites in adherent MDCK cells and a novel strategy for identifying intracellular metabolic markers for use as cell amount indicators in data normalization. *Anal. Bioanal. Chem.* **2011**, *400*, 2983–2993. [CrossRef] [PubMed]

26. Ritter, J.B.; Genzel, Y.; Reichl, U. Simultaneous extraction of several metabolites of energy metabolism and related substances in mammalian cells: Optimization using experimental design. *Anal. Biochem.* **2008**, *373*, 349–369. [CrossRef]

27. Dunn, W.B.; Wilson, I.D.; Nicholls, A.W.; Broadhurst, D. The importance of experimental design and QC samples in large-scale and MS-driven untargeted metabolomic studies of humans. *Bioanalysis* **2012**, *4*, 2249–2264. [CrossRef]

28. Bi, H.; Krausz, K.W.; Manna, S.K.; Li, F.; Johnson, C.H.; Gonzalez, F.J. Optimization of harvesting, extraction, and analytical protocols for UPLC-ESI-MS-based metabolomic analysis of adherent mammalian cancer cells. *Anal. Bioanal. Chem.* **2013**, *405*, 5279–5289. [CrossRef]

29. Beckonert, O.; Keun, H.C.; Ebbels, T.M.; Bundy, J.; Holmes, E.; Lindon, J.C.; Nicholson, J.K. Metabolic profiling, metabolomic and metabonomic procedures for NMR spectroscopy of urine, plasma, serum and tissue extracts. *Nat. Protoc.* **2007**, *2*, 2692–2703. [CrossRef]

30. Cuperlovic-Culf, M.; Barnett, D.A.; Culf, A.S.; Chute, I. Cell culture metabolomics: Applications and future directions. *Drug Discov. Today* **2010**, *15*, 610–621. [CrossRef]

31. Matyash, V.; Liebisch, G.; Kurzchalia, T.V.; Shevchenko, A.; Schwudke, D. Lipid extraction by methyl-tert-butyl ether for high-throughput lipidomics. *J. Lipid Res.* **2008**, *49*, 1137–1146. [CrossRef] [PubMed]

32. Muschet, C.; Moller, G.; Prehn, C.; de Angelis, M.H.; Adamski, J.; Tokarz, J. Removing the bottlenecks of cell culture metabolomics: Fast normalization procedure, correlation of metabolites to cell number, and impact of the cell harvesting method. *Metabolomics* **2016**, *12*, 151. [CrossRef] [PubMed]

33. Blankenship, D.; Niemi, J.; Hilow, E.; Karl, M.; Sundararajan, S. Oral pioglitazone reduces infarction volume and improves neurologic function following MCAO in rats. *Adv. Exp. Med. Biol.* **2011**, *701*, 157–162. [CrossRef] [PubMed]

 *metabolites*

Review

Metabolomics Applied to the Study of Extracellular Vesicles

Charles Williams [1,2], Mari Palviainen [3,4], Niels-Christian Reichardt [2,5], Pia R.-M. Siljander [3,4] and Juan M. Falcón-Pérez [1,6,7,8,*]

[1] Exosomes Laboratory, CIC bioGUNE, Bizkaia Technology Park, 48160 Derio, Spain; cwilliams@cicbiomagune.es

[2] Glycotechnology Laboratory, CIC biomaGUNE, Paseo Miramón 182, 20014 San Sebastián, Spain; nreichardt@cicbiomagune.es

[3] EV Group, Molecular and Integrative Biosciences Research Programme, Faculty of Biological and Environmental Sciences, and CURED, Drug Research Program, Faculty of Pharmacy, Division of Pharmaceutical Biosciences, University of Helsinki, Viikinkaari 9, 00790 Helsinki, Finland; mari.palviainen@helsinki.fi (M.P.); pia.siljander@helsinki.fi (P.R.-M.S.)

[4] EV-Core, University of Helsinki, Viikinkaari 9, 00790 Helsinki, Finland

[5] CIBER-BBN, Paseo Miramón 182, 20014 San Sebastián, Spain

[6] Metabolomics Platform, CIC bioGUNE, Bizkaia Technology Park, 48160 Derio, Spain

[7] Centro de Investigación Biomédica en Red de Enfermedades Hepáticas y Digestivas (CIBERehd), 28029 Madrid, Spain

[8] IKERBASQUE Basque Foundation for Science, Bilbao, 48013 Bizkaia, Spain

[*] Correspondence: jfalcon@cicbiogune.es

Received: 4 October 2019; Accepted: 7 November 2019; Published: 12 November 2019

Abstract: Cell-secreted extracellular vesicles (EVs) have rapidly gained prominence as sources of biomarkers for non-invasive biopsies, owing to their ubiquity across human biofluids and physiological stability. There are many characterisation studies directed towards their protein, nucleic acid, lipid and glycan content, but more recently the metabolomic analysis of EV content has also gained traction. Several EV metabolite biomarker candidates have been identified across a range of diseases, including liver disease and cancers of the prostate and pancreas. Beyond clinical applications, metabolomics has also elucidated possible mechanisms of action underlying EV function, such as the arginase-mediated relaxation of pulmonary arteries or the delivery of nutrients to tumours by vesicles. However, whilst the value of EV metabolomics is clear, there are challenges inherent to working with these entities—particularly in relation to sample production and preparation. The biomolecular composition of EVs is known to change drastically depending on the isolation method used, and recent evidence has demonstrated that changes in cell culture systems impact upon the metabolome of the resulting EVs. This review aims to collect recent advances in the EV metabolomics field whilst also introducing researchers interested in this area to practical pitfalls in applying metabolomics to EV studies.

Keywords: extracellular vesicles; exosomes; microvesicles; biomarkers; diagnostics; metabolic pathways

---

1. Introduction

Extracellular vesicles (EVs) are bioactive nanosized vesicles that are secreted by cells that mediate intercellular communication through the transmission of functional biomolecules [1]. The term "EV" encompasses multiple designations as determined by the biogenetic pathway in question or biophysical characteristics of EVs [2]. Multivesicular-body-derived exosomes, for example, are created via the inward budding of endosomes, with the resulting intraluminal vesicles released through fusion of the endosome with the cell membrane. Elsewise, microvesicles directly bud off from the cell membrane and apoptotic bodies are the products of fragmented, dying cells. In addition to these three known classes of

EV, uncharacterised subtypes are likely to exist differing in their composition, ultrastructure, or size [3,4]. Since the discovery in the mid-2000s that EVs can transfer functional mRNA [5,6], there have been increasing reports ascribing biological functions to EVs across an astounding range of contexts [1]. From developmental biology [7] and cardiovascular homeostasis [8] through to immunity [9] and angiogenesis [10], there appears to be a role for EVs in every instance where biological material is exchanged between cells. Correspondingly, the pathological dysregulation of these processes has also been reported in cancer [11], neurodegeneration [12], obesity [13] and inflammatory diseases [14], whilst the deployment of EVs by pathogens and microbiota is informing new paradigms on how we interact with microorganisms [15–17].

Consisting essentially of a lipid bilayer, EVs also comprise a repertoire of proteins, glycans and nucleic acids [1]. Some of these components are incorporated into vesicles during biogenesis and are thought to be characteristic markers of EVs. Notable examples include CD9, CD63 and CD81 of the tetraspanin protein family [18], high phosphatidylserine lipid content [19] and a conserved glycosylation signature that has been described for mammalian EVs [20]. Depending on the EV class in question, markers differ and guidelines from the Minimal information for studies of extracellular vesicles (MISEV) working group recommend that transmembrane/glycosylphosphatidylinositol-anchored(GPI-anchored) proteins and cytosolic proteins be analysed in all bulk EV preparations in order to demonstrate the presence of EVs as well as non-EV structural proteins for purity control [21]. However, the comparatively limited repertoire of molecular markers cannot account for the diversity of functions ascribed to EVs, and online databases of EV omics studies show a vast assortment of EV-associated compounds beyond these [22,23], signifying other means of cargo packaging. Molecules may be trafficked directly to EVs through interactions with other biomolecules or indirectly packaged during vesicle formation processes that encapsulate pockets of cytosol [24]. Moreover, EV content is dynamic and shifts in response to perturbations in culture conditions [25]. As such, EVs present as reflections of their cellular source, and the analysis of vesicle content provides a nanoscopic window into the physiopathological state of said cells. There are various examples of this concept in action with the identification of EV protein, lipid and nucleic acid biomarkers for several diseases [26–28]. Additionally, EVs have been isolated from many human biofluids [1], engendering a huge interest in the use of EVs for non-invasive liquid biopsies, and the EV diagnostics sector is projected to reach a value of $100 million by 2021 [29]. In recent years, several examples of metabolite biomarkers have been published [30,31] and EV metabolomics holds great promise against this backdrop.

Metabolites are described as any biologically relevant molecule <2 kDa in size. This comprises a large range of molecular species, from steroid hormones and lipids to the metabolic intermediates of nutrient anabolism and the monomers of the major biopolymer classes. Considering that all cellular processes involve metabolites in some form, assaying of metabolomes can provide information on any dysregulation to these in a manner analogous to using EVs to study cellular status. For example, UHPLC-MS analysis of urinary EVs from prostate cancer patients identified abnormal levels of the steroid hormone dehydroepiandrosterone sulphate (DHEAS), suggesting this metabolite as a potential marker of prostate cancer and also as a means of monitoring disease progression [32]. Beyond providing disease biomarkers, metabolomics can also illuminate fundamental EV biology. It was shown that cancer-derived fibroblasts can supply nutrients to cancer cells through EVs, which further illuminates the mechanisms of the tumour niche [33]. Elsewise, EVs can also act as "metabolically active machines", as proven by the metabolomic analysis of serum samples mixed with hepatocyte EVs. Therein, alterations in the levels of over 90 metabolites were detected, signifying the presence of functional enzymes in EVs [34].

Whilst EVs present exciting opportunities for metabolomics researchers, there are intractable practical difficulties inherent to working with EVs that must be considered from the outset of any potential projects—difficulties that mainly relate to the different classes of EVs and the heterogeneity within these. For example, whilst isolation protocols can be targeted towards exosomes, the physical and biomolecular properties of exosomes, microvesicles and apoptotic bodies overlap, meaning that any exosome-targeted preparation likely also contains other EVs as contaminants and it is difficult to assign vesicle designation with full certainty [21,35]. Even then, supposing totally pure exosome purification, the stochastic nature of vesicle biogenesis gives each exosome a potentially unique composition [2]. Any subsequent omics characterisation thus provides information for the bulk vesicle population and ignores the possible contribution of relevant subpopulations. EV subpopulations are also problematic for sample preparation, as multiple reports have shown how different isolation methods generate different results during downstream characterization [36–39]. The matrix from which EVs are obtained may partially dictate the isolation method, but the choice is further complicated by the increasing number of commercial and published EV isolation methods. Finally, production methods can also bias EV composition. This has been demonstrated for EVs from two prostate cancer cell lines produced from either standard culture flatware or from bioreactors, with the metabolic signature of these models shifting depending on the cell culture conditions [40]. Fortunately, despite these complexities, the actual metabolomic analysis of a purified EV sample is straightforward and does not present any sample-specific issues. In our hands, established techniques of metabolite extraction and analysis have proven sufficient in generating sufficient data. A workflow of possible methodologies for an EV metabolomics study is presented below (Figure 1).

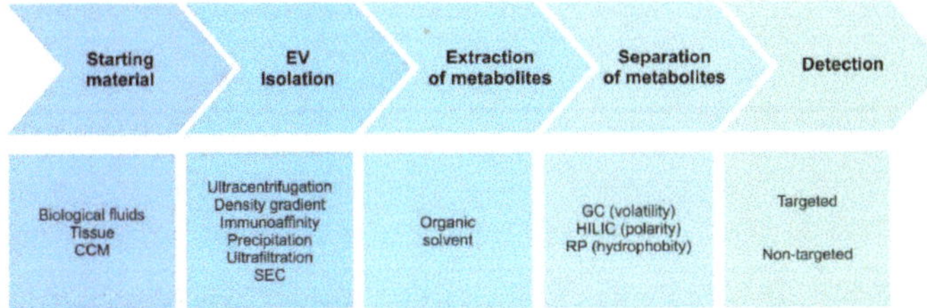

Figure 1. Key stages of the metabolomic analysis workflow for extracellular vesicle (EV) samples. Methodologies of EV isolation can impact the metabolome and should be carefully considered when comparing results from different studies. (CCM, cell conditioned medium; SEC, size exclusion chromatography; GC, gas chromatography; HILIC, hydrophilic interaction chromatography; RP, reversed-phase chromatography; Targeted analysis, detailed analysis of a predefined subset of the metabolome; Non-targeted analysis, maximum metabolite coverage).

Herein, we aim to cover the recent findings from EV metabolomics studies whilst placing an emphasis on the EV isolation and analysis methodologies used to highlight the best practices. The key studies so far and their salient information have been collected in Table 1 for easy reference. Consideration will also be given to the EV origins within studies, how the wider body of EV research can inform future EV metabolomics studies and whether a gold standard of EV sample preparation can ever be achieved. We will also cover how metabolomics of other sample types has been useful in understanding the role of EVs as metabolically active machines. It should be noted that there is crossover between EV metabolomics and lipidomics, as the size of biologically relevant lipids categorises them as metabolites. Since EV lipidomics is a more established field and has been excellently reviewed elsewhere [41,42], we have focused on metabolites as small polar molecules involved in cellular processes of metabolism.

Table 1. Primary studies comprising metabolite analysis of EV samples.

Research Description	Sample Type, Source	EV Isolation Method	Metabolomics Workflow	Significant Metabolites
EVs secreted by cancerous cells contain a suite of metabolites that can be received and metabolised by cancer cells [33]	CD63+ EVs from cancer-associated fibroblasts	Total exosome isolation reagent	Methanol–chloroform fractionation of EVs followed by GC-MS or UHPLC	Pyruvate, citrate, glutamine, arginine, palmate, stearate
Plasma EVs from adenocarcinoma patients and PANC 1 cell culture shown to possess broad metabolomes [43]	CD63+/CD9+/TSG101+ EVs from human plasma EVs from PANC 1 cell culture	$100,000 \times g$ ultracentrifugation	Methanol–chloroform fractionation of EVs followed by UHPLC-ESI-MS	Amino acids, substituted sugars
Metabolomics analysis of urinary EVs revealed potential prostate cancer biomarkers [32]	EVs from urine of either benign prostatic hyperplasia or prostate cancer patients	$100,000 \times g$ ultracentrifugation	Methanol–chloroform fractionation of EVs followed by UHPLC-MS	Dehydroepiandrosterone sulphates, other androsterone sulphate isomers
Identified vesicular hexanal as a candidate chemoattractant for *Anopheles gambiae* vectors of malaria [44]	EVs from patient-derived red blood cell culture	$110,000 \times g$ ultracentrifugation	GC-MS with headspace solid phase microextraction of EV samples	Hexanal, pentane2,2,4-trimethyl-pentane, 1,2,3-propanetriol diacetate
Comparative metabolomics of EVs from cells cultured with either conventional flatware or bioreactors revealed significant differences [40]	Large and small CD91+/CD9+/TSG101+ EVs from PC-3 and VCaP cell culture	$20,000 \times g$ and $110,000 \times g$ ultracentrifugation	Acetonitrile dissolution of EVs followed by UHPLC/Q-TOF-MS with separation by either reverse phase or hydrophilic interaction	Amino acids, phosphatidylcholines, phosphatidylethanolamines, sphingomyelins
Metabolomics of urinary and platelet-derived EVs show enrichment of specific molecules compared to sample matrices [45]	CD9+/CD63+/TSG101+ EVs from urine of prostate cancer patients and platelet-derived EVs from matched serum samples	$100,000 \times g$ ultracentrifugation for urinary EVs (uEVs) and $110,000 \times g$ for platelet-derived EVs	Acetonitrile dissolution of EVs followed by UHPLC-MS-MS	Spermidine, ornithine, carnitine derivatives, nicotinamide adenine dinucleotide, amino acids

Table 1. *Cont.*

Research Description	Sample Type, Source	EV Isolation Method	Metabolomics Workflow	Significant Metabolites
Enrichment of certain metabolite classes detected in EVs after irradiation of rhesus macaques [46]	Plasma-derived CD63+ EVs from rhesus macaques	120,000× *g* ultracentrifugation	Acetonitrile dissolution of EVs followed by UHPLC/Q-TOF-MS	Carnitines, sphingomyelins, amino acids, 5-methycytosine, nonic acids
Outer membrane vesicles from toxigenic and nontoxigenic *Bacteroides fragilis* spp. exhibit different metabolomes [47]	Outer membrane vesicles from *Bacteroides fragilis* spp. culture	100,000× *g* ultracentrifugation	Cold methanol extraction of EV metabolites followed by UHPLC-MS or GC-MS	Creatinine, creatine, glycerate-2P, fumarate, malate, amino acids
Large EVs from atherosclerotic plaques present taurine enrichment [48]	EVs from human carotid atherosclerotic plaques	20,500× *g* ultracentrifugation	Proton nuclear magnetic resonance spectroscopy	Taurine, lactate, glycerophosphocholine
Change in serum EV metabolome after chemotherapy [49]	Serum exosomes from pancreatic cancer patients before and after chemotherapy	100,000× *g* ultracentrifugation	50% methanol and freeze–thaw cycle for extraction followed by LC-Q-TOF-MS	Alanyl-histidine, 6-dimethylaminopurine, leucyl-proline, and methionine sulfoxide
EVs secreted by mesenchymal stem cells contain a suite of metabolites that can be received and metabolised by other cells [50]	EVs derived from mesenchymal stem cell culture	Ultrafiltration followed by 110,000× *g* ultracentrifugation	Methanol extraction followed by CE-UV and HPLC-MS/MS	Diacylglycerols, sphingomyelins, lactate, glutamate

2. Metabolomics of Patient-Derived EVs

Considering the potential of EVs for non-invasive liquid biopsies, the identification of robust metabolite biomarkers in EVs is a key goal. Such studies necessitate the use of EVs extracted from clinical samples for greater impact and translation. An instructive example comes from a metabolomics study of urinary EVs (uEVs) in the context of prostate cancer [32]. Therein, 50 mL of urine from patients with either prostate cancer (PCa) or benign prostatic hyperplasia (BPH) was sterile filtered and frozen at −80 °C immediately following collection. Subsequent EV isolation from the urine was performed by differential ultracentrifugation—first centrifuged for 30 min at 10,000 × g to remove the majority of larger vesicles before pelleting of small uEVs at 100,000 × g for 75 min. Processing of all uEV samples was performed by fractionation with methanol and chloroform into different phases of metabolites for UHPLC-MS analysis. Through this methodology, researchers detected differential metabolite content in the patient-derived uEVs and highlighted 76 of these as statistically significantly different between the PBH and PCa samples. Crucially, 3beta-hydroxyandors-5-en-17-one-3-sulphate (dehydroepiandrosterone sulphate, DHEAS) was raised in the uEVs from PCa patients compared to BHP patients, potentially enabling clinicians to distinguish between these often-confused diagnoses. The study continued with bioinformatics analysis of published gene expression data to identify a decrease in steroid sulphatase in the PCa samples, correlating with disease progression. DHEAS is a substrate for this enzyme, and its loss in advanced PCa suggests that DHEAS accumulation could also be used to stratify patients according to disease progression. The question remains as to whether a baseline DHEAS level could be established to show meaningful DHEAS deviations as a biomarker for initial PCa identification. Furthermore, the pathological relevance of EV-associated DHEAS is potentially interesting. No more than 10% of DHEAS is produced from the gonads of healthy males, instead primarily originating from the adrenal glands [51]. It is possible that the dysregulation of the steroid sulphatase gene in the cancer context could lead to more constitutive DHEAS signalling, exacerbating prostate tumour severity. In all, this study highlights how a combinatorial approach of metabolomics in combination with other methods can add value to diagnostics and provide new insights to disease.

Another proof-of-concept study for extracting EV metabolite biomarkers from biofluids was established with the plasma of endometrial adenocarcinoma (EAC) patients [43]. Plasma was obtained from 10 mL samples of blood from both EAC patients and control subjects before storage at −80 °C. EVs were again isolated through ultracentrifugation, although with a first centrifugation at 16,500× g and the second at 100,000× g for 120 min. Methanol–chloroform fractionation was employed to release metabolites and UPLC-ESI-MS was used for data acquisition. Due to inconsistencies between the data and metabolite databases, accurate mass-based identification of the detected metabolites could not be performed, beyond a few amino acids and substituted sugars. However, principal component analysis of metabolites revealed a clear separation between the plasma of healthy control subjects and those of the cancer patients, indicating the potential for even incomplete EV metabolomics in disease identification. Both these studies used ultracentrifugation for EV purification, albeit with some differences to their protocols. Previously described as the "gold standard" of EV isolation [52], this approach is straightforward and well suited to generating proof-of-concept data. However, ultracentrifugation is not appropriate for clinical diagnostics due to specific equipment requirements [53]. Other isolation methods are also available, such as size exclusion chromatography, immunoaffinity and precipitation-based methods. Each has their own advantages and drawbacks in terms of selectivity and yield, and these have been described with detail in a recent review [54]. In the case of urinary EVs, different clinically relevant EV purification methods were compared against ultracentrifugation, selected on the basis of minimal equipment requirement [37]. Said methodologies comprised several commercially available EV isolation kits working on precipitation principals and a method of lectin-based EV capture [55]. The protein and RNA composition of purified EVs was found to be heavily influenced by the isolation method, with EV markers CD9 and CD63 nearly absent in some purified samples. Inter-donor samples were also found to vary wildly in marker content when the same isolation method was used, highlighting the difficulty of establishing biomarker baselines. Considering this evidence, it is essential that this comparative approach is extended to EV metabolomics in order to ascertain whether there are similar method-dependent effects on EV metabolite content.

3. Metabolomics of EVs from Cell Culture

Whilst extracting EVs directly from biofluids has direct clinical prospects, there are advantages to culturing patient-derived cells in vitro for understanding disease. Specifically, the EVs of cancer-associated fibroblasts have been examined and a potential role found for their EVs in modulating the metabolism of cells from the cancerous PC3 line [33]. This study took fibroblasts originally isolated from cancer patient tumours and cultured these in either normal conditions or in media supplemented with various ^{13}C-labeled molecules including glucose, pyruvate, and lysine. Therein, the commercial Total Exosome Isolation Reagent from Thermo Fisher was used to isolate EVs directly from the cell culture media and the study employed complementary metabolomics approaches to characterise the cargo of these EVs. GC-MS (gas chromatography) highlighted a high vesicular content of the nutrients pyruvate and citrate, whilst UHPLC identified a number of essential amino acids, with glutamine and arginine particularly enriched. Downstream functional studies proved that these metabolites, which were supplied by the EVs produced in the presence of ^{13}C-labeled molecules, could be taken up and utilised by the receiving PC3 cells. The EVs were found to be within the size range for exosomes (30–100nm) according to nanoparticle tracking analysis, and were positive for CD63.

Beyond considerations in EV sample preparation, some forethought must also be given to the types of molecules one wishes to analyse. No single method is able to cover the entire spectrum of the metabolome. This is well exemplified with the EVs from red blood cells in relation to malaria and the transmission of the *Plasmodium falciparum* parasite by *Anopheles* spp. [44]. Understanding the processes of mosquito attraction to humans requires the study of volatile organic compounds as the transmitters of human scent over distances. Red blood cells from infected and non-infected volunteers were cultured ex vivo and the resulting conditioned media subjected to differential ultracentrifugation for 30 min at 15,000× *g* and 110,000× *g* for 70 min for EV purification. Metabolite extraction was performed concurrently with GC-MS using the headspace solid phase microextraction (HS-SPME) method. Briefly, EV samples were sealed in GC vials in the presence of a divinylbenzene/carboxen/polydimethylsiloxane fibre for 12 h at 37 °C before injection to the mass spectrometer. The HS-SPME GC-MS method was chosen to select for volatile organic compounds, and 18 of such were identified from EVs, of which diacetin was found to be increased in the EVs derived from infected red blood cells. The ultracentrifugation supernatants were also analysed, and in turn, hexanal was found solely in infected supernatants. Together, these are possible mosquito chemoattractants upregulated after infection with *P. falciparum*. which were discovered by metabolomics.

Importantly, the method of cell culture can also impact upon the metabolite composition of EVs, and may need to be considered when results of different studies are compared. This issue was highlighted by a recent article comparing the EVs of two prostate cancer cell lines produced either in conventional cell culture flatware or bioreactor culture [40]. Large and small EVs from either 20,000× *g* or 110,000× *g* ultracentrifugation steps were collected from both the VCaP and PC-3 lines cultured in either condition. EV samples were disassembled in acetonitrile and subjected to LC-ESI-MS with separation with either reversed-phase or hydrophilic columns. EVs were found to possess broadly similar physical characteristics regardless of the culture method, but significantly different levels of 459 metabolites were seen across both cell lines. Indeed, some molecules were unique to bioreactor or flatware culture, indicating different pathways of metabolic activity in the producing cells altering the identity of the resulting EVs. Whilst illustrative of the issues in EV sample generation, the implications of this work go beyond the differential metabolite results. There is significant interest in translating EVs as therapeutics to the clinic which will necessitate production scale-up to bioreactor levels from the flatware of standard research laboratories. Functional metabolite cargoes have already been proven in the case of cancer nutrient supply [33] and depending on the treatment modality of the candidate therapy, their efficacy may be impacted upon changes in metabolite content. This study should serve as a warning to therapy developers to characterise their products at all levels during process development. Moreover, understanding how the production conditions influence the metabolite content of the EVs may highlight novel methods to modulate their innate content.

4. Enrichment of Metabolites in EVs vs. EV Source

Metabolite enrichment in exosomes can sometimes facilitate the detection of certain low-abundance disease markers which are below the detection limit using the sample matrix alone. However, careful quantification of metabolite concentrations in both exosomes and matrices is required to validate enrichment. Failure to actively prove enrichment can have varied consequences. In the best-case scenario, it may be that a raw matrix shows biomarkers at sufficient concentrations to enable diagnosis, forgoing the need for advanced EV-targeted methods. However, it is more likely that a putative EV biomarker is not significantly enriched when compared to the biofluid, invalidating EV-associated biomarker application. More concerning outcomes are also possible. For the EV field, this is best exemplified by the contamination of EV preparations by exogenous nucleic acids from foetal bovine serum (FBS) media additives in a number of high-profile studies [56,57]. Prior to these cases, it was assumed that the ultracentrifugation-mediated EV-depletion of FBS was sufficient to remove excipient contaminants. Now, the findings of many publications have been called into question—a warning to all not to take EV sample preparation pitfalls lightly. Similarly, the recent identification of "exomeres" as distinct, functional nanoparticulate entities that can copurify with EV preparations may also present issues in future [58,59].

Therefore, it is important to always analyse the EV source matrix alongside the processed EV sample (e.g., the cell culture media direct from the bottle and without exposure to cells). For ultracentrifugation-based EV studies, retention and analysis of the supernatant can also suffice to determine any carryover from the cell culture media or biofluids, depending on the project. Alternatively, commercial isolation kits typically function by precipitation methods, and the work and supernatants from these should also be tested. A practical example comes from a comparative study on uEVs and platelet-derived EVs (pEVs) [45]. Therein, the metabolite content of uEVs purified by ultracentrifugation was measured by UHPLC-MS and compared to that of the originating urine filtrate, alongside pEVs compared with the parental platelets. In both systems, there was a high degree of metabolite overlap between the EVs and the source material, but certain classes of metabolites were enriched in the EV samples, suggesting an active recruitment of these molecules into the EVs. The largest changes in uEVs were observed for spermidine, ornithine and nicotinamide adenine dinucleotide (NAD), each enriched >600-fold compared to urine. Nucleotide and amino acid pathway metabolites were also significantly enriched. On the other hand, pEVs contained 11 unique molecules compared to the source platelets which only had one, but the enrichments were less dramatic for common metabolites. For example, a 250-fold upregulation of adenosine was the largest of these, followed by carnitine and various carnitine derivatives with 10- to 50-fold enrichments. Finally, inter-sample comparison of uEVs showed larger differences between the prostate cancer patients and the healthy controls than in the comparison between urine samples. Further comparison of the metabolite profiles from uEVs against pEVs revealed some common metabolites but also a broad range of unique molecules, highlighting the point that specific biofluids will be better starting points for specific diseases. Together, these data support the isolation of EVs as carriers of metabolites to identify otherwise missable biomarkers.

5. Metabolomics of Non-EV Samples

It is interesting to note how the metabolomics toolkit can be applied to questions of EV biology without direct EV analysis. This approach is well exemplified by complementary studies proving EVs as metabolically active entities by assaying the blood metabolome after incubation with EVs from rat primary hepatocyte culture [34,60]. Therein, small volumes of rat serum were incubated for 1 h with the EVs collected from 110,000× *g* ultracentrifugation and the metabolites then extracted through methanol–chloroform fractionation for UHPLC-MS analysis. In both studies, blood metabolome shifts were observed after incubation with EVs, specifically molecules of the arginine biosynthesis pathway, demonstrating that EVs are capable of actively modulating metabolites with their enzymatic content. This provides a novel concept of EVs as metabolic machines. These findings were further developed for arginase-positive EVs in an assay of pulmonary endothelial induction [34], based on the action of arginine as a nitric oxide precursor. The EVs inhibited the relaxation of isolated pulmonary arteries, providing direct proof that EVs can effect physiological changes through metabolome alteration. In related stories of active enzyme content, the presence of functional asparaginase, α2-6-sialyltransferase 1 (ST6Gal-I) and extracellular nicotinamide phosphoribosyltransferase (eNAMPT) have all been demonstrated in EVs [59,61,62]. eNAMPT is of particular note for its role in nicotinamide adenine dinucleotide metabolism, heavily associated with mechanisms of aging. Given the rapid rate of publication for these recent discoveries, it is exceedingly likely that metabolomics will reveal more such metabolically active EVs in future [63].

6. Conclusions and Future Directions

EV metabolomics is a nascent field but one with great potential, as evidenced by the interesting and varied findings presented in this review. The issues of EV sample preparation will always be inherent but are not insurmountable, as long as methods are clearly reported alongside sufficient controls and a suite of vesicle characterisation tests. That said, understanding in the field is continually evolving, as exemplified by the issues of contaminating nucleic acids from FBS media [57] or the disputed use of previously accepted EV markers [64]. To better facilitate inter-study comparison now and in posterity, full reporting is essential. It is worth consulting the MISEV and EV-TRACK (Transparent Reporting and Centralizing Knowledge in Extracellular Vesicle Research) guidelines in this regard [21,65]. These initiatives detail the minimal reporting requirements that should be included in a reproducible, quality study. However, there are no specific instructions for metabolomics reporting as yet, and these also need to be established or perhaps adopted from previous metabolomics standardisation initiatives [66]. EVpedia has allowances for metabolomics datasets [22], and we encourage researchers to fully utilise this resource and upload any findings to facilitate future bioinformatics analyses.

There is a great interest in using EVs for diagnostic purposes. However, the validation of biomarkers of any sort is a lengthy pursuit with many regulatory hurdles where failure is possible [67]. In these early days of EV metabolomics, it is worth pre-empting these challenges as much as possible through careful methodological planning. In this regard, it may be necessary to further standardise working practices. For example, the amount of EVs used for metabolomics varies across studies. One such work gives a vesicle count of 10^{10} particles for their experiments [45], and a similar number should perhaps be adopted by the field to ensure consistency between datasets. Moreover, the vast majority of studies presented here have used ultracentrifugation and the supremacy of this method as the "gold standard" has been challenged in recent years. With the ever-increasing amount of published methods, it is unlikely that a single best method will emerge and instead the choice of isolation method will be determined by the end application. Toward this end, we recommend that clinically relevant methods be adopted as soon as possible for biomarker studies [37,53]. Finally, the practical challenge of detecting biomarker shifts in a given patient is also worth considering. These may be too subtle to detect without previous patient sample data, necessitating a program of routine EV metabolite analysis that is unfeasible in practice. Future efforts should be made to establish a baseline EV metabolome for healthy humans from which deviations can be inferred.

EV metabolomics is useful for understanding certain EV functions, but questions remain regarding fundamental EV biology. For example, the investigation of EV subpopulation metabolomes has not been described, but may prove valuable to this area of high interest. This is a challenging task to undertake due the already low yield of total EV samples, particularly with the highly selective purification methods of marker-targeted affinity isolation or ultracentrifugation in combination with further chromatography steps [21]. Nonetheless, inevitable improvements in the sensitivity of metabolomics, such as nanoflow liquid chromatography-nanoelectrospray ionization (nLC-nESI) [68], along with more refined purification techniques, will hopefully enable these questions to be answered.

Despite the encouraging results collected herein, the full power of EV metabolomics is yet to be realized. We have used examples from biological pathologies, but the metabolome changes from physical stimuli can also be detected—as in the case of primates exposed to damaging radiation [46], which could potentially prove useful in monitoring exposure levels of radiotherapy technicians. Moreover, the metabolomics of pathogen EVs may further the understanding of infection from bacterial species [47]. Elsewise, the studies with patient-derived EVs presented here have focused on cancer, but metabolite biomarkers have already been suggested in neurodegeneration and even psychiatric disorders [69]. Such pathologies present different challenges in identifying valid patient–control cohorts but there is no reason that EV-targeted metabolomics may not prove useful in enriching the scanty biomarkers in these contexts. We hope that having read this review, metabolomics and EV researchers are encouraged to seek EV metabolite-related projects and contribute materially to this promising field.

Author Contributions: C.W. and M.P. were equally responsible for the writing—original draft preparation, N.-C.R., P.R.-M.S. and J.M.F.-P. were equally responsible for the writing—review and editing.

Funding: This work has been supported by Health Basque Government (2015111149 to JMF-P), Ramón Areces Foundation to JMF-P, Spanish Ministry of Economy and Competitiveness MINECO (SAF2015-66312 and RTI2018-094969-B-I00 to JMF-P), REDIEX (Excellence Network on Exosomes funded by MINECO) and the Severo Ochoa Excellence Accreditation (SEV-2016-0644), Spanish Ministry of Economy and Competitiveness MINECO (CTQ2017-90039-R to NCR), CIC biomaGUNE and CIC bioGUNE joint fellowship program (to CW). Funding for open access charge: Severo Ochoa Excellence Accreditation (SEV-2016-0644). All of them were co-financed by ERDF (FEDER) Funds from the European Commission, "A way of making Europe". This work was performed under the Maria de Maeztu Units of Excellence Program from the Spanish State Research Agency – Grant No. MDM-2017-0720. This work has also been supported by the Academy of Finland program grant no. 287089 (PS) and 1315227 (MP), The Medicinska Understödsföreningen Liv och Hälsa r.f. (PS) and Magnus Ehrnrooth Foundation (PS).

Conflicts of Interest: The authors declare no conflicts of interest pertaining to this submission.

References

1. Yáñez-Mó, M.; Siljander, P.R.-M.; Andreu, Z.; Zavec, A.B.; Borràs, F.E.; Buzas, E.I.; Buzas, K.; Casal, E.; Cappello, F.; Carvalho, J.; et al. Biological properties of extracellular vesicles and their physiological functions. *J. Extracell. Vesicles* **2015**, *4*, 27066. [CrossRef] [PubMed]

2. Willms, E.; Cabañas, C.; Mäger, I.; Wood, M.J.A.; Vader, P. Extracellular Vesicle Heterogeneity: Subpopulations, Isolation Techniques, and Diverse Functions in Cancer Progression. *Front. Immunol.* **2018**, *9*, 738. [CrossRef] [PubMed]

3. Van Niel, G.; D'Angelo, G.; Raposo, G. Shedding light on the cell biology of extracellular vesicles. *Nat. Rev. Mol. Cell Biol.* **2018**, *19*, 213–228. [CrossRef] [PubMed]

4. Höög, J.L.; Lötvall, J. Diversity of extracellular vesicles in human ejaculates revealed by cryo-electron microscopy. *J. Extracell. Vesicles* **2015**, *4*, 28680. [CrossRef] [PubMed]

5. Valadi, H.; Ekström, K.; Bossios, A.; Sjöstrand, M.; Lee, J.J.; Lötvall, J.O. Exosome-mediated transfer of mRNAs and microRNAs is a novel mechanism of genetic exchange between cells. *Nat. Cell Biol.* **2007**, *9*, 654–659. [CrossRef] [PubMed]

6. Ratajczak, J.; Miekus, K.; Kucia, M.; Zhang, J.; Reca, R.; Dvorak, P.; Ratajczak, M.Z. Embryonic stem cell-derived microvesicles reprogram hematopoietic progenitors: Evidence for horizontal transfer of mRNA and protein delivery. *Leukemia* **2006**, *20*, 847–856. [CrossRef] [PubMed]

7. Cruz, L.; Romero, J.A.A.; Iglesia, R.P.; Lopes, M.H. Extracellular Vesicles: Decoding a New Language for Cellular Communication in Early Embryonic Development. *Front. Cell Dev. Biol.* **2018**, *6*, 94. [CrossRef] [PubMed]

8. Stahl, P.D.; Raposo, G. Extracellular Vesicles: Exosomes and Microvesicles, Integrators of Homeostasis. *Physiology (Bethesda)* **2019**, *34*, 169–177. [CrossRef] [PubMed]

9. Robbins, P.D.; Morelli, A.E. Regulation of immune responses by extracellular vesicles. *Nat. Rev. Immunol.* **2014**, *14*, 195–208. [CrossRef] [PubMed]

10. Todorova, D.; Simoncini, S.; Lacroix, R.; Sabatier, F.; Dignat-George, F. Extracellular Vesicles in Angiogenesis. *Circ. Res.* **2017**, *120*, 1658–1673. [CrossRef] [PubMed]

11. Xu, R.; Rai, A.; Chen, M.; Suwakulsiri, W.; Greening, D.W.; Simpson, R.J. Extracellular vesicles in cancer—Implications for future improvements in cancer care. *Nat. Rev. Clin. Oncol.* **2018**, *15*, 617–638. [CrossRef] [PubMed]

12. Croese, T.; Furlan, R. Extracellular vesicles in neurodegenerative diseases. *Mol. Asp. Med.* **2018**, *60*, 52–61. [CrossRef] [PubMed]

13. Pardo, F.; Villalobos-Labra, R.; Sobrevia, B.; Toledo, F.; Sobrevia, L. Extracellular vesicles in obesity and diabetes mellitus. *Mol. Asp. Med.* **2018**, *60*, 81–91. [CrossRef] [PubMed]

14. Buzas, E.I.; György, B.; Nagy, G.; Falus, A.; Gay, S. Emerging role of extracellular vesicles in inflammatory diseases. *Nat. Rev. Rheumatol.* **2014**, *10*, 356–364. [CrossRef] [PubMed]

15. Kuipers, M.E.; Hokke, C.H.; Smits, H.H.; Nolte, E.N. Pathogen-Derived Extracellular Vesicle-Associated Molecules That Affect the Host Immune System: An Overview. *Front. Microbiol.* **2018**, *9*, 2182. [CrossRef] [PubMed]

16. Rodrigues, M.; Fan, J.; Lyon, C.; Wan, M.; Hu, Y. Role of Extracellular Vesicles in Viral and Bacterial Infections: Pathogenesis, Diagnostics, and Therapeutics. *Theranostics* **2018**, *8*, 2709–2721. [CrossRef] [PubMed]

17. Ahmadi Badi, S.; Moshiri, A.; Fateh, A.; Jamnani, F.R.; Sarshar, M.; Vaziri, F.; Siadat, S.D. Microbiota-Derived Extracellular Vesicles as New Systemic Regulators. *Front. Microbiol.* **2017**, *8*, 1610. [CrossRef] [PubMed]

18. Andreu, Z.; Yáñez-Mó, M. Tetraspanins in Extracellular Vesicle Formation and Function. *Front. Immunol.* **2014**, *5*, 442. [CrossRef] [PubMed]

19. Murphy, D.E.; de Jong, O.G.; Brouwer, M.; Wood, M.J.; Lavieu, G.; Schiffelers, R.M.; Vader, P. Extracellular vesicle-based therapeutics: Natural versus engineered targeting and trafficking. *Exp. Mol. Med.* **2019**, *51*, 32. [CrossRef] [PubMed]

20. Williams, C.; Royo, F.; Aizpurua-Olaizola, O.; Pazos, R.; Boons, G.-J.; Reichardt, N.-C.; Falcon-Perez, J.M. Glycosylation of extracellular vesicles: Current knowledge, tools and clinical perspectives. *J. Extracell. Vesicles* **2018**, *7*, 1442985. [CrossRef] [PubMed]

21. Théry, C.; Witwer, K.W.; Aikawa, E.; Alcaraz, M.J.; Anderson, J.D.; Andriantsitohaina, R.; Antoniou, A.; Arab, T.; Archer, F.; Atkin-Smith, G.K.; et al. Minimal information for studies of extracellular vesicles 2018 (MISEV2018): A position statement of the International Society for Extracellular Vesicles and update of the MISEV2014 guidelines. *J. Extracell. Vesicles* **2018**, *7*, 1535750. [CrossRef] [PubMed]

22. Kim, D.-K.; Lee, J.; Kim, S.R.; Choi, D.-S.; Yoon, Y.J.; Kim, J.H.; Go, G.; Nhung, D.; Hong, K.; Jang, S.C.; et al. EVpedia: A community web portal for extracellular vesicles research. *Bioinformatics* **2015**, *31*, 933–939. [CrossRef] [PubMed]

23. Kalra, H.; Simpson, R.J.; Ji, H.; Aikawa, E.; Altevogt, P.; Askenase, P.; Bond, V.C.; Borràs, F.E.; Breakefield, X.; Budnik, V.; et al. Vesiclepedia: A compendium for extracellular vesicles with continuous community annotation. *PLoS Biol.* **2012**, *10*, e1001450. [CrossRef] [PubMed]

24. Abels, E.R.; Breakefield, X.O. Introduction to Extracellular Vesicles: Biogenesis, RNA Cargo Selection, Content, Release, and Uptake. *Cell. Mol. Neurobiol.* **2016**, *36*, 301–312. [CrossRef] [PubMed]

25. de Jong, O.G.; Verhaar, M.C.; Chen, Y.; Vader, P.; Gremmels, H.; Posthuma, G.; Schiffelers, R.M.; Gucek, M.; van Balkom, B.W.M. Cellular stress conditions are reflected in the protein and RNA content of endothelial cell-derived exosomes. *J. Extracell. Vesicles* **2012**, *1*, 18396. [CrossRef] [PubMed]

26. Dhondt, B.; Van Deun, J.; Vermaerke, S.; de Marco, A.; Lumen, N.; De Wever, O.; Hendrix, A. Urinary extracellular vesicle biomarkers in urological cancers: From discovery towards clinical implementation. *Int. J. Biochem. Cell Biol.* **2018**, *99*, 236–256. [CrossRef] [PubMed]

27. Dickhout, A.; Koenen, R.R. Extracellular Vesicles as Biomarkers in Cardiovascular Disease; Chances and Risks. *Front. Cardiovasc. Med.* **2018**, *5*, 113. [CrossRef] [PubMed]

28. Lane, R.E.; Korbie, D.; Hill, M.M.; Trau, M. Extracellular vesicles as circulating cancer biomarkers: Opportunities and challenges. *Clin. Transl. Med.* **2018**, *7*, 14. [CrossRef] [PubMed]

29. Roy, S.; Hochberg, F.H.; Jones, P.S. Extracellular vesicles: The growth as diagnostics and therapeutics; a survey. *J. Extracell. Vesicles* **2018**, *7*, 1438720. [CrossRef] [PubMed]

30. Davison, J.; O'Gorman, A.; Brennan, L.; Cotter, D.R. A systematic review of metabolite biomarkers of schizophrenia. *Schizophr. Res.* **2018**, *195*, 32–50. [CrossRef] [PubMed]

31. Voge, N.V.; Perera, R.; Mahapatra, S.; Gresh, L.; Balmaseda, A.; Loroño-Pino, M.A.; Hopf-Jannasch, A.S.; Belisle, J.T.; Harris, E.; Blair, C.D.; et al. Metabolomics-Based Discovery of Small Molecule Biomarkers in Serum Associated with Dengue Virus Infections and Disease Outcomes. *PLoS Negl. Trop. Dis.* **2016**, *10*, e0004449. [CrossRef] [PubMed]

32. Clos-Garcia, M.; Loizaga-Iriarte, A.; Zuñiga-Garcia, P.; Sánchez-Mosquera, P.; Rosa Cortazar, A.; González, E.; Torrano, V.; Alonso, C.; Pérez-Cormenzana, M.; Ugalde-Olano, A.; et al. Metabolic alterations in urine extracellular vesicles are associated to prostate cancer pathogenesis and progression. *J. Extracell. Vesicles* **2018**, *7*, 1470442. [CrossRef] [PubMed]

33. Zhao, H.; Yang, L.; Baddour, J.; Achreja, A.; Bernard, V.; Moss, T.; Marini, J.C.; Tudawe, T.; Seviour, E.G.; San Lucas, F.A.; et al. Tumor microenvironment derived exosomes pleiotropically modulate cancer cell metabolism. *Elife* **2016**, *5*, e10250. [CrossRef] [PubMed]

34. Royo, F.; Moreno, L.; Mleczko, J.; Palomo, L.; Gonzalez, E.; Cabrera, D.; Cogolludo, A.; Vizcaino, F.P.; van-Liempd, S.; Falcon-Perez, J.M. Hepatocyte-secreted extracellular vesicles modify blood metabolome and endothelial function by an arginase-dependent mechanism. *Sci. Rep.* **2017**, *7*, 42798. [CrossRef] [PubMed]

35. Ramirez, M.I.; Amorim, M.G.; Gadelha, C.; Milic, I.; Welsh, J.A.; Freitas, V.M.; Nawaz, M.; Akbar, N.; Couch, Y.; Makin, L.; et al. Technical challenges of working with extracellular vesicles. *Nanoscale* **2018**, *10*, 881–906. [CrossRef] [PubMed]

36. Li, P.; Kaslan, M.; Lee, S.H.; Yao, J.; Gao, Z. Progress in Exosome Isolation Techniques. *Theranostics* **2017**, *7*, 789–804. [CrossRef] [PubMed]

37. Royo, F.; Zuñiga-Garcia, P.; Sanchez-Mosquera, P.; Egia, A.; Perez, A.; Loizaga, A.; Arceo, R.; Lacasa, I.; Rabade, A.; Arrieta, E.; et al. Different EV enrichment methods suitable for clinical settings yield different subpopulations of urinary extracellular vesicles from human samples. *J. Extracell. Vesicles* **2016**, *5*, 29497. [CrossRef] [PubMed]

38. Tang, Y.-T.; Huang, Y.-Y.; Zheng, L.; Qin, S.-H.; Xu, X.-P.; An, T.-X.; Xu, Y.; Wu, Y.-S.; Hu, X.-M.; Ping, B.-H.; et al. Comparison of isolation methods of exosomes and exosomal RNA from cell culture medium and serum. *Int. J. Mol. Med.* **2017**, *40*, 834–844. [CrossRef] [PubMed]

39. Freitas, D.; Balmaña, M.; Poças, J.; Campos, D.; Osório, H.; Konstantinidi, A.; Vakhrushev, S.Y.; Magalhães, A.; Reis, C.A. Different isolation approaches lead to diverse glycosylated extracellular vesicle populations. *J. Extracell. Vesicles* **2019**, *8*, 1621131. [CrossRef] [PubMed]

40. Palviainen, M.; Saari, H.; Kärkkäinen, O.; Pekkinen, J.; Auriola, S.; Yliperttula, M.; Puhka, M.; Hanhineva, K.; Siljander, P.R.-M. Metabolic signature of extracellular vesicles depends on the cell culture conditions. *J. Extracell. Vesicles* **2019**, *8*, 1596669. [CrossRef] [PubMed]

41. Skotland, T.; Sandvig, K.; Llorente, A. Lipids in exosomes: Current knowledge and the way forward. *Prog. Lipid Res.* **2017**, *66*, 30–41. [CrossRef] [PubMed]

42. Skotland, T.; Hessvik, N.P.; Sandvig, K.; Llorente, A. Exosomal lipid composition and the role of ether lipids and phosphoinositides in exosome biology. *J. Lipid Res.* **2019**, *60*, 9–18. [CrossRef] [PubMed]

43. Altadill, T.; Campoy, I.; Lanau, L.; Gill, K.; Rigau, M.; Gil-Moreno, A.; Reventos, J.; Byers, S.; Colas, E.; Cheema, A.K. Enabling Metabolomics Based Biomarker Discovery Studies Using Molecular Phenotyping of Exosome-Like Vesicles. *PLoS ONE* **2016**, *11*, e0151339. [CrossRef] [PubMed]

44. Correa, R.; Coronado, L.M.; Garrido, A.C.; Durant-Archibold, A.A.; Spadafora, C. Volatile organic compounds associated with Plasmodium falciparum infection in vitro. *Parasites Vectors* **2017**, *10*, 215. [CrossRef] [PubMed]

45. Puhka, M.; Takatalo, M.; Nordberg, M.-E.; Valkonen, S.; Nandania, J.; Aatonen, M.; Yliperttula, M.; Laitinen, S.; Velagapudi, V.; Mirtti, T.; et al. Metabolomic Profiling of Extracellular Vesicles and Alternative Normalization Methods Reveal Enriched Metabolites and Strategies to Study Prostate Cancer-Related Changes. *Theranostics* **2017**, *7*, 3824–3841. [CrossRef] [PubMed]

46. Cheema, A.K.; Hinzman, C.P.; Mehta, K.Y.; Hanlon, B.K.; Garcia, M.; Fatanmi, O.O.; Singh, V.K. Plasma Derived Exosomal Biomarkers of Exposure to Ionizing Radiation in Nonhuman Primates. *Int. J. Mol. Sci.* **2018**, *19*, 3427. [CrossRef] [PubMed]

47. Zakharzhevskaya, N.B.; Vanyushkina, A.A.; Altukhov, I.A.; Shavarda, A.L.; Butenko, I.O.; Rakitina, D.V.; Nikitina, A.S.; Manolov, A.I.; Egorova, A.N.; Kulikov, E.E.; et al. Outer membrane vesicles secreted by pathogenic and nonpathogenic Bacteroides fragilis represent different metabolic activities. *Sci. Rep.* **2017**, *7*, 5008. [CrossRef] [PubMed]

48. Mayr, M.; Grainger, D.; Mayr, U.; Leroyer, A.S.; Leseche, G.; Sidibe, A.; Herbin, O.; Yin, X.; Gomes, A.; Madhu, B.; et al. Proteomics, metabolomics, and immunomics on microparticles derived from human atherosclerotic plaques. *Circ. Cardiovasc. Genet.* **2009**, *2*, 379–388. [CrossRef] [PubMed]

49. Luo, X.; An, M.; Cuneo, K.C.; Lubman, D.M.; Li, L. High-Performance Chemical Isotope Labeling Liquid Chromatography Mass Spectrometry for Exosome Metabolomics. *Anal. Chem.* **2018**, *90*, 8314–8319. [CrossRef] [PubMed]

50. Vallabhaneni, K.C.; Penfornis, P.; Dhule, S.; Guillonneau, F.; Adams, K.V.; Mo, Y.Y.; Xu, R.; Liu, Y.; Watabe, K.; Vemuri, M.C.; et al. Extracellular vesicles from bone marrow mesenchymal stem/stromal cells transport tumor regulatory microRNA, proteins, and metabolites. *Oncotarget* **2015**, *6*, 4953–4967. [CrossRef] [PubMed]

51. Labrie, F.; Luu-The, V.; Labrie, C.; Bélanger, A.; Simard, J.; Lin, S.-X.; Pelletier, G. Endocrine and intracrine sources of androgens in women: Inhibition of breast cancer and other roles of androgens and their precursor dehydroepiandrosterone. *Endocr. Rev.* **2003**, *24*, 152–182. [CrossRef] [PubMed]

52. Momen-Heravi, F.; Balaj, L.; Alian, S.; Mantel, P.-Y.; Halleck, A.E.; Trachtenberg, A.J.; Soria, C.E.; Oquin, S.; Bonebreak, C.M.; Saracoglu, E.; et al. Current methods for the isolation of extracellular vesicles. *Biol. Chem.* **2013**, *394*, 1253–1262. [CrossRef] [PubMed]

53. Sáenz-Cuesta, M.; Arbelaiz, A.; Oregi, A.; Irizar, H.; Osorio-Querejeta, I.; Muñoz-Culla, M.; Banales, J.M.; Falcón-Pérez, J.M.; Olascoaga, J.; Otaegui, D. Methods for extracellular vesicles isolation in a hospital setting. *Front. Immunol.* **2015**, *6*, 50. [CrossRef] [PubMed]

54. Monguió-Tortajada, M.; Gálvez-Montón, C.; Bayes-Genis, A.; Roura, S.; Borràs, F.E. Extracellular vesicle isolation methods: Rising impact of size-exclusion chromatography. *Cell. Mol. Life Sci.* **2019**, *76*, 2369–2382. [CrossRef] [PubMed]

55. Echevarria, J.; Royo, F.; Pazos, R.; Salazar, L.; Falcon-Perez, J.M.; Reichardt, N.-C. Microarray-based identification of lectins for the purification of human urinary extracellular vesicles directly from urine samples. *Chembiochem* **2014**, *15*, 1621–1626. [CrossRef] [PubMed]

56. Wei, Z.; Batagov, A.O.; Carter, D.R.F.; Krichevsky, A.M. Fetal Bovine Serum RNA Interferes with the Cell Culture derived Extracellular RNA. *Sci. Rep.* **2016**, *6*, 31175. [CrossRef] [PubMed]

57. Tosar, J.P.; Cayota, A.; Eitan, E.; Halushka, M.K.; Witwer, K.W. Ribonucleic artefacts: Are some extracellular RNA discoveries driven by cell culture medium components? *J. Extracell. Vesicles* **2017**, *6*, 1272832. [CrossRef] [PubMed]

58. Zhang, H.; Freitas, D.; Kim, H.S.; Fabijanic, K.; Li, Z.; Chen, H.; Mark, M.T.; Molina, H.; Martin, A.B.; Bojmar, L.; et al. Identification of distinct nanoparticles and subsets of extracellular vesicles by asymmetric flow field-flow fractionation. *Nat. Cell Biol.* **2018**, *20*, 332–343. [CrossRef] [PubMed]

59. Zhang, Q.; Higginbotham, J.N.; Jeppesen, D.K.; Yang, Y.-P.; Li, W.; McKinley, E.T.; Graves-Deal, R.; Ping, J.; Britain, C.M.; Dorsett, K.A.; et al. Transfer of Functional Cargo in Exomeres. *Cell Rep.* **2019**, *27*, 940–954.e6. [CrossRef] [PubMed]

60. Royo, F.; Palomo, L.; Mleczko, J.; Gonzalez, E.; Alonso, C.; Martínez, I.; Pérez-Cormenzana, M.; Castro, A.; Falcon-Perez, J.M. Metabolically active extracellular vesicles released from hepatocytes under drug-induced liver-damaging conditions modify serum metabolome and might affect different pathophysiological processes. *Eur. J. Pharm. Sci.* **2017**, *98*, 51–57. [CrossRef] [PubMed]

61. Iraci, N.; Gaude, E.; Leonardi, T.; Costa, A.S.H.; Cossetti, C.; Peruzzotti-Jametti, L.; Bernstock, J.D.; Saini, H.K.; Gelati, M.; Vescovi, A.L.; et al. Extracellular vesicles are independent metabolic units with asparaginase activity. *Nat. Chem. Biol.* **2017**, *13*, 951–955. [CrossRef] [PubMed]

62. Yoshida, M.; Satoh, A.; Lin, J.B.; Mills, K.F.; Sasaki, Y.; Rensing, N.; Wong, M.; Apte, R.S.; Imai, S.-I. Extracellular Vesicle-Contained eNAMPT Delays Aging and Extends Lifespan in Mice. *Cell Metab.* **2019**, *30*, 329–342.e5. [CrossRef] [PubMed]

63. Angulo, M.A.; Royo, F.; Falcón-Pérez, J.M. Metabolic Nano-Machines: Extracellular Vesicles Containing Active Enzymes and Their Contribution to Liver Diseases. In *Current Pathobiology Reports*; Springer: Berlin, Germany, 2019.

64. Liao, Z.; Jaular, L.M.; Soueidi, E.; Jouve, M.; Muth, D.C.; Schøyen, T.H.; Seale, T.; Haughey, N.J.; Ostrowski, M.; Théry, C.; et al. Acetylcholinesterase is not a generic marker of extracellular vesicles. *J. Extracell. Vesicles* **2019**, *8*, 1628592. [CrossRef] [PubMed]

65. Van Deun, J.; Mestdagh, P.; Agostinis, P.; Akay, Ö.; Anand, S.; Anckaert, J.; Martinez, Z.A.; Baetens, T.; Beghein, E.; Bertier, L.; et al. EV-TRACK: Transparent reporting and centralizing knowledge in extracellular vesicle research. *Nat. Methods* **2017**, *14*, 228–232. [CrossRef] [PubMed]

66. Sumner, L.W.; Amberg, A.; Barrett, D.; Beale, M.H.; Beger, R.; Daykin, C.A.; Fan, T.W.-M.; Fiehn, O.; Goodacre, R.; Griffin, J.L.; et al. Proposed minimum reporting standards for chemical analysis Chemical Analysis Working Group (CAWG) Metabolomics Standards Initiative (MSI). *Metabolomics* **2007**, *3*, 211–221. [CrossRef] [PubMed]

67. Kraus, V.B. Biomarkers as drug development tools: Discovery, validation, qualification and use. *Nat. Rev. Rheumatol.* **2018**, *14*, 354–362. [CrossRef] [PubMed]

68. Chetwynd, A.J.; David, A. A review of nanoscale LC-ESI for metabolomics and its potential to enhance the metabolome coverage. *Talanta* **2018**, *182*, 380–390. [CrossRef] [PubMed]

69. Sethi, S.; Brietzke, E. Omics-Based Biomarkers: Application of Metabolomics in Neuropsychiatric Disorders. *Int. J. Neuropsychopharmacol.* **2015**, *19*, pyv096. [CrossRef] [PubMed]

 *metabolites*

Review

Current Practice in Untargeted Human Milk Metabolomics

Isabel Ten-Doménech [1,†], Victoria Ramos-Garcia [1,†], José David Piñeiro-Ramos [1], María Gormaz [1,2], Anna Parra-Llorca [1], Máximo Vento [1,2], Julia Kuligowski [1,*] and Guillermo Quintás [3,4]

1   Neonatal Research Unit, Health Research Institute La Fe, Avenida Fernando Abril Martorell 106, 46026 Valencia, Spain; isabel_ten@iislafe.es (I.T.-D.); victoria_ramos@iislafe.es (V.R.-G.); jose_pineiro@iislafe.es (J.D.P.-R.); gormaz_mar@gva.es (M.G.); annaparrallorca@gmail.com (A.P.-L.); maximo.vento@uv.es (M.V.)
2   Division of Neonatology, University & Polytechnic Hospital La Fe, Avenida Fernando Abril Martorell 106, 46026 Valencia, Spain
3   Health and Biomedicine, Leitat Technological Center, Carrer de la Innovació, 2, 08225 Terrassa, Spain; gquintas@leitat.org
4   Unidad Analítica, Health Research Institute La Fe, Avenida Fernando Abril Martorell 106, 46026 Valencia, Spain
*   Correspondence: julia.kuligowski@uv.es; Tel.: +34-961-246-661
†   Both authors contributed equally.

Received: 29 November 2019; Accepted: 19 January 2020; Published: 22 January 2020

Abstract: Human milk (HM) is considered the gold standard for infant nutrition. HM contains macro- and micronutrients, as well as a range of bioactive compounds (hormones, growth factors, cell debris, etc.). The analysis of the complex and dynamic composition of HM has been a permanent challenge for researchers. The use of novel, cutting-edge techniques involving different metabolomics platforms has permitted to expand knowledge on the variable composition of HM. This review aims to present the state-of-the-art in untargeted metabolomic studies of HM, with emphasis on sampling, extraction and analysis steps. Workflows available from the literature have been critically revised and compared, including a comprehensive assessment of the achievable metabolome coverage. Based on the scientific evidence available, recommendations for future untargeted HM metabolomics studies are included.

Keywords: human milk; metabolome; sampling; extraction; liquid chromatography–mass spectrometry (LC-MS); nuclear magnetic resonance (NMR); gas chromatography–mass spectrometry (GC-MS); capillary electrophoresis—mass spectrometry (CE-MS)

---

1. Introduction

Human milk (HM) has been markedly established as the optimal way of providing infants with the necessary nutrients and bioactive factors for their early development. Many health associations and organisms, including World Health Organization, recommend exclusive breastfeeding for the first six months of life [1]. Health benefits of HM for infants include reduced mortality and morbidity, including sepsis, respiratory diseases, otitis media, gastroenteritis, and urinary tract infections, among others [2]. In addition, studies reporting on long-term benefits of HM consumption such as lower risk of suffering from type 1 diabetes and inflammatory bowel disease or overweight in adulthood emerged [3]. HM may also be associated with a slightly improved neurological outcome as cohort studies report [4], especially in preterm infants [5], although potential confounders must be accounted for [6].

HM composition is dynamic and influenced by several factors including genetics, gestational and infant's age, circadian rhythm, maternal nutrition, or ethnicity. It provides a series of nutrients such as lipids, proteins, carbohydrates, and vitamins, jointly with a number of bioactive factors that contribute to several physiological activities in the newborn infant as well as to short- and long-term outcomes [7,8]. Living cells including stem cells, hormones, growth factors, enzymes, microbiota, and even genetic material are part of this vast array of HM components with impact in early development, particularly the immune system [9]. In addition, HM appears to be one of the richest sources of microRNAs [10]. On the other hand, because of the maternal environmental exposure and lifestyle, the presence of some contaminants such as persistent organic pollutants or pharmacologically active substances in HM has been described [11,12].

Due to its complex composition, the analysis of HM is not straightforward. While the advent of "omics" approaches has offered valuable insights into the composition of this unique biofluid, untargeted metabolomic and lipidomic studies have only recently been applied to HM [13]. The comprehensive study of the HM metabolome, which includes the intermediate and end products of metabolism, can shed light on maternal status or phenotype [14,15]. The generation, analysis, and integration of large and complex data sets obtained in metabolomic studies go hand in hand with the following challenges: (i) the intrinsic complexity of the sample: a rich variety of jointly present, structurally heterogeneous compounds at concentrations that strongly vary covering several orders of magnitude; (ii) pre-analytical steps related to sampling, storage, and pre-processing (e.g., extraction, clean-up); and (iii) the diversity of platforms currently available including nuclear magnetic resonance (NMR), as well as gas chromatography (GC), liquid chromatography (LC), and capillary electrophoresis (CE) coupled to mass spectrometry (MS). The analysis of the HM metabolome has been approached employing a variety of extraction and analytical techniques to respond to a spectrum of clinically relevant questions. Several studies have compared HM metabolome with formula milk [13,16–20] or with milk from other mammalian species including monkey [21], donkey [17], and cow [18], whereas others have made efforts in defining the metabolome of preterm milk [13,16,22–26] and the evaluation of the HM metabolome during the course of lactation [15,23,27–30]. Furthermore, the influence of maternal diet [14,15,31], phenotype [14,32], obesity [30], or atopy status [33], as well as geographical location [33,34], time of the day [29,35], chemotherapy [36], or preeclampsia during pregnancy [31] on the HM metabolome have been reported.

Recent review articles that address the HM metabolome or lipidome [12,37–41] are available. For information on the compounds and compound families typically found in HM and their function the reader is referred to [37–40]. Readers with a particular interest in HM lipidomics are referred to a recent compilation study [41]. Technical aspects of HM analysis when performing metabolomics studies in HM have been recently described [12]. This review article gathers recent literature available on metabolomic analysis of HM, particularly focusing on untargeted approaches as indicated in Figure 1, to provide an up-to-date overview of the key factors that may influence HM metabolome coverage. Based on the information provided within the available literature, recommendations to guide study design and analytical method development of untargeted HM metabolomics assays were developed.

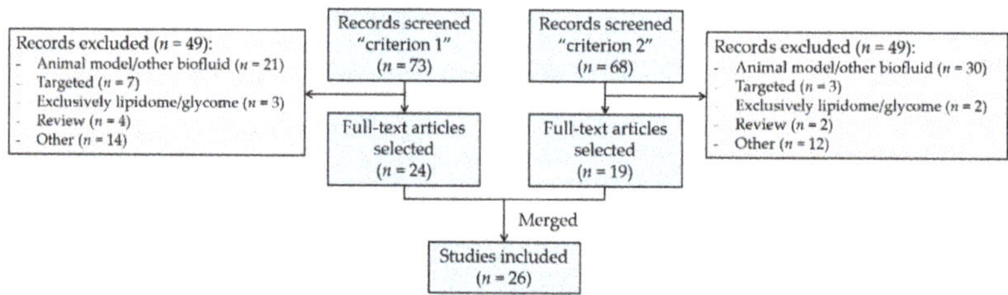

Figure 1. Flow diagram of literature selection and review process. Search "criterion 1": term ("human milk" OR "breast milk"), AND "metabolom*", AND "infant"; only articles. Search "criterion 2": term ("human milk" OR "breast milk"), AND "metabolom*", AND ("GC" OR "LC" OR "NMR" OR "CE"); only articles. Web of Science database was employed for literature search.

2. Considerations Regarding the Study Design

HM is a biofluid characterized by a dynamically varying composition according to several factors including lactation time, time of the day, throughout each feed, maternal status, and the environmental exposure. Although compositional variations have been mainly studied regarding the protein content of HM [42], changes of other compound classes such as fat or vitamins have been also reported [43,44]. Considering the intrinsic variability of HM, the complexity of obtaining representative HM samples is not negligible. Sources of variation related to sample manipulation and compositional variation can be minimized using standard operational procedures (SOPs). SOPs are fundamental to maintain quality assurance (QA) and quality control (QC) process and facilitate repeatable and reproducible research within and across laboratories. However, biologically meaningful results across studies will only be obtained if several key factors during the sample collection process are successfully controlled. This is of special importance in untargeted approaches, where the interpretation of results is especially challenging, and confounding factors introduced by a non-exhaustive sampling protocol can be wrongly attributed to differences between subjects of a studied population. Conversely, biologically meaningful information can be missed or remain unnoticed due to unwanted bias introduced during sample collection.

The information regarding study design provided in HM metabolomics studies varies considerably [13–24,26–32,34–36,45–47] as shown in Figure 2. Repeatedly reported factors have been grouped into three categories and are discussed in detail in the following sections: (1) maternal-infant-related factors (blue bars), (2) time-related factors (green bars), and (3) HM collection-related factors (orange bars). It should be noted that, although the importance of each factor might vary with the scientific question of each study, the authors encourage (i) the use of SOPs employed during sample collection to assure homogenous and representative sampling and (ii) the reporting of all documented factors in order to enhance comparability between results of metabolomic studies on HM. In case of HM, samples are typically collected, handled and sometimes temporary stored and transported by the mothers and not, such as it is the case for other biofluids (e.g., plasma or serum), by health professionals. During study design it is therefore very important to assure that mothers receive detailed instructions and/or training for the correct handling of collected samples. In addition, one should keep in mind that sampling protocols should neither interfere with infant feeding nor negatively impact on the mother-baby bonding. Hence, the collection of transitional and mature milk is usually preferred over colostrum, especially in studies involving mothers of preterm infants, where colostrum is usually kept exclusively for the infant's supply.

2.1. Maternal-Infant-Related Factors

In HM metabolomic studies, gestational age is frequently reported (see Figure 2), although the impact of this factor on the HM metabolome has not been fully characterized. Studies focused on preterm milk showed that, analogously to full-term milk, its composition is dynamic throughout the first month of lactation [13,16,22]. However, after 5–7 weeks, metabolite composition of HM from mothers of preterm infants resembled that collected from mothers of full-term infants [23]. On the other hand, Marincola et al. [13] observed that HM from mothers of early preterm infants (26 weeks of gestation) differentiated from milk samples from term infants. However, the low number of samples involved in the study ($n = 20$ and $n = 3$ mothers of preterm and term infants, respectively) hindered the assessment of the statistical significance of the impact of gestational age on the milk metabolite composition. Sundekilde et al. [23] carried out a longitudinal study on milk from mothers of preterm and full-term infants covering similar lactation periods (3–14 weeks and 3–26 weeks after birth, respectively) and showed that some metabolites were present at significantly different levels in full-term milk compared to preterm milk. On the contrary, Longini et al. [16] did not observe significant differences between preterm and full-term milk within the first week after delivery, only being able to discriminate milk samples from early preterm infants (<29 weeks of gestation). It is worth noting that the effect of gestational age on the HM metabolome has been mainly studied employing NMR platforms [13,16,22,23], in which metabolite coverage is limited (see Figure 5) and some metabolite classes (e.g., lipids) are barely accessible. For this reason, and in order to further evaluate the impact of gestational age on the milk metabolome, we warrant more comprehensive metabolomic studies employing complementary analytical platforms.

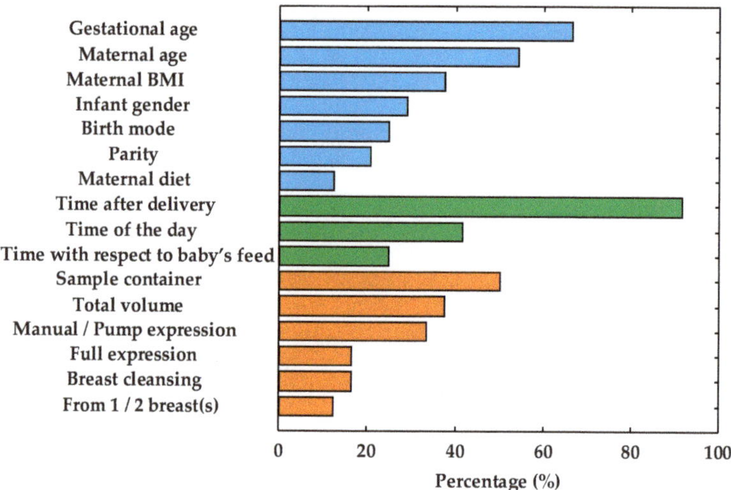

Figure 2. Reporting frequency of factors relevant to the human milk (HM) sampling process: Maternal-infant-related factors (blue bars), time-related factors (green bars), and HM collection-related factors (orange bars). Note: BMI = body mass index.

Other potentially relevant, miscellaneous information about the studied population of mother-infant pairs such as infant gender, parity, and birth mode, have been frequently reported in metabolomic studies (see Figure 2), although these characteristics often remain in the background since the studies focus on other aspects. The influence of these factors on the metabolite composition of HM has not been elucidated yet, and this might be addressed in forthcoming studies. An additional factor that is not typically reported in HM metabolomics studies is maternal secretor status. Significant differences in the oligosaccharides profile of milk between so-called secretors (Se$^+$), which are those mothers that provide a functional *FUT2* gene, and non-secretors (Se$^-$) have been reported [48]. Secretor status is mainly established based on the presence (Se$^+$) or absence (Se$^-$) of 2′-fucosyllactose, with a prevalence rate of approximately 80% of secretors over non-secretors [14,22,23,26,32,49]. Maternal secretor status is therefore usually determined *a posteriori* during data processing and analysis. Oligosaccharides are polar compounds that are present at concentrations in the mM range that will likely be preserved during sample extraction procedures employed for metabolomics studies. As their presence/absence might potentially affect clustering of milk based on maternal secretor status [24], to provide this information, when available, might be of interest.

2.2. Time-Related Factors

HM undergoes significant changes over time, having established three differentiated lactation stages: colostrum, transitional milk and mature milk. As can be seen in Figure 2, lactation time is reported in the vast majority of metabolomic studies. In particular, several studies have focused on the HM metabolome throughout lactation [15,23,27–30], all of them concluding that significant differences in the metabolic profile over time exist. Therefore, it seems reasonable to report this factor. On the other hand, although it has been demonstrated that diurnal variation affects HM fat content [50], its effect on the overall metabolite composition is a controversial issue which has not yet been adequately addressed in the available literature. Whereas no significant changes in some lipids and small polar metabolites have been observed [29,35], differences in some micronutrients (e.g., vitamins) could be evidenced [44]. The use of a pool of a 24-h expression of HM should compensate for changes due to diurnal variation, thus, obtaining more representative samples [13,24,25] in longitudinal studies. However, the drawback is that this practice is incompatible with breastfeeding of the infant, which, in turn, might raise severe ethical concerns. A feasible compromise for ameliorating diurnal variations is the use of pooled morning and evening samples [29,35].

Regarding time of collection with respect to baby's feed, the influence of this variable has not been studied to date, but given the differences found between fore- and hindmilk [51], it seems reasonable to assume that this factor might be potentially relevant.

2.3. HM Collection-Related Factors

Any uncontrolled variable within an experiment can result in a potential source of bias. In this sense, although less attention has been payed to other factors related to the expression and storage of HM (see Figure 2, orange bars), they may be relevant to the outcomes of metabolomic studies. As can be seen, the type of sample container is indicated in 50% of the studies, whereas other specifications regarding HM expression are included scarcely. The latter factor deserves some special attention, since differences in the milk fat content between foremilk (initial milk of a feed) and hindmilk (last milk of a feed) have been reported [51]. Therefore, full expression of breast(s) is desirable in order to obtain a representative HM aliquot [52]. The influence of all other factors, to date, remains unstudied.

2.4. Pasteurization and Storage

HM banks rely on stringent protocols in which pasteurization, indispensable for minimizing the potential to transmit infectious agents, as well as freezing and long-term storage procedures are established. The pasteurization process affects some of the nutritional and biological properties of HM [53–55]. In this review, three studies that use milk from HM banks are included [16,20,23], but only one specifies whether or not HM has undergone pasteurization [23]. Variability of the metabolite profile of HM caused by pasteurization has not been comprehensively explored to date. Future studies focused on the systematic exploration of the effect of thermal treatment on HM are warranted.

HM is usually stored frozen employing −20 °C and −80 °C for short- and long-term storage, respectively. However, duration of storage and the effect of repeated freeze-thaw cycles are identified as additional factors with potential impact on HM composition that are missing in most published studies. In lipidomic studies, the integrity of HM samples is preserved by subjecting HM to inactivation of endogenous enzymes such as lipases in order to minimize lipolysis and lipogenesis. In this sense, immediate storage at −80 °C is advisable [41]. Particularly for metabolite composition analysis, storage at −80 °C is widespread [13,15,18,19,24–26,28,31,45,46], sometimes with a prior short-term storage at −20 °C [14,21,22,27,29,33,34]. Wu et al. [29] investigated the effect of storage conditions by keeping samples for different times at −20 °C and then transferring them to −80 °C versus storing samples directly at −80 °C. Variations in duration of storage at −20 °C versus −80 °C showed no detectable effect on the metabolites considered (e.g., lactose and other carbohydrates, choline and its derivatives, and a variety of amino acids) by visual inspection of sample clusters in principal component analysis scores plots. However, analysis of variance evidenced differences in butyrate, caprate, and acetate contents. However, time of storage considered in this study was limited to two weeks, which is not representative for conditions employed in clinical studies or standard home routines. It is therefore clear that further studies are required in this regard.

On the other hand, HM employed for research studies is typically stored in small aliquots. When working with raw milk, this procedure might introduce bias due to phase separation prior to the preparation of the aliquots. Hence, the homogenization of HM with a disruptor, resulting in a stable emulsion with reduced size of milk fat globules [56], as employed prior to the quantitation of macronutrients with HM analyzers, might be advisable.

3. Metabolite Extraction from HM

For metabolite extraction from HM, an array of methods has been reported. An overview of the employed approaches is shown in Figure 3. The selection of the extraction method is conditioned by the study objective and the subsequent analysis method. As in other untargeted metabolomics workflows, for HM metabolomics, the selected sample preparation approach should enable a high degree of metabolome coverage while making the sample matrix compatible with the analytical platform. Other considerations might include the available amount of sample volume and the use of one sample extraction procedure for subsequent analysis by multiple, complementary analytical platforms [13,27,28].

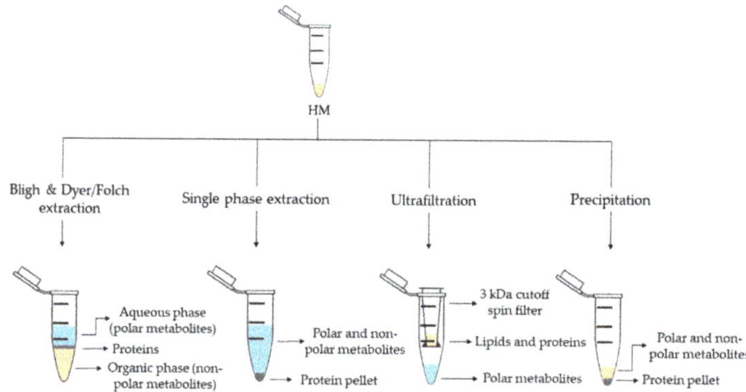

Figure 3. Sample preparation approaches employed in human milk (HM) metabolomics.

Liquid-liquid extraction (LLE) is the classical extraction method employed in metabolomics and lipidomics. This method, developed by Folch et al. [57] in 1957, uses a chloroform-methanol mixture (2:1, *v/v*), which results in two differentiate phases: an upper phase containing polar metabolites and a lower phase containing nonpolar metabolites. Subsequently, in 1959 Bligh and Dyer [58] developed a modified method using a miscible chloroform-methanol-water mixture and later separated into two phases by adding chloroform or water. Both approaches enable the separation of polar and nonpolar metabolites, thus, allowing the analysis of a wide range of metabolites and making them compatible with several analytical platforms. While the use of Bligh and Dyer LLE is widely extended for HM metabolomics studies (see Table 1) [13,16–19,24,25,29,32], only Andreas et al. [28] used a modified Folch extraction protocol for processing HM samples.

Methyl tert-butyl ether (MTBE) in combination with methanol has recently been proposed for single-phase extraction [27]. MTBE is a nontoxic and noncarcinogenic solvent and it is therefore considered a safe and environmentally friendly alternative to harmful solvents employed in traditional LLE methods, such as chloroform, which is a suspected human carcinogen. In this extraction method, a unique phase containing both, polar and nonpolar metabolites is obtained with a protein pellet at the bottom (see Figure 3). Thus, the simultaneous analysis of lipidome and metabolome in a very small amount of biological sample is achievable. This method has been successfully employed to determine polar metabolites and fatty acids (FAs) in HM by GC-MS [27,28], as well as lipids and polar metabolites by LC-MS [15,27,28], thus, increasing the metabolome coverage by the combined use of complementary analytical platforms.

Table 1. Sample preparation steps and platforms employed in untargeted analysis of HM metabolome.

Sample Preparation (1st. step)	Sample Preparation (2nd. step)	Compound Class	Platform	Column/Capillary	References
Bligh & Dyer extraction	Deuterated solvent addition to aqueous phase	Polar metabolites	^{1}H-NMR	-	[13,16,29,32]
	Derivatization of aqueous phase: methoximation and silylation	Polar metabolites and FAs	GC-MS	DB-5ms	[17–19]
	Derivatization of organic phase: methylation	FAs	GC-MS	DB-5ms	[13]
	Direct injection of aqueous phase	Polar metabolites	LC-QTOF-MS (+)	HILIC	[35]
	Redissolution of aqueous phase in H_2O:ACN (95:5)	Polar metabolites	LC-Orbitrap-MS (+, −)	C18	[24]
	Redissolution of organic phase in (ACN:IPA:H_2O (65:30:5)	Lipidic metabolites	LC-Orbitrap-MS (+,−)	C18	[25]
Folch extraction	Deuterated solvent addition to aqueous and organic phases	Hydrophobic and polar metabolites	^{1}H-NMR	-	
	Redissolution of aqueous phase in formic acid and centrifugation	Polar metabolites (amino acids)	CE-TOF-MS (+)	60 m × 50 μm I.D.	[28]
	Redissolution of organic phase in IPA:H_2O:ACN (2:1:1) and centrifugation	Lipidic metabolites	UPLC-QTOF-MS (+,−)	C18	
Single phase extraction	Derivatization: methoximation and silylation	Polar metabolites and FAs	GC-MS	DB-5ms	[27,28]
	Direct injection	Lipidic (and polar) metabolites	LC-QTOF-MS (+,−)	C8	[27,28]
			UPLC-QTOF-MS (+)	C18	[15]
Fat extraction with n-hexane/IPA	Deuterated solvent addition	TGs	^{13}C-NMR; ^{1}H-NMR	-	[20]
Filtration 3 kDa cutoff spin filter	Deuterated solvent addition	Polar metabolites	^{1}H-NMR	-	[14,21,22,29,33]

Table 1. *Cont.*

Sample Preparation (1st. step)	Sample Preparation (2nd. step)	Compound Class	Platform	Column/Capillary	References
Protein precipitation	Derivatization: methoximation and silylation	Polar metabolites	GC-MS	DB-5ms	[36]
	Hybrid SPE-Phospholipid extraction and redissolution in diluted organic phase of Bligh & Dyer extraction	Lipidic metabolites	LC-QTOF-MS (+)	C8	[35]
	Fat removal with CH_2Cl_2 and dansylation of aqueous phase	Polar metabolites (amine/phenol submetabolome)	Chemical isotope labelling LC-QTOF-MS (+)	C18	[45,46]
	Direct injection	Polar metabolites and FAs	UPLC-QTOF-MS (+,−)	C18	[18]
Fat removal by centrifugation	Two additional centrifugations and deuterated solvent addition	Polar metabolites	^{1}H-NMR	-	[34]
	Filtration 10 kDa cutoff spin filter and deuterated solvent addition	Polar metabolites	^{1}H-NMR	-	[23,26]
Homogenization	Deuterated solvent addition	Polar metabolites	^{1}H-NMR	-	[31]
H_2O-dilution	$NaBH_4$-reduction and PGC cartridge	Oligosaccharides	UPLC-TQD-MS (+)	Hypercarb®	[24]

CE, capillary electrophoresis; FAs, fatty acids; GC, gas chromatography; HILIC, hydrophilic interaction liquid chromatography; IPA, 2-propanol; I.D., inner diameter; LC, liquid chromatography; MS, mass spectrometry; ^{13}C-NMR, carbon-13 nuclear magnetic resonance; ^{1}H-NMR, proton nuclear magnetic resonance; PGC, porous graphitic carbon; QTOF, quadrupole time of flight; TGs, triacylglycerols; TQD, triple quadrupole; UPLC, ultraperformance liquid chromatography; +, positive ionization mode; -, negative ionization mode.

Ultrafiltration makes use of centrifugal molecular weight cutoff filters. Different molecular weight cut-off filters are commercially available for this purpose and repeated centrifugation steps might be employed to remove proteins and lipids (see Table 1). Unlike single-phase extraction, ultrafiltration allows to separate polar metabolites from the HM without dilution [14,21,22,29], however this method does not have the capacity to study the global metabolome of HM. At present, this extraction method has only been used in combination with NMR analyses [14,21,22,29].

Precipitation with organic solvents separates the polar and nonpolar metabolites of the proteins that settle at the bottom of the tube which can then be easily removed by centrifugation. This simple method has been employed for the analysis of polar metabolites by GC-MS after derivatization [36] as well as for the analysis of polar and nonpolar metabolites by LC-MS without further pre-processing [18]. Furthermore, this approach has been implemented in more sophisticated workflows as recently shown by Hewelt-Belka et al. [35]. Here, the authors combined LLE and a protein precipitation and solid-phase extraction (SPE) procedure to prepare HM samples, thereby, enabling the detection of high- and low-abundant lipid species (e.g., glycerolipids and phospholipids) in one LC-MS run.

4. Analytical Platforms Employed in HM Metabolomics

As reflected in Table 1, the use of all analytical platforms that are commonly employed in untargeted metabolomics studies, such as LC-MS, GC-MS, NMR, and, to a lesser extent, CE-MS, has been reported for performing HM metabolomics. ^{1}H-NMR is the most frequently used technique [13,14, 16,20–22,26,28,29,31–33] for both, the analysis of polar and hydrophobic metabolites in HM. ^{13}C-NMR has been reported for the detection of triacylglycerols [20]. NMR is a highly reproducible technique that allows a straightforward library matching after spectral alignment, while at the same time supporting structural elucidation of detected metabolites. However, it presents lower sensitivity, and hence, the achievable metabolome coverage is low in comparison to other analytical platforms.

All other analytical platforms rely on the use of MS detection. LC-MS provides high sensitivity and is characterized by a huge versatility due to the availability of (i) a large selection of chromatographic columns with a variety of stationary phases, that in combination with appropriate mobile phases achieve compound separation based on different retention mechanisms and (ii) an array of different instrumental configurations (i.e. different ion sources and mass analyzers). For example, reversed phase (C8, C18) LC-quadrupole time of flight MS (LC-QTOF-MS) [15,18,27,28,35,46] and LC-Orbitrap-MS [24,25] have been reported for the detection of both, polar and lipidic metabolites in HM; and hydrophilic interaction LC (HILIC)-QTOF-MS for polar metabolite detection [35]. On the other hand, for the successful screening of HM oligosaccharides, a porous graphitic carbon column installed on a LC-triple quadrupole (TQD)-MS instrument was used [24].

GC-MS is the most suitable platform for measuring volatile compounds, while other non-volatile compounds must be derivatized prior to analysis. For HM analysis, methoximation followed by silylation or methylation are commonly employed (see Table 1). The most frequently used column is the DB-5ms column for both polar and FA detection [13,17–19], in some cases with an integrated 10 m pre-column (deactivated fused silica) [27,28,36].

Regarding CE-MS, only one study has been reported for polar metabolite detection in HM [28]. CE provides a series of advantages over other techniques, mainly due to the small sample volumes employed and the efficient separation of polar compounds that is difficult to be achieved by LC columns. However, issues with poor reproducibility, matrix effects and sensitivity may be hindering a widely extended use of this technique for the analysis of complex biological samples such as HM.

Due to the diversified composition of HM, no single analytic technique can resolve the entire HM metabolome. Only multiplatform approaches enable a comprehensive characterization providing a high metabolome coverage including polar and nonpolar metabolites present in HM. In this sense, two studies performing a multiplatform approach were found in the literature combining LC-MS and GC-MS [18,27], and only one study that performed analysis using four different techniques (LC-MS, GC-MS, NMR, and CE-MS) was reported [28].

The use of high-end analytical platforms requires the implementation of QA and QC processes to improve data quality, repeatability, and reproducibility, especially in untargeted metabolomics. For practical guidelines on the use of QC measures in untargeted, MS-based assays the reader is referred to [59]. Pooled QC samples are prepared by mixing small aliquots of the study samples, and therefore, they are considered representative in terms of matrix composition and concentration ranges of the metabolites present in the study samples. QC samples are analyzed repeatedly throughout the analytical sequence alongside the study samples. The signal of each feature detected in QC samples can be used to model and correct systematic changes in the instrument response during the analytical sequence. Additionally, the obtained data can be used to perform intra-study reproducibility assessments and to correct for systematic variation across batches. In HM metabolomics, Smilowitz et al. [14], Andreas et al. [28], and Gay et al. [33] used QC samples for NMR studies, while Villaseñor et al. [27], Mung et al. [46], Hewelt-Belka et al. [35], and Alexandre-Gouabau et al. [24,25] used pooled HM samples for QC purposes in LC-MS-based assays. Considering the highly complex sample matrix of HM, the authors strongly recommend the implementation of QC measures, including the analysis of QC samples, to increase reproducibility and facilitate the joint analysis of data from different studies.

5. The HM Metabolome: Compound Annotation and Coverage

As in other areas of metabolomic research, compound identification is still a major bottleneck in data analysis and interpretation. The Metabolomics Standards Initiative's (MSI) defines four levels of metabolite identification, which include: identified metabolites (level 1); putatively annotated compounds (level 2); putatively annotated compound classes (level 3); and unknown compounds (level 4) [60]. Due to the limited availability of pure analytical standards required to reach level 1, biological databanks and spectral databases are the most important resources for metabolite annotation (levels 2 and 3). A large number of databases are available today, providing different levels of information and complementary data on chemical structures, physicochemical properties, biological functions, and pathway mapping of metabolites [61]. The metabolomics community classifies these resources in several categories: (i) chemical databases; (ii) spectral libraries; (iii) pathway databases; (iv) knowledge databases; and (v) references repositories [62].

Regarding HM metabolomics, the most frequently used databases and libraries are: Human Metabolome Database (HMDB) [63], Metabolite and Chemical Entity Database (METLIN) [64], National Institute of Science and Technology (NIST) library, Fiehn RTL Library [65], LipidMAPS Structure Database (LMSD) [66], Milk Metabolome Database (MCDB) [67,68], Kyoto Encyclopedia of Genes and Genomes (KEGG) [69], MycompoundID with the evidence-based metabolome library (EML) [70], Chenomx NMR Suite Profiles and other online university databases, such as CEU-mass mediator [71,72].

Metabolite assignment in NMR spectra has been performed based on literature data and commercial resonance databases, such as Chenomx NMR Suite Profiles. Metabolite annotation was contrasted with in-house libraries containing pure compound spectra. Some of the proposed assignments were confirmed by two-dimensional NMR spectra, such as Correlation Spectroscopy (COSY) [13,29,31,32], Homonuclear Correlation Spectroscopy (TOCSY) [13,31,32,34], Diffusion-Ordered Spectroscopy (DOSY) [32], Heteronuclear Single Quantum Coherence Spectroscopy (HSQC) [32,34], and Heteronuclear Multiple Bond Correlation (HMBC) [32].

In LC-MS and CE-MS-based studies of the HM metabolome, tentative metabolite annotation has been carried out by matching of accurate masses, isotopic profiles, and/or fragmentation patterns to candidate metabolites in online databases such as KEGG, METLIN, LipidMAPS, and HMDB [18, 24,25,27,28,35]. In-house built databases generated by the analysis of commercial standards are also commonly employed [24,25]. In GC-MS, retention index (RI) corrections are made by analyzing a fatty acid methyl ester (FAME) mixture standard solution and assigning a match score between the experimental FAME mixture and theoretical RI values based on the values contained in the Fiehn RTL library. Furthermore, metabolites were complementarily annotated by comparing their mass fragmentation patterns with those available in Fiehn RTL and NIST libraries [13,17–19,27,28,36].

A comprehensive list of annotated and/or identified metabolites in HM from untargeted metabolomics studies [14,15,17–19,21–29,31–36] is reported in Table S1. This table contains information about the metabolites reported in each reference, such as their molecular formula, IDs (LipidMAPS and/or HMDB IDs), the extraction procedure performed, the analytical platform used, and the detected metabolite class. Readers can select metabolites dynamically by filtering data according to the latter information. A total of 1187, 111, and 128 metabolites were reported using LC-MS, GC-MS, and NMR, respectively (see Figure 4). As shown in the Venn diagram, LC-MS and GC-MS allowed the detection of 36 common metabolites (mainly carbohydrates and FAs); a total of 29 metabolites overlapped between LC-MS and NMR (principally oligosaccharides); and 21 metabolites (predominantly amino acids and organic acids) were commonly reported in GC-MS and NMR based studies. Only 13 metabolites were reported by all three platforms, i.e., creatine, tyrosine, arabinose, galactose, glucose, lactose, maltose, capric acid/caprate, caprylic acid/ caprylate, citric acid/citrate, pyruvic acid/pyruvate, hippuric acid/hippurate, and myo-inositol. These metabolites were assigned to different classes including amino acids, carbohydrates, FAs, and organic acids.

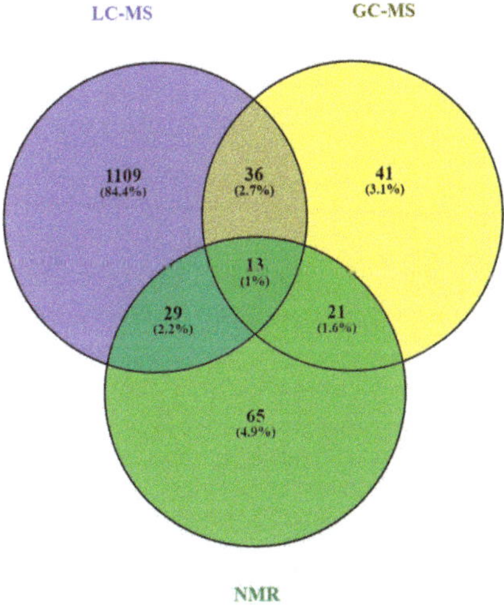

Figure 4. Venn diagram of metabolites reported in human milk (HM) according to the technique in [73]. Note: GC-MS, gas chromatography—mass spectrometry; LC-MS, liquid chromatography—mass spectrometry; NMR, nuclear magnetic resonance.

Based on the available data from the literature, the distribution of metabolite classes present in HM according to each technique was assessed. As can be seen in Figure 5, the difference in detected metabolite classes as observed by LC-MS in comparison to GC-MS and NMR is evident. Using GC-MS and NMR, carbohydrates are the most reported metabolites in HM, followed by amino acids, organic acids, organooxygen compounds, and organoheterocyclic compounds, with all these metabolite classes being certainly less abundant in LC-MS studies. In the case of NMR, organonitrogen compounds have also been reported, as well as nucleosides and nucleotides on a smaller scale. In the case of lipid classes, fatty acyls have been identified by LC-MS and GC-MS with similar incidence and in lesser extent by NMR. It is indubitable that lipid classes are more comprehensively studied by LC-MS assays, where glycerophospholipids, glycerolipids, and fatty acyls are detected at relatively high abundances, followed by sphingolipids, sterol lipids, and, to a lesser extent, prenol lipids.

Table 2 shows a list of metabolites reported in > 80% of studies employing either LC-MS, GC-MS, or NMR-based assays. This table is intended to aid method development of future untargeted metabolomics workflows tailored to the study of the HM metabolome, as it shows a shortlist of metabolites that should be detected by each platform regardless of the instrumental settings employed. It should be noted that due to the high versatility of LC-MS, there is a greater variation in metabolites recorded and in return, the list of consistently reported metabolites in HM across studies is shorter than for NMR and GC-MS, where differences in experimental conditions and variations between the employed detection parameters and instruments are smaller. Again, this table represents the high orthogonality between the detected metabolites using NMR and LC-MS. While the use of LC-MS is clearly of advantage for the measurement of different lipids, NMR provides information on amino acids and small organic acids. Metabolome coverage provided by GC-MS falls in-between the other two platforms, consistently providing information on lipids, sugars, amino acids, and organic acids.

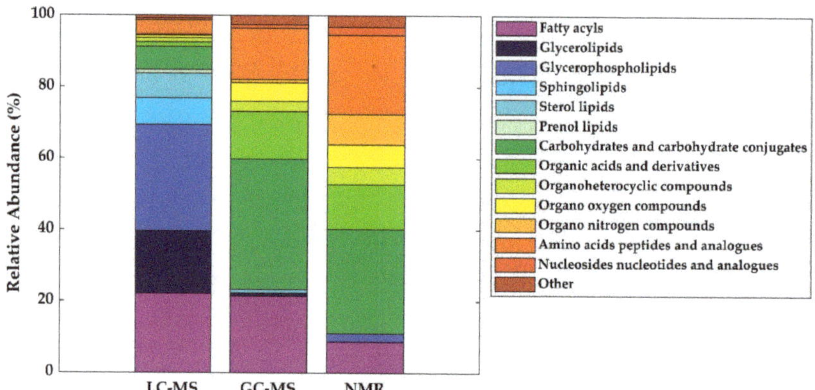

Figure 5. Distribution of metabolite classes annotated and/or identified in HM according to technique. Note: GC-MS, gas chromatography—mass spectrometry; LC-MS, liquid chromatography—mass spectrometry; NMR, nuclear magnetic resonance.

Table 2. Most frequently reported metabolites (>80% of studies) according to technique.

Metabolite class	LC-MS	GC-MS	NMR
Fatty acyls	Linoleic acid (C18:2) Oleic acid (C18:1) Palmitoleic acid (C16:1)	Oleic acid (C18:1) Palmitic acid (C16:0) Stearic acid (C18:0)	-
Glycerolipids	DG (36:1)	-	-
Glycerophospholipids	LysoPC (16:0)	-	-
Carbohydrates and carbohydrate conjugates	-	Fructose Fucose Ribose	Lactose
Organic acids and derivatives	-	Malic acid Urea	Acetate Citrate Lactate
Organo nitrogen compounds	-	-	Choline
Amino acids, peptides, and analogues	-	Alanine Glutamate Glycine Pyroglutamic acid Serine Valine	Alanine Creatine Glutamate Glutamine Isoleucine Leucine Tyrosine Valine

GC-MS, gas chromatography—mass spectrometry; LC-MS, liquid chromatography—mass spectrometry; NMR, nuclear magnetic resonance; DG, diacylglycerol; PC, phosphatidylcholine.

6. Conclusions and Future Perspectives

In less than a decade, 26 research papers have been published trying to shed light on the complex and dynamic composition of HM and the feasibility of different options for sample extraction and metabolite detection has been demonstrated. Due to the many factors that influence HM composition, a thorough study design including SOPs for milk extraction, collection, and storage is indispensable for obtaining biologically meaningful results. Multi-platform approaches are encouraged for providing adequate metabolome coverage, as the diversity of compounds contained in HM will not be properly reflected using one single assay. In line with metabolomics workflows tailored to other sample types, the reproducibility of HM metabolomics studies will benefit from the implementation of QA/QC procedures. Automated metabolite annotation and identification with pure chemical standards is warranted and the authors encourage the use of publicly accessible platforms for enabling the exchange of raw data for comparison between studies.

Supplementary Materials: The following are available online at http://www.mdpi.com/2218-1989/10/2/43/s1, Table S1: List of metabolites annotated and/or identified in HM metabolomic studies.

Author Contributions: Conceptualization, G.Q., J.K., M.G., and M.V.; methodology, A.P.-L., I.T.-D., J.D.P.-R. and V.R.-G.; software, G.Q., I.T.-D. and J.K., and validation, G.Q. and J.K.; formal analysis, I.T.-D and V.R.-G.; investigation, I.T.-D. and V.R.-G.; resources, J.K.; data curation, I.T.-D. and V.R.-G.; writing—original draft preparation, A.P.-L., I.T.-D., J.D.P.-R and V.R.-G.; writing—review and editing, all authors; visualization, I.T.-D. and V.R.-G.; supervision, J.K. and M.V; project administration, G.Q. and J.K.; funding acquisition, J.K. All authors have read and agreed to the published version of the manuscript.

Funding: A.P.-L., I.T.-D. and J.K. received salary support by the *Instituto de Salud Carlos III* (Ministry of Economy and Competitiveness, Spain), grant numbers CD19/00176, CM18/00165, and CP16/00034, respectively. This project has received funding from the European Union's Horizon 2020 research and innovation programme under grant agreement No. 818110.

Conflicts of Interest: The authors declare no conflict of interest.

References

1. World Health Organization Breastfeeding. Available online: https://www.who.int/topics/breastfeeding/en/ (accessed on 2 July 2019).
2. Geddes, D.; Perrella, S. Breastfeeding and human lactation. *Nutrients* **2019**, *11*, 802–806. [CrossRef]
3. Owen, C.G.; Martin, R.M.; Whincup, P.H.; Davey Smith, G.; Cook, D.G. Does breastfeeding influence risk of type 2 diabetes in later life? A quantitative analysis of published evidence. *Am. J. Clin. Nutr.* **2006**, *84*, 1043–1054. [CrossRef]
4. Der, G.; Batty, G.D.; Deary, I.J. Effect of breast feeding on intelligence in children: Prospective study, sibling pairs analysis, and meta-analysis. *Br. Med. J.* **2006**, *333*, 945–948. [CrossRef] [PubMed]
5. Rozé, J.C.; Darmaun, D.; Boquien, C.Y.; Flamant, C.; Picaud, J.C.; Savagner, C.; Claris, O.; Lapillonne, A.; Mitanchez, D.; Branger, B.; et al. The apparent breastfeeding paradox in very preterm infants: Relationship between breast feeding, early weight gain and neurodevelopment based on results from two cohorts, EPIPAGE and LIFT. *BMJ Open* **2012**, *2*, 1–9. [CrossRef] [PubMed]
6. Horta, B.L.; Loret De Mola, C.; Victora, C.G. Breastfeeding and intelligence: A systematic review and meta-analysis. *Acta Paediatr. Int. J. Paediatr.* **2015**, *104*, 14–19. [CrossRef] [PubMed]
7. Lönnerdal, B. Bioactive proteins in human milk: Mechanisms of action. *J. Pediatr.* **2010**, *156*, S26–S30. [CrossRef]
8. Musilova, S.; Rada, V.; Vlkova, E.; Bunesova, V. Beneficial effects of human milk oligosaccharides on gut microbiota. *Benef. Microbes* **2014**, *5*, 273–283. [CrossRef]
9. Ballard, O.; Morrow, A.L. Human milk composition: Nutrients and bioactive factors. *Pediatr. Clin. N. Am.* **2013**, *60*, 49–74. [CrossRef]
10. Alsaweed, M.; Hartmann, P.E.; Geddes, D.T.; Kakulas, F. MicroRNAs in Breastmilk and the Lactating Breast: Potential Immunoprotectors and Developmental Regulators for the Infant and the Mother. *Int. J. Environ. Res. Public Health* **2015**, *12*, 13981–14020. [CrossRef]

11. Van den Berg, M.; Kypke, K.; Kotz, A.; Tritscher, A.; Lee, S.Y.; Magulova, K.; Fiedler, H.; Malisch, R. WHO/UNEP global surveys of PCDDs, PCDFs, PCBs and DDTs in human milk and benefit–risk evaluation of breastfeeding. *Arch. Toxicol.* **2017**, *91*, 83–96. [CrossRef]

12. Garwolińska, D.; Namieśnik, J.; Kot-Wasik, A.; Hewelt-Belka, W. State of the art in sample preparation for human breast milk metabolomics—Merits and limitations. *TrAC Trends Anal. Chem.* **2019**, *114*, 1–10. [CrossRef]

13. Marincola, F.C.; Noto, A.; Caboni, P.; Reali, A.; Barberini, L.; Lussu, M.; Murgia, F.; Santoru, M.L.; Atzori, L.; Fanos, V. A metabolomic study of preterm human and formula milk by high resolution NMR and GC/MS analysis: Preliminary results. *J. Matern.-Fetal Neonatal Med.* **2012**, *25*, 62–67. [CrossRef] [PubMed]

14. Smilowitz, J.T.; Sullivan, A.O.Õ.; Barile, D.; German, J.B.; Lo, B. The human milk metabolome reveals diverse oligosaccharide profiles. *J. Nutr.* **2013**, *143*, 1709–1718. [CrossRef] [PubMed]

15. Li, K.; Jiang, J.; Xiao, H.; Wu, K.; Qi, C.; Sund, J.; Li, D. Changes in metabolites profile of breast milk over lactation stages and their relationship with dietary intake in Chinese: HPLC-QTOFMS based metabolomic analysis. *Food Funct.* **2018**, *9*, 5189–5197. [CrossRef] [PubMed]

16. Longini, M.; Tataranno, M.L.; Proietti, F.; Tortoriello, M.; Belvisi, E.; Vivi, A.; Tassini, M.; Perrone, S.; Buonocore, G. A metabolomic study of preterm and term human and formula milk by proton MRS analysis: Preliminary results. *J. Matern.-Fetal Neonatal Med.* **2014**, *7058*, 27–33. [CrossRef]

17. Murgia, A.; Scano, P.; Contu, M.; Ibba, I.; Altea, M.; Demuru, M.; Porcu, A.; Caboni, P. Characterization of donkey milk and metabolite profile comparison with human milk and formula milk. *LWT* **2016**, *74*, 427–433. [CrossRef]

18. Qian, L.; Zhao, A.; Zhang, Y.; Chen, T.; Zeisel, S.H.; Jia, W.; Cai, W. Metabolomic approaches to explore chemical diversity of human breast-milk, formula milk and bovine milk. *Int. J. Mol. Sci.* **2016**, *17*, 2128–2143. [CrossRef]

19. Scano, P.; Murgia, A.; Demuru, M.; Consonni, R.; Caboni, P. Metabolite profiles of formula milk compared to breast milk. *Food Res. Int.* **2016**, *87*, 76–82. [CrossRef]

20. Lopes, T.I.B.; Cañedo, M.C.; Oliveira, F.M.P.; Ancantara, G.B. Toward precision nutrition: Commercial infant formulas and human milk compared for stereospecific distribution of fatty acids using metabolomics. *Omics J. Integr. Biol.* **2018**, *22*, 484–492. [CrossRef]

21. O'Sullivan, A.; He, X.; McNiven, E.M.S.; Hinde, K.; Haggarty, N.W.; Lönnerdal, B.; Slupsky, C.M. Metabolomic phenotyping validates the infant rhesus monkey as a model of human infant metabolism. *J. Pediatr. Gastroenterol. Nutr.* **2013**, *56*, 355–363. [CrossRef]

22. Spevacek, A.R.; Smilowitz, J.T.; Chin, E.L.; Underwood, M.A.; German, J.B.; Slupsky, C.M. Infant maturity at birth reveals minor differences in the maternal milk metabolome in the first month of lactation. *J. Nutr.* **2015**, *145*, 1698–1708. [CrossRef] [PubMed]

23. Sundekilde, U.K.; Downey, E.; Mahony, J.A.O.; Shea, C.O.; Ryan, C.A.; Kelly, A.L.; Bertram, H.C. The effect of gestational and lactational age on the human milk metabolome. *Nutrients* **2016**, *8*, 304–318. [CrossRef] [PubMed]

24. Alexandre-Gouabau, M.-C.; Moyon, T.; David-Sochard, A.; Fenaille, F.; Cholet, S.; Royer, A.-L.; Guitton, Y.; Billard, H.; Darmaun, D.; Rozé, J.-C.; et al. Comprehensive preterm breast milk metabotype associated with optimal infant early growth pattern. *Nutrients* **2019**, *11*, 528–553. [CrossRef] [PubMed]

25. Alexandre-Gouabau, M.C.; Moyon, T.; Cariou, V.; Antignac, J.P.; Qannari, E.M.; Croyal, M.; Soumah, M.; Guitton, Y.; David-Sochard, A.; Billard, H.; et al. Breast milk lipidome is associated with early growth trajectory in preterm infants. *Nutrients* **2018**, *10*, 164–192. [CrossRef]

26. Dessì, A.; Briana, D.; Corbu, S.; Gavrili, S.; Marincola, F.C.; Georgantzi, S.; Pintus, R.; Fanos, V.; Malamitsi-Puchner, A. Metabolomics of breast milk: The importance of phenotypes. *Metabolites* **2018**, *8*, 79–88. [CrossRef]

27. Villaseñor, A.; Garcia-Perez, I.; Garcia, A.; Posma, J.M.; Fernández-Lópes, M.; Nicholas, A.J.; Modi, N.; Holmes, E.; Barbas, C. Breast milk metabolome characterization in a single-phase extraction, multiplatform analytical approach. *Anal. Chem.* **2014**, *86*, 8245–8252. [CrossRef]

28. Andreas, N.J.; Hyde, M.J.; Gomez-romero, M.; Lopez-Gonzalvez, M.A.; Villaseñor, A.; Wijeyesekera, A.; Barbas, C.; Modi, N.; Holmes, E.; Garcia-Perez, I. Multiplatform characterization of dynamic changes in breast milk during lactation. *Electrophoresis* **2015**, *36*, 2269–2285. [CrossRef]

29. Wu, J.; Domellöf, M.; Zivkovic, A.M.; Larsson, G.; Öhman, A.; Nording, M.L. NMR-based metabolite profiling of human milk: A pilot study of methods for investigating compositional changes during lactation. *Biochem. Biophys. Res. Commun.* **2016**, *469*, 626–632. [CrossRef]

30. Isganaitis, E.; Venditti, S.; Matthews, T.J.; Lerin, C.; Demerath, E.W.; Fields, D.A. Maternal obesity and the human milk metabolome: Associations with infant body composition and postnatal weight gain. *Am. J. Clin. Nutr.* **2019**, *110*, 111–120. [CrossRef]

31. Dangat, K.; Upadhyay, D.; Kilari, A.; Sharma, U.; Kemse, N.; Mehendale, S.; Lalwani, S.; Wagh, G.; Joshi, S.; Jagannathan, N.R. Altered breast milk components in preeclampsia; An in-vitro proton NMR spectroscopy study. *Clin. Chim. Acta* **2016**, *463*, 75–83. [CrossRef]

32. Praticò, G.; Capuani, G.; Tomassini, A.; Baldassarre, E.; Delfini, M.; Miccheli, A. Exploring human breast milk composition by NMR-based metabolomics. *Nat. Prod. Res.* **2013**, *28*, 95–101. [CrossRef] [PubMed]

33. Gay, M.C.L.; Koleva, P.T.; Slupsky, C.M.; Toit, E.; Eggesbo, M.; Johnson, C.C.; Wegienka, G.; Shimojo, N. Worldwide variation in human milk metabolome: Indicators of breast physiology and maternal lifestyle? *Nutrients* **2018**, *10*, 1151–1162. [CrossRef] [PubMed]

34. Gómez-Gallego, C.; Morales, J.M.; Monleón, D.; du Toit, E.; Kumar, H.; Linderborg, K.M.; Zhang, Y.; Yang, B.; Isolauri, E.; Salminen, S.; et al. Human breast milk NMR metabolomic profile across specific geographical locations and its association with the milk microbiota. *Nutrients* **2018**, *10*, 1355–1375. [CrossRef] [PubMed]

35. Hewelt-Belka, W.; Garwolińska, D.; Belka, M.; Bączek, T.; Namieśnik, J.; Kot-Wasik, A. A new dilution-enrichment sample preparation strategy for expanded metabolome monitoring of human breast milk that overcomes the simultaneous presence of low- and high-abundance lipid species. *Food Chem.* **2019**, *288*, 154–161. [CrossRef]

36. Urbaniak, C.; Mcmillan, A.; Angelini, M.; Gloor, G.B.; Sumarah, M.; Burton, J.P.; Reid, G. Effect of chemotherapy on the microbiota and metabolome of human milk, a case report. *Microbiome* **2014**, *2*, 24–35. [CrossRef]

37. Demmelmair, H.; Koletzko, B. Variation of metabolite and hormone contents in human milk. *Clin. Perinatol.* **2017**, *44*, 151–164. [CrossRef]

38. Marincola, F.C.; Dessì, A.; Corbu, S.; Reali, A.; Fanos, V. Clinical impact of human breast milk metabolomics. *Clin. Chim. Acta* **2015**, *451*, 103–106. [CrossRef]

39. Slupsky, C.M. Metabolomics in human milk research. *Nestle Nutr. Inst. Workshop Ser.* **2019**, *90*, 179–190.

40. Bardanzellu, F.; Fanos, V.; Reali, A. "Omics" in human colostrum and mature milk: Looking to old data with new eyes. *Nutrients* **2017**, *9*, 843–867. [CrossRef]

41. George, A.D.; Gay, M.C.L.; Trengove, R.D.; Geddes, D.T. Human milk lipidomics: Current techniques and methodologies. *Nutrients* **2018**, *10*, 1169–1179. [CrossRef]

42. Lönnerdal, B.; Erdmann, P.; Thakkar, S.K.; Sauser, J.; Destaillats, F. Longitudinal evolution of true protein, amino acids and bioactive proteins in breast milk: A developmental perspective. *J. Nutr. Biochem.* **2017**, *41*, 1–11. [CrossRef] [PubMed]

43. Gidrewicz, D.A.; Fenton, T.R. A systematic review and meta-analysis of the nutrient content of preterm and term breast milk. *BMC Pediatr.* **2014**, *14*, 216–230. [CrossRef] [PubMed]

44. Hampel, D.; Shahab-Ferdows, S.; Islam, M.M.; Peerson, J.M.; Allen, L.H. Vitamin concentrations in human milk vary with time within feed, circadian rhythm, and single-dose supplementation. *J. Nutr.* **2017**, *147*, 603–611. [CrossRef] [PubMed]

45. Mung, D.; Li, L. Development of chemical isotope labeling LC-MS for milk metabolomics: Comprehensive and quantitative profiling of the amine/phenol submetabolome. *Anal. Chem.* **2017**, *89*, 4435–4443. [CrossRef] [PubMed]

46. Mung, D.; Li, L. Applying quantitative metabolomics based on chemical isotope labeling LC-MS for detecting potential milk adulterant in human milk. *Anal. Chim. Acta* **2018**, *1001*, 78–85. [CrossRef] [PubMed]

47. Garcia, C.; Duan, R.D.; Brévaut-Malaty, V.; Gire, C.; Millet, V.; Simeoni, U.; Bernard, M.; Armand, M. Bioactive compounds in human milk and intestinal health and maturity in preterm newborn: An overview. *Cell. Mol. Biol.* **2013**, *59*, 108–131.

48. Garwolińska, D.; Namieśnik, J.; Kot-Wasik, A.; Hewelt-Belka, W. Chemistry of human breast milk—A comprehensive review of the composition and role of milk metabolites in child development. *J. Agric. Food Chem.* **2018**, *66*, 11881–11896. [CrossRef]

49. Totten, S.M.; Zivkovic, A.M.; Wu, S.; Ngyuen, U.; Freeman, S.L.; Ruhaak, L.R.; Darboe, M.K.; German, J.B.; Prentice, A.M.; Lebrilla, C.B. Comprehensive profiles of human milk oligosaccharides yield highly sensitive and specific markers for determining secretor status in lactating mothers. *J. Proteome Res.* **2012**, *11*, 6124–6133. [CrossRef]

50. Lubetzky, R.; Littner, Y.; Mimouni, F.B.; Dollberg, S.; Mandel, D.; Lubetzky, R.; Lubetzky, R.; Littner, Y.; Mimouni, F.B.; Dollberg, S.; et al. Circadian variations in fat content of expressed breast milk from mothers of preterm infants. *J. Am. Coll. Nutr.* **2006**, *25*, 151–154. [CrossRef]

51. Saarela, T.; Kokkonen, J.; Koivisto, M. Macronutrient and energy contents of human milk fractions during the first six months of lactation. *Acta Paediatr.* **2005**, *94*, 1176–1181. [CrossRef]

52. Jensen, R.G.; Lammi-Keefe, C.J.; Koletzko, B. Representative sampling of human milk and the extraction of fat for analysis of environmental lipophilic contaminants. *Toxicol. Environ. Chem.* **1997**, *62*, 229–247. [CrossRef]

53. Bertino, E.; Peila, C.; Cresi, F.; Maggiora, E.; Sottemano, S.; Gazzolo, D.; Arslanoglu, S.; Coscia, A. Donor human milk: Effects of storage and heat treatment on oxidative stress markers. *Front. Pediatr.* **2018**, *6*, 1–5. [CrossRef] [PubMed]

54. Peila, C.; Moro, G.E.; Bertino, E.; Cavallarin, L.; Giribaldi, M.; Giuliani, F.; Cresi, F.; Coscia, A. The effect of holder pasteurization on nutrients and biologically-active components in donor human milk: A review. *Nutrients* **2016**, *8*, 447. [CrossRef] [PubMed]

55. García-Lara, N.R.; Vieco, D.E.; De La Cruz-Bértolo, J.; Lora-Pablos, D.; Velasco, N.U.; Pallás-Alonso, C.R. Effect of holder pasteurization and frozen storage on macronutrients and energy content of breast milk. *J. Pediatr. Gastroenterol. Nutr.* **2013**, *57*, 377–382. [CrossRef]

56. Billard, H.; Simon, L.; Desnots, E.; Sochard, A.; Boscher, C.; Riaublanc, A.; Alexandre-Gouabau, M.C.; Boquien, C.Y. Calibration adjustment of the mid-infrared analyzer for an accurate determination of the macronutrient composition of human milk. *J. Hum. Lact.* **2016**, *32*, NP19–NP27. [CrossRef]

57. Folch, J.; Lees, M.; Sloane Stantley, G.H. A simple method for the isolation and purification of total lipides from animal tissues. *J. Biol. Chem.* **1957**, *266*, 497–509.

58. Bligh, E.G.; Dyer, W.J. A rapid method of total lipid extraction and purification. *Can. J. Biochem. Physiol.* **1959**, *37*, 911–917. [CrossRef]

59. Broadhurst, D.; Goodacre, R.; Reinke, S.N.; Kuligowski, J.; Wilson, I.D.; Lewis, M.R.; Dunn, W.B. Guidelines and considerations for the use of system suitability and quality control samples in mass spectrometry assays applied in untargeted clinical metabolomic studies. *Metabolomics* **2018**, *14*, 72–89. [CrossRef]

60. Sumner, L.W.; Samuel, T.; Noble, R.; Gmbh, S.D.; Barrett, D.; Beale, M.H.; Hardy, N. Proposed minimum reporting standards for chemical analysis Chemical Analysis Working Group (CAWG) Metabolomics Standards Initiative (MSI). *Metabolomics* **2007**, *3*, 211–221. [CrossRef]

61. Vinaixa, M.; Schymanski, E.L.; Neumann, S.; Navarro, M.; Salek, R.M.; Yanes, O. Mass spectral databases for LC/MS- and GC/MS-based metabolomics: State of the field and future prospects. *TrAC Trends Anal. Chem.* **2016**, *78*, 23–35. [CrossRef]

62. Fiehn, O.; Barupal, D.K.; Kind, T. Extending biochemical databases by metabolomic surveys. *J. Biol. Chem.* **2011**, *286*, 23637–23643. [CrossRef] [PubMed]

63. Wishart, D.S.; Feunang, Y.D.; Marcu, A.; Guo, A.C.; Liang, K.; Vázquez-Fresno, R.; Sajed, T.; Johnson, D.; Li, C.; Karu, N.; et al. HMDB 4.0: The human metabolome database for 2018. *Nucleic Acids Res.* **2018**, *46*, D608–D617. [CrossRef] [PubMed]

64. Smith, C.A.; O'Maille, G.; Want, E.J.; Qin, C.; Trauger, S.A.; Brandon, T.R.; Custodio, D.E.; Abagyan, R.; Siuzdak, G. METLIN: A metabolite mass spectral database. *Ther. Drug Monit.* **2005**, *27*, 747–751. [CrossRef] [PubMed]

65. Kind, T.; Wohlgemuth, G.; Lee, D.Y.; Lu, Y.; Palazoglu, M.; Shahbaz, S.; Fiehn, O. FiehnLib—Mass spectral and retention index libraries for metabolomics based on quadrupole and time-of-flight gas chromatography/mass spectrometry. *Anal. Chem.* **2009**, *81*, 10038–10048. [CrossRef]

66. Cardiff University; Babraham Institute; University of California, S.D. LIPID MAPS Lipidomics Gateway. Available online: http://www.lipidmaps.org/ (accessed on 8 November 2019).

67. Foroutan, A.; Guo, A.C.; Vazquez-fresno, R.; Lipfert, M.; Zhang, L.; Zheng, J.; Badran, H.; Budinski, Z.; Mandal, R.; Ametaj, B.N.; et al. Chemical composition of commercial cow's milk. *J. Agric. Food Chem.* **2019**, *67*, 4897–4914. [CrossRef]

68. Milk Composition Database. Available online: http://www.mcdb.ca/ (accessed on 5 November 2019).

69. KEGG PATHWAY Database. Available online: https://www.kegg.jp/kegg/pathway.html (accessed on 8 November 2019).

70. Li, L.; Li, R.; Zhou, J.; Zuniga, A.; Stanislaus, A.E.; Wu, Y.; Huan, T.; Zheng, J.; Shi, Y.; Wishart, D.S.; et al. MyCompoundID: Using an evidence-based metabolome library for metabolite identification. *Anal. Chem.* **2013**, *85*, 3401–3408. [CrossRef]

71. Gil de la Fuente, A.; Godzien, J.; Saugar, S.; Garcia-Carmona, R.; Badran, H.; Wishart, D.S.; Barbas, C.; Otero, A. CEU Mass Mediator 3.0: A Metabolite Annotation Tool. *J. Proteome Res.* **2019**, *18*, 797–802. [CrossRef]

72. CEU Mass Mediator. Available online: http://ceumass.eps.uspceu.es/mediator/ (accessed on 5 November 2019).

73. Oliveros, J.C. Venny. An Interactive Tool for Comparing Lists with Venn's Diagrams. Available online: http://bioinfogp.cnb.csic.es/tools/venny/index.html (accessed on 4 November 2019).

Review

Metabolomics in the Context of Plant Natural Products Research: From Sample Preparation to Metabolite Analysis

Mohamed A. Salem [1,*], **Leonardo Perez de Souza** [2], **Ahmed Serag** [3], **Alisdair R. Fernie** [2,4], **Mohamed A. Farag** [5,6], **Shahira M. Ezzat** [5,7] **and Saleh Alseekh** [2,4,*]

[1] Department of Pharmacognosy, Faculty of Pharmacy, Menoufia University, Gamal Abd El Nasr st., Shibin Elkom, Menoufia 32511, Egypt

[2] Max Planck Institute of Molecular Plant Physiology, Am Mühlenberg 1, 14476 Potsdam-Golm, Germany; LPerez@mpimp-golm.mpg.de (L.P.d.S.); Fernie@mpimp-golm.mpg.de (A.R.F.)

[3] Pharmaceutical Analytical Chemistry Department, Faculty of Pharmacy, Al-Azhar University, Cairo 11751, Egypt; Ahmedserag777@azhar.edu.eg

[4] Center of Plant Systems Biology and Biotechnology (CPSBB), Plovdiv 4000, Bulgaria

[5] Pharmacognosy Department, Faculty of Pharmacy, Cairo University, Cairo 11562, Egypt; Mohamed.farag@pharma.cu.edu.eg (M.A.F.); shahira.ezzat@pharma.cu.edu.eg (S.M.E.)

[6] Chemistry Department, School of Sciences & Engineering, The American University in Cairo, New Cairo 11835, Egypt

[7] Department of Pharmacognosy, Faculty of Pharmacy, October University for Modern Sciences and Arts (MSA), Giza 11787, Egypt

[*] Correspondence: mohamed.salem@phrm.menofia.edu.eg (M.A.S.); Alseekh@mpimp-golm.mpg.de (S.A.); Tel.: +49-(0)331-567-8211 (S.A.)

Received: 25 November 2019; Accepted: 11 January 2020; Published: 15 January 2020

Abstract: Plant-derived natural products have long been considered a valuable source of lead compounds for drug development. Natural extracts are usually composed of hundreds to thousands of metabolites, whereby the bioactivity of natural extracts can be represented by synergism between several metabolites. However, isolating every single compound from a natural extract is not always possible due to the complex chemistry and presence of most secondary metabolites at very low levels. Metabolomics has emerged in recent years as an indispensable tool for the analysis of thousands of metabolites from crude natural extracts, leading to a paradigm shift in natural products drug research. Analytical methods such as mass spectrometry (MS) and nuclear magnetic resonance (NMR) are used to comprehensively annotate the constituents of plant natural products for screening, drug discovery as well as for quality control purposes such as those required for phytomedicine. In this review, the current advancements in plant sample preparation, sample measurements, and data analysis are presented alongside a few case studies of the successful applications of these processes in plant natural product drug discovery.

Keywords: metabolomics; plant natural products; drug discovery; metabolite extraction; liquid chromatography; gas chromatography; mass spectrometry; NMR

---

1. Introduction

Nature provides a rich source of numerous bioactive compounds that have been extensively employed in traditional medicine since time immemorial [1]. In recent years, the Food and Drug Administration (FDA) has approved an impressive number of modern drugs that are also natural products or directly derived therefrom [2]. These mostly constitute compounds belonging to secondary metabolic pathways with prominent examples including taxol from *Taxus brevifolia*, vinblastine from

Catharanthus roseus, doxorubicin from *Streptomyces peucetius,* and cyclosporine from *Tolypocladium inflatum* [3,4]. The first step in the discovery of lead compounds from natural sources is the release of bioactive metabolites from their biomass through various extraction techniques viz., supercritical fluid extraction [5,6], microwave-assisted and ultrasonic-assisted extraction [7], molecular distillation methods [8], and membrane separation technology [9]. Moreover, bioassay-guided fractionation using chromatographic methods such as preparative high performance liquid chromatography (HPLC) is applied for the isolation and purification of active metabolites from their crude extracts [10]. Further technologies such as nuclear magnetic resonance spectroscopy (NMR), mass spectrometry (MS), and ultraviolet-visible spectroscopy (UV–Vis) have allowed the detailed characterization and ultimately, the structural elucidation of these agents [11]. Finally, the bioactivity (effects in cell lines, animal models, and human volunteers) is investigated for assessing the pharmacological potential of the candidate compounds. Nevertheless, some pitfalls were observed when employing this classical approach for lead compound discovery, where degradation or chemical modification of the bioactive compounds during the process of isolation and purification often occurs. Furthermore, important biological information that was present in the original extract might be lost during activity-guided fractionation as the samples are not fully analyzed [12]. Moreover, better therapeutic effects were reported when using the whole extracts, as practiced in traditional medicine, rather than a single-compound based remedy. This effect could be attributed to the synergy between bioactive components (e.g., studies showed the synergistic effects of different plant extracts and doxorubicin in cancer treatment [13], Apocynaceae plants and antibiotics against *Acinetobacter baumannii* [14], and catechin and resveratrol as antioxidants [15]).

Unlike the classical approach in natural products research, metabolomics experiments offer an improved expedited route for drug discovery [16]. The basic goal of metabolomics is to provide a comprehensive qualitative and/or quantitative analysis of all metabolites present in a living system [17]. Interestingly, this concept could be extended in natural product drug discovery via studying the relationship between the whole metabolome of natural-derived remedies and their biological effects [18]. Implementing such an approach not only overcomes the aforementioned pitfalls of the classical techniques used in natural product research, but also provides a broader insight of the biochemical status and gene functions of the studied organisms. Furthermore, the signature between specific compounds in the metabolome and their bioactivity could also be revealed, aided by advanced bioinformatics tools, a process that can be highly useful in pharmacological standardization and biological fingerprinting of natural extracts [19]. The workflow of metabolomics experiments involves an efficient extraction of these endogenous metabolites to be subjected for qualitative and quantitative analysis. However, unlike other omics technologies, no single analytical platform is capable of analyzing all metabolites simultaneously due to their extreme complexity and huge chemical diversity. Recent developments in analytical chemistry platforms such as hyphenating mass spectrometry with gas chromatography (GC), liquid chromatography (LC) or capillary electrophoresis (CE), and nuclear magnetic resonance (NMR) spectroscopy have led to a highly efficient set up for metabolome analysis [20], however, these do not yet reach comprehensibility [21]. Although huge datasets are generated from these instruments, the evolution of chemometrics and multivariate data analysis algorithms provides a powerful tool for extracting useful information from such high dimensionality results [22]. They enable, for example, the detection of compounds that correlate to the medicinal efficacy in test animal or human systems based on their analytical spectral fingerprints [23]. Moreover, pattern recognition and classification algorithms have also allowed the implementation of metabolomics as an effective tool for the quality control of herbal medicinal products [24,25]. Nevertheless, metabolite identification remains the most challenging aspect of metabolomics experiments [26]. Large mass spectral and NMR spectral databases have been created to untangle such problems [27]. Furthermore, bioinformatics tools based on molecular networking such as GNPS [28] and MetGem [29] have been implemented not only to assign known metabolites from their complex mixtures, but also to elucidate the chemical structures of novel compounds of interest.

This review will discuss the recent developments of metabolomics in the context of plant natural product drug discovery including current advances in sample preparation techniques and analytical profiling platforms. In addition, computational tools employed for metabolomics data processing and metabolites identification are reviewed. Finally, some successful applications of metabolomics in the identification of bioactive agents and in the quality control of natural products are outlined, and an outlook for the increasing use of metabolomics in these fields is provided.

2. Sample Preparation for Metabolomic Studies

In metabolomics studies, biological samples are collected, extracted, measured, and finally, the resulting data are analyzed (Figure 1). Sample preparation is a crucial step in metabolomics as it greatly affects the reliability of the metabolomics results. Minor changes in the sample collection, extraction, or storage greatly affect metabolite stability and hence can lead to major changes in the observed metabolome. Metabolomics samples have to be collected uniformly and rapidly to avoid changes due to the fast enzymatic turnover rate [30]. The ultimate aim is to minimize the biologically-irrelevant changes resulting from sample processing. Improper handling of biological samples is the most likely source of bias in metabolomic studies [30]. In order to validate plant metabolomics studies, the minimum parameters related to experimental design, sample extraction to data analysis should ideally follow the Metabolomics Standards Initiative (MSI) [31].

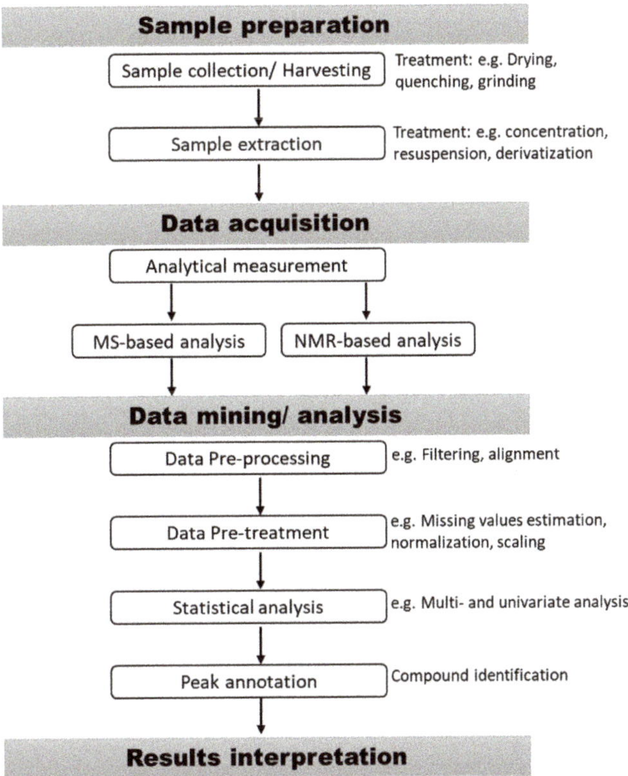

Figure 1. Key steps of a plant metabolomics study. Additional steps can be performed or modified in certain approaches based on the analytical methods or point of interest.

2.1. Sample Collection

A wide range of biological samples have been extensively reported in metabolomics studies. This includes tissues (animal-and plant-derived tissues), fluids (such as urine, whole blood, serum, sweat, cerebrospinal fluid, breast milk, amniotic fluid, saliva, etc.), and cell cultures (human, animal, plant, algae, etc.) [32,33]. Such complex samples have different matrices, and therefore, require different sample preparation protocols. Here, we focus on natural products that are mainly derived from plant origin, while also giving a general overview of other tissues. Depending on the natural abundance of some metabolites and the level of detection, various amounts of biological material per sample have to be considered. Usually, 1–100 mg of tissue, 10–250 μL of fluids, or 10^5-10^7 cells per each biological replicate is required [33]. A minimum number of 3–5 biological replicates is recommended for each condition in metabolomic studies [31]. Biological replicates (parallel measurements of samples from different individuals) rather than technical replicates (repeated measurements of the same sample) are to be considered [34].

2.2. Harvesting Methods

Sample freezing methods such as using dry ice or liquid nitrogen are highly recommended during the harvesting of fresh samples to avoid enzyme-induced metabolic changes [31]. Removing unwanted components such as soil particles is also recommended before collection. Long-term storage of samples prior to extraction should, however, be avoided. For short term storage (e.g., for few days up to two weeks), samples can be kept in liquid nitrogen, dry ice, or a -80 °C freezer [31]. Prior to extraction, the harvested samples can be exposed to processing methods such as lyophilization, cell lysis, and/or grinding, depending on the biological material. The conditions related to cultivation parameters, collected tissue type, seasonality, developmental stage, harvesting time, and sample processing should be reported for each condition since metabolites are greatly affected by such parameters [35], with environmental aspects being reported to induce both qualitative and quantitative variations in the metabolite composition of both plant primary and secondary metabolites [36].

2.3. Sample Extraction

Unlike the genome and proteome, which can be captured using a single extraction protocol, the metabolome is difficult to capture within a single solvent due to the diverse chemistry of metabolites [37]. Therefore, complex biological samples, which as stated above, have different matrices, and therefore, require different extraction protocols. Generally, metabolites are preferentially extracted with the solvent following the rule of thumb "like dissolves like". Polar and semi-polar metabolites can be extracted with hydrophilic solvents such as hydro-alcoholic solutions, while, lipids can be extracted with more hydrophobic solvents. Several protocols for metabolome extraction have been developed and extensively reviewed [32,33,35,38–40]. The selected method must be rapid and efficient, whilst at the same time, cover a wide range of target metabolites, thus maintaining a high level of precision and accuracy.

Liquid–liquid fractionation provides significant simplifying steps compared to single extraction methods [41]. Partial purification of each fraction is achieved via removing interfering compounds such as hydrophobic molecules that are enriched in lipophilic solvents and more polar metabolites in hydrophilic solvents [42]. While different extraction protocols are varied with regard to the selected solvent(s) and the amount that is to be used, most protocols include a deproteinization step. The presence of proteins in the analyzed samples can severely affect the instrument's accuracy, precision, and lifetime [43]. Extraction efficiency can usually be enhanced by selecting an extraction solvent mixture consisting of one or two steps compared to single extraction solvent [33]. Liquid–liquid extraction methods were based on the so called gold standard extraction protocols of the 'Folch' and 'Bligh and Dyer' methods utilizing chloroform/methanol mixture in different proportions [44,45]. Liquid–liquid extraction using chloroform/methanol/water have been used for the analysis of lipids and polar metabolites [46–49]. As a cleaner and safer solvent, methyl *tert*-butyl ether (MTBE) has also been used as an alternative to chloroform for liquid–liquid extraction [50], particularly for the recovery of metabolites and lipids from bacterial, plant, algal, flies, and diverse mammalian samples [41,42,51–60]. The use of isotope-labeled internal standards during sample preparation is recommended to assess matrix effects [39,61]. For the selection of suitable solvent(s) and an extraction method that is suitable for different metabolite classes, readers are referred to the available extraction protocol reviews [32,35,39,40,62,63].

2.4. Extract Concentration, Dilution, Enrichment, and Re-Suspension

In addition to liquid–liquid extraction, other methods such as organic solvent precipitation as well as single step or multiple step solid-phase extraction (SPE) have also been used to partially purify samples prior to analytical measurement [64]. After sample extraction, solvents are evaporated to concentrate the metabolites. Samples are typically concentrated in a vacuum concentrator at room temperature without heating. The use a nitrogen flow evaporator is recommended in cases where metabolites are sensitive to oxidative modifications as, for example, some lipids. Finally, the dry extracts are typically resuspended in an analysis-compatible solvent [43]. Dried samples can be subjected to further steps such as derivatization before GC/MS analysis [49]. Short term storage of extracted liquid samples, even at low temperature (-20 °C), is not recommended, however, samples can be stored, if necessary, in a dry state for a short time prior to analysis [65].

3. Analytical Methods for Metabolites Analysis

The diversity of metabolites in the plant kingdom is staggering; a commonly quoted estimate is that plants produce somewhere between 100,000 and one million metabolites [66,67]. Metabolite measurements have been carried out for decades because of the fundamental regulatory importance of metabolites as components of biochemical pathways, the importance of certain metabolites in the human diet, and their use as diagnostic markers for a wide range of biological conditions including disease and response to chemical treatment [37]. There are wide ranges of metabolomics approaches, which ultimately aim to measure the entire small molecule complement of the cell. Current metabolomics strategies are mainly reliant on four major approaches: gas chromatography-mass spectrometry (GC-MS), liquid chromatography–mass spectrometry (LC-MS), capillary electrophoresis–mass spectrometry (CE-MS), and nuclear magnetic resonance (NMR) spectroscopy. A number of detailed protocols [49,63,68,69], in addition to several excellent technical reviews [70,71] regarding the utilization of these analytical tools in metabolomics experiments, have been published, however, a brief technical overview of these four major methods is provided herein.

3.1. Gas Chromatography-Mass Spectrometry (GC-MS)

GC-MS has been one of the most popular metabolomics techniques to measure the levels of volatile and semi-volatile organic compounds in a wide variety of samples. In GC-MS, polar metabolites are derivitized to render them volatile and then separated by GC. Various derivatization methods such as alkylation, acylation, methoximation, trimethylsilylation, and silylation can be used [72,73]. Electron impact ionization results in highly reproducible fragmentation patterns that are essential for large-scale experiments [37]. A great advantage of GC-MS is that it is both relatively sensitive and highly robust, and can routinely and reproducibly measure hundreds of analytes across thousands of samples [72,74–76]. That said, a technical challenge in GC-MS profiling is to separate each metabolite signal from overlapping peaks in the raw GC-MS chromatogram [77]. Several peak-picking software packages equipped with sophisticated peak deconvolution functions are available to deal with a relatively high-throughput data of thousands of samples [77]. In addition, various databases have been developed to aid in assigning structures to spectral peaks observed in metabolomics experiments [78–80]. Furthermore, limitations of GC-MS have recently been improved by the development of two-dimensional GC × GC coupled with high resolution mass spectrometry [81–84]. In this approach, two columns with different properties (non-polar vs. polar) are connected through a modulator, allowing further separation of compounds that co-elute from the first column, thereby giving rise to enhanced resolution and peak capacity [85,86]. The GC-MS method has had myriad applications in plant, pharmacological, and medical metabolomics studies, and is considered as one of most suitable techniques of the accurate determination of primary metabolites, however, it is severely compromised in measuring some secondary metabolites. Several classes of volatile and non-volatile metabolites such as phenolics, alkaloids, and terpenoids have been analyzed and identified by GC-MS [79,87].

3.2. Liquid Chromatography-Mass Spectrometry (LC-MS)

Liquid chromatography-mass spectrometry (LC-MS) has become the most comprehensive technique to measure a wide range of diverse metabolites. Unlike GC-MS, it does not require prior sample treatment, and crude extracts obtained by simple extraction can be introduced directly to the LC-MS. The choice of columns including reversed phase, ion exchange, and hydrophobic interaction provides metabolite separation on the basis of differential chemical properties. Nowadays, reversed-phase columns such as C18 or C8 are the most commonly used for LC gradient separation [63]. The development of ultra-performance LC rendered the technique even more powerful with regard to resolution, sensitivity, and throughput. LC-MS is a unique method for measuring plant secondary metabolites such as flavonoids and alkaloids [88–90], membrane lipids (lipidomics) [60,91,92], and primary metabolites such as amino acids [93]. Despite the fact that LC-MS is the most comprehensive technique in hand, LC-MS by no means approaches the metabolic complement of a typical plant cell [21]. This is mainly due to the poor availability of standard compounds for secondary metabolites [94]. In the past years, a considerable improvement in the number of metabolites that can be measured and annotated has been realized, which is mainly due to the improved machine performance afforded by the development of Ultra-Performance Liquid Chromatography (UPLC) coupled with high-resolution mass analysis methods such as time-of-flight (TOF) MS, Fourier transform (FT) MS, and Orbitrap-based MS [88,90,95,96]. In addition, this improvement has relied on increased efforts in the collection of standard compounds and sharing of reference extracts [97–99] for use in peak annotation authentication and by the increased sophistication of computational approaches for compound annotation (see the web resources listed in [27,100]).

3.3. Capillary Electrophoresis-Mass Spectrometry (CE-MS)

Capillary electrophoresis (CE) separates polar and charged compounds on the basis of their charge-to-mass ratio [74]. CE offers fast and high-resolution separation of charged analytes from small injection volumes. Coupled to mass spectrometry (MS), it represents a powerful analytical technique providing (exact) mass information and enables molecular characterization based on fragmentation. Although exquisitely sensitive and able to readily capture metabolite classes that the other platforms do not, including nucleotides and highly charged metabolites [101], CE is time consuming and is further hampered by the fact that it covers a range of metabolites that have highly diverse extraction requirements. Therefore, the use of CE-MS in plant metabolism remains relatively rare [74,102]. In addition, CE has poor migration time reproducibility and a lack of reference libraries, which may only be partially overcome by the prediction of migration time [103,104]. However, CE has some distinct advantages over other instruments employed for metabolomics: mainly the facts that it utilizes low separation volume, which is particularly suitable for the study of biological fluids in small experimental animals, and operates under a homogeneous separation environment [105].

3.4. Nuclear Magnetic Resonance (NMR)

NMR spectroscopy is a powerful analytical tool that has traditionally occupied a vital position for structure elucidation of natural products [106–110]. Unlike MS, NMR is not discriminatory and is less biased as the results does not rely on the type of ionization condition or the preferences of the used instruments [111]. Hence, this technique allows for the simultaneous detection of the abundant primary metabolites (organic acids, amino acids, and sugars) alongside secondary metabolites (alkaloids, terpenoids, and flavonoids) as typically found in plant natural extracts [112]. In addition, NMR is a very useful technique for the structure elucidation of novel and/or unexpected compounds including those with identical masses and/or different isotopomer distributions [12,113]. Moreover, NMR-based methods are highly reproducible, noninvasive, nondestructive, and require minimal sample preparation as the sample does not get in physical contact with the device as in MS [114]. Another major strength in NMR spectroscopy lies in the direct proportionality between the NMR spectrum signals and the corresponding real molar levels of the detected metabolites, making absolute quantification of all detected metabolites possible without the need for calibration curves of individual analytes, posing it as a powerful tool in quality control purposes of drug extracts [115]. However, a five times delay of the spin–lattice relaxation times (T1) of the slowest relaxing nuclei in the extract is essential in such quantitative experiments to avoid baseline distortions and increase the accuracy of signal integration. This will significantly increase the duration of these experiments compared to other routine NMR metabolomics methods.

Nevertheless, NMR also has a number of disadvantages in metabolomics analysis [116]. The technique's major disadvantage is the low sensitivity as the polarization rate of the active NMR nuclei obeys the Boltzmann distribution law. However, recent developments in NMR hardware have at least partially improved this drawback to some extent. The use of superconducting magnets that are commonly operated in metabolomics experiments at field resonance of 600 MHz can increase the sensitivity and resolution of the detected signals [117]. Moreover, the development of cryogenic probes in which the electronics are cooled down to a very low temperature (e.g., 20 K) has resulted in a better sensitivity than standard probes via keeping the noise generated due to random thermal motion of electrons at a minimum [118]. There are currently two types of cryoprobe available, the first is cooled by a closed cycle helium cooler system, which yields a signal-to-noise enhancement up to a factor of five. The second type is a liquid nitrogen-cooled system, commonly used in metabolomics experiments, that is able to provide a sensitivity enhancement of up to a factor of three, but with the capability to be cooled down and warmed up relatively quickly [119]. However, such probes may be of limited relevance for samples with high ionic and/or dielectric conductivity such as those with salts or polar solvents, commonly encountered in plant-derived samples [120]. That said, the low concentrations of natural products remains a limitation of NMR analysis. To partially compensate for this, special probes known as microprobes have been also developed in which the analysis of a few microliters of samples could be possible, thus overcoming the limited availability of the samples [121] as is typical in natural product studies. Dynamic nuclear polarization (DNP) is another interesting advancement that has also been employed to enhance the sensitivity of NMR signals. The fundamental principle behind DNP involves transferring the polarization of the electron spins to the nuclei of interest via microwave irradiation close to the electron Larmor frequency, resulting in a temporary hyperpolarization in this spin-active nuclei (usually naturally abundant ^{13}C, ^{15}N, and ^{1}H). After the polarization transfer step is completed, the sample is transferred to the NMR spectrometer to collect the enhanced (>1000-fold) NMR signals [122].

3.4.1. 2D NMR Based Metabolomics Strategies

Although NMR methods have largely been employed primarily to identify compounds from pure samples as in classical natural product research, they have recently played a larger role in the analysis of mixtures of unfractionated natural extracts [123]. The bottleneck in such complicated analysis is the congregation of the ^{1}H NMR spectra, which could be untangled using two main strategies. The first strategy aims to simplify the abundant ^{1}H NMR spectra through the use of diffusion and/or relaxation filters or through creating a projection of the ^{1}H decoupled spectrum like in 2D ^{1}H J-resolved NMR experiments. Relaxation filters are pulse sequences that selectively attenuate, or even remove, signals of components with shorter T1 or spin–spin relaxation times, T2 [124]. Hence, they could be utilized to simplify the ^{1}H NMR spectra through selective elimination of signals belonging to high molecular weight or rigid structure (quickly relaxing) compounds. They have also been implemented to attenuate water signals using the water attenuation T2 relaxation Carr-Purcell-Meiboom-Gill (WATR-CPMG) pulse sequence [125]. On the other hand, diffusion filter based experiments such as diffusion ordered spectroscopy (DOSY) commonly simplify ^{1}H NMR spectra based on the difference of the translational diffusion coefficients between molecules of different molecular size, achieving what has be described as "NMR chromatography" [126]. The DOSY experiments can be performed in 1D mode, where two or more ^{1}H NMR experiments are acquired for compounds possessing different diffusion coefficients or in 2D mode (2D DOSY), where the proton spectrum is depicted in one dimension and the diffusion coefficient is spread in an orthogonal dimension [127]. Interestingly, the extension of 2D DOSY to 3D DOSY is gaining much popularity in metabolomics analysis via incorporating another proton dimension as in 3D COSY-DOSY and 3D TOCSY-DOSY, or another carbon dimension as in 3D DOSY-HSQC [128]. Such DOSY methods have been recently applied in quality control and metabolic fingerprinting of commercial natural extracts [128,129]. Furthermore, they have also been used to identify novel metabolites from natural resources, for example, novel bromopyrrole alkaloids, and agesamides A and B were identified in the crude bromopyrrole fractions from Okinawan marine sponges using the 2D DOSY method [130]. 2D ^{1}H J-resolved NMR experiments represent another approach that simplify ^{1}H NMR spectra where the overlapping resonance of the ^{1}H dimension is resolved through representing the coupling constant of each signal (J value) in a second dimension [131]. 2D ^{1}H J-resolved NMR has been widely used in metabolomics classifications studies [24,132], however, prior knowledge of the chemical composition of the studied samples is required, which make this experiment of limited utility in the analysis of novel compound discovery [108]. On the other hand, a newer approach called pure-shift ^{1}H NMR spectroscopy has emerged in which better removal of homonuclear J couplings has been achieved [133]. In this experiment, J coupling multiplicities are collapsed into single lines via implementing specific pulse sequences that result in a broadband homo-decoupled spectrum. This method has been applied recently to investigate the metabolome of *Physalis peruviana* fruit aqueous extracts [134]. The simplified data allowed for the identification of glutamic acid, a metabolite not observed in previous studies of the same extract due to the heavy overlap of its NMR signals.

The second strategy that can reliably handle ^{1}H NMR overlapping peaks is the spreading of the crowded ^{1}H NMR peaks in a second frequency dimension via magnetization transfer [135]. This transfer could be between nuclei of the same type as in H–H correlated spectroscopy (COSY) and H–H total correlated spectroscopy (TOCSY), nuclei of different types such as heteronuclear single quantum coherence spectroscopy (HSQC) and heteronuclear multiple bond correlation (HMBC) experiments, or through space as in nuclear overhauser effect spectroscopy (NOESY) and rotating frame overhauser enhancement spectroscopy (ROESY) experiments [135]. Full details of these common 2D NMR methods that have been used extensively in metabolomics analysis of natural products, and are thus beyond the scope of this review, but can be found elsewhere [12,108,135,136]. It is noteworthy to mention that spreading the crowded signals in indirect proton dimensions such as in COSY and TOCSY results in a relatively short acquisition time due to the high natural abundance of protons, however, some overlap might still exist as a region of only 10 ppm is employed. In contrast, using carbon as a second dimension such as in HSQC and HMBC will require a much longer acquisition time, but allow the resolution of the overlapping signals and the indirect dimension to be increased to more than 200 ppm units. Nevertheless, it should be noted that acquiring data using HSQC or HMBC can aid in the detection of less abundant compounds compared to the direct acquisition from the carbon channel as it is typical in ^{13}C-NMR. It is worth mentioning that different pulse sequences have been developed for quantitative HSQC experiments. For example, a new phase modulated pulse sequence called the quantitative, offset-compensated, CPMG-adjusted HSQC (Q-OCCAHSQC) experiment has been proposed to enable the use of HSQC spectra for the quantitative determination of many primary metabolites [137]. However, a combination of different 2D NMR experiments is favored for better analytical performance and enhanced identification strategies (e.g., COSY, (1H,13C)-HMQC, (1H,13C)-HMBC, and NOESY) were used to analyze the unfractionated defensive secretion of the walking stick *Parectatosoma mocquerysi* identifying the novel monoterpene parectadial [138]. COSY, HMQC, and HMBC have also aided in the identification of six terpenoids/sterol secondary metabolites in the crude extract of *Sarcophyton* marine coral [139]. Similarly, a combination of ^{1}H, ^{1}H COSY, HSQC, and HMBC have identified and quantified structurally related α- and β-bitter acids and their degradation products in hop resin [140] using run times comparable to those used in HPLC. Interestingly, in the same study, HMBC spectra coupled with multivariate techniques were utilized as a novel approach for the classification of 13 commercial hop cultivars. Application of 2D-NMR metabolomics for drug discovery is considerably easier in cases where the secondary metabolites are naturally abundant as typical in the case of hop resin, which contains much lower levels of primary metabolites [54]. In an interesting application of 2D NMR methods, differential analyses of NMR spectra (DANS) has been oriented for natural products drug discovery and comparative metabolomics. DANS uses a simple algorithm that graphically compares the 2D NMR spectra of different biological states to spot the peaks that are discriminatory among this set of spectra. For example, 2D double quantum filtered (DQF)-COSY was employed for DANS analysis of seven different culturing protocols of the filamentous fungus, *Tolypocladium cylindrosporum*, leading to the detection and identification of two novel indole alkaloids, TC-705A and TC-705B, in the unfractionated extracts [141]. Furthermore, DQF-COSY experiments have been applied for the identification of signaling molecules in the model organism *Caenorhabditis elegans* [142] and polyene antibiotic "bacillaene" from *Bacillus subtilis* [143]. On the other hand, multidimensional NMR experiments have already been used to increase the resolution of the newly developed benchtop NMR instruments [144]. Benchtop NMR provides inexpensive and direct NMR access within any metabolomics laboratory, however, they suffer from decreased chemical shift dispersion and peaks overlapping as they are typically operated in the range of 40–80 MHz proton resonance frequency [145]. Benchtop NMR maintain the same J-coupling frequency as the high field instruments since it is independent of the Larmor frequency, making it suitable for some metabolomics and quality control applications, particularly in 2D modes [146].

3.4.2. Solid State NMR Based Metabolomics

Another exciting application of NMR spectroscopy is the analysis of semi-solid samples such as fresh plant leaf or intact tissues via the implementation of high resolution solid state magic angle spinning (HR-MAS) NMR [147]. In this technique, the sample is mixed with a minimal volume of solvent. Next, an NMR spectrum is acquired using a special HR-MAS probe rotated at the angle of 54.74° (the 'magic angle') and a spinning rate of 4 kHz. HR-MAS NMR provides spectra with a similar resolution to that of classical liquid-state NMR techniques due to the elimination of chemical anisotropies and dipolar coupling that hamper analysis of the solid states samples. This technique requires few sample preparation steps, thus it is gaining much attention in the quality control of herbal medicine and food analysis [148,149]. HR-MAS NMR has been also reported for the metabolic profiling of red algae [150], carrot [151], and Arabidopsis [152] tissues. However, the method is not fully quantitative as the local microenvironment in tissues or cells (e.g., pH) could affect the metabolite chemical shifts, thus hindering its potential, as does the difficulty in incorporating a reference compound [153].

3.4.3. Hyphenated NMR-Based Metabolomics

Whereas standalone NMR spectroscopy has unquestionable merits, on-line hyphenation of separation techniques and NMR spectroscopy (HPLC-NMR) can provide another powerful workflow for de novo identification of novo natural products in crude extracts [154,155]. In hyphenated HPLC-NMR, a flow cell matching the HPLC module is used instead of conventional NMR probes [156]. The separated analytes are eluted from the HPLC column to the crossed flow cell, and their on-line—^{1}H NMR spectra are recorded. Some drawbacks were observed in this setup, particularly in the case of applying a gradient mobile phase [157]. Hence, most instrumental setups have changed from on-flow cell to stopped flow HPLC-NMR, and further to the loop/cartridge storage devices that showed wider application in natural product chemistry [158,159] and secondary metabolite identification [160]. Another major development in hyphenated NMR techniques is high-performance liquid chromatography-solid phase extraction-NMR spectroscopy (HPLC-SPE-NMR)where a post-column SPE cartridge is used as an analyte enrichment device prior to NMR acquisition [161]. In this setup, the HPLC eluted component is allowed to be captured onto a solid adsorbent, allowing the HPLC mobile phase to be removed. The adsorbed analyte is then transferred to the NMR probe using the minimal volume of solvents with sufficient elution power (e.g., CD_3OD, CD_3CN). The decrease in analyte chromatographic peak volume via the use of minimal solvent confers increased sensitivity of this technique over conventional HPLC-NMR, in addition to significant cost reduction since employing a deuterated HPLC mobile phase is no longer mandatory [162]. However, careful optimization of SPE trapping is necessary from an analytical point of view, as some metabolites may not be trapped onto the SPE stationary phase. In addition, the use of a small volume of deuterated NMR solvents for flushing the required analyte from the SPE stationary phase is a critical parameter that needs to be carefully optimized. HPLC-SPE-NMR has been applied in the structural characterization of secondary metabolites belonging to different classes including flavonoids [163], terpenoids [164], steroids [165], aromatic alkaloids [166], diarylheptanoids [167], iridoids [168], and saponins [169] and has great potential in fields such as chemical ecology in which targeted metabolites are often signaling molecules present at low levels [170]. Furthermore, multiple complementary hyphenated approaches including PDA and MS have been integrated with HPLC-SPE-NMR as versatile platforms for authentication and the structural identification of secondary metabolites directly from extracts [171,172]. The sample is split after the initial HPLC separation with a small portion being routed for PDA and MS analysis and the remainder to the SPE cartridges for collection [173]. HPLC–PDA–MS–SPE–NMR has been used for the dereplication of six novel quinolinone alkaloids, named haplacutine A–F from the *Haplophyllum acutifolium* crude extract [174] and 23 coumarins including six new compounds from the crude ethyl acetate extract of *Coleonema album* leaves [175] and even minor metabolites could also be detected and fully characterized using this advanced platform [176]. These developments thus suggest that NMR, despite its lower sensitivity will retain a role in role in drug discovery in the decades to come.

4. Metabolomic Data Processing and Interpretation

GC-MS, LC-MS, and NMR are arguably the most relevant techniques within the context of natural product discovery. GC-MS is a robust platform with great peak resolution in the chromatographic dimension, capable of providing both stable separation and mass spectra fragmentation via its most common ionization source, namely electron ionization (EI). This renders data processing as well as metabolite annotation relatively facile for GC-MS. Well established methods for obtaining robust retention indexes based on series of standards [49], together with the great stability of EI spectra, result in an outstanding machine independent reproducibility of the results obtained and facilitate the use of standard databases for the identification of analytes such as the Golm Metabolome Database (GMD) and MassBank of North America (MoNA), among others [27]. The high resolution of the chromatographic separation and stability of ionization also facilitates the deconvolution of mass features into compound mass spectra and several relatively user-friendly tools are available that are capable of performing all steps from the raw data to peak area and assessment and metabolite annotation matching against databases [177–179]. However, the portion of the metabolome that can be covered by GC is significantly limited by the necessary properties of its analytes, namely, being volatile and stable at the high temperatures of analysis. Several compounds can be derivatized to fit such properties, this is unfortunately not the case for most natural products. Moreover, compound annotation relies almost exclusively in matching against known compounds in databases. De novo annotation of compounds based solely on an analyte mass spectra and chromatographic behavior is uncommon. This is because the interpretation of EI spectra is very complex due to the high degree of fragmentation, which can be partially mitigated by the adoption of a milder ionization source such as chemical ionization. Finally, GC has a disadvantage in relation to LC as it cannot be easily translated into a purification technique for further characterization of unknowns by more powerful structural techniques such as NMR. Still, a few interesting attempts have been made in expanding GC-MS structural elucidation such as the work by Matsuo et al. [180]. Here, the authors achieved considerable improvements over traditional spectra and retention index matching approaches by integrating multiple cheminformatics procedures including the use of quality controls to reduce mass spectra background noise and remove artifact peaks, principal component analysis for the selection of biologically relevant variables, EI-MS spectral search in databases, and both retention index filtering and predictions [180]. A further peculiarity of GC-MS is the lack of linearity with respect to quantitation across peaks, requiring quantitative analysis to be made on an analyte by analyte basis [181,182].

LC-MS is the most versatile technique in terms of metabolome coverage and certainly the most relevant for rapid dereplication of natural products in complex mixtures. Its high sensitivity combined with the multitude of stationary phases with different chemistries and ionization sources covering a broad spectra of different compounds allows it to be optimized for nearly every class of natural product. The greatest challenges faced by LC-MS are related to the susceptibility of the results to the many factors influencing separation and ionization efficiency such as changes in solvent composition, column stability, electrospray formation, ion suppression, and other so-called matrix effects. These factors negatively affect spectra and retention reproducibility. The resulting shifts of compound retention times across multiple runs are hard to predict or account for with procedures such as the calculation of retention indexes for GC. The lack of reproducibility in both retention and spectra make the identification of known compounds based on database matches significantly less efficient than for GC. Data processing is also a more challenging task since the lack of ionization stability combined with the lower resolution of LC separation complicates combining mass features into compound spectra, resulting in much more complex datasets with an overwhelming amount of mass features. Despite all of these challenges, the great flexibility of LC-MS can lead to intensive developments for this technique within the context of metabolomics and several pipelines are available to process, analyze, and interpret LC-MS based metabolomics data [32].

Considering that the focus of this review is on the discovery of plant natural products, we mainly describe the processing of untargeted metabolomics experiments, defined as the unbiased processing of all signals within a dataset. Several excellent freeware are currently available for the processing of mass chromatograms with some popular options including XCMS [183], MZmine 2 [184], OpenMS [185], and MS-DIAL [179], all of which have been extensively used for diverse sets of metabolomics data. All of these software work on similar principles, identifying *m/z* signals above a certain threshold as well as the boundaries confining the chromatographic peak representing these signals over time (often referred as features) and returning the respective peak areas/heights. After peak detection, they are usually aligned across multiple samples prior to data analysis. One of the first challenges in processing the resulting data matrixes is the sheer number of features resulting from this process, usually in the order of tens of thousands, much of which are uninformative or redundant, being attributed, for example, to background noise, contaminations and in source fragmentation. Several practices are commonly included in experimental design to assist in the elimination of these signals in the initial steps of data analysis. The inclusion of extraction blanks and technical replicates of quality controls are good examples of such practices and have been shown to be useful for the identification of features that do not originate from the samples or that exhibit low reproducibility [186]. This data cleaning process can significantly facilitate further steps of data analysis, particularly for the identification of features that are most relevant for the separation of experimental groups, for instance, in activity guided fractionation of complex extracts [187]. Another challenging and essential aspect of processing untargeted metabolomics experiments is the identification of features representing fragments, adducts, and isotopes of a same compound. As mentioned before, this process is much more challenging for LC-MS in relation to GC-MS, however, there are a few tools such as CAMERA [188], which can assign putative identity for features based on their retention time and correlation in relation to one another as well as mass differences matching common adducts and in source fragmentation patterns in electrospray ionization.

Particularly relevant for the identification of compounds by mass spectrometry are the experiments including tandem MS fragmentation of either specific ions in a data-dependent acquisition mode (DDA), or of all ions in a data-independent acquisition mode (DIA). Both types of experiment provide second order MS spectra (MS/MS), which are more informative and suitable for compound identification based on matching spectra signals against databases. In DDA experiments, specific ions are isolated and fragmented, therefore all signals in the MS/MS spectra correspond to fragments of the isolated ions. In DIA experiments, larger isolation windows are combined to fragment all ions in first order spectra, therefore there is no link between parental and daughter ions but with the advantage of a comprehensive fragmentation. Several specialized tools as well as extensions to the tools previously mentioned were developed to process such MS/MS data, a few examples including MetDIA [189], which is based on algorithms from XCMS, and MS-DIAL [179], which was developed with a particular focus on processing DIA data despite also being able to process DDA and GC-MS data.

MS/MS spectra are currently the starting point for most mass spectrometry based structural elucidation approaches. The initial procedure usually consists of using the experimental data to search matches through different databases. Several relevant databases for natural product discovery were recently reviewed by Wolfender et al. [190]. We highlight here some of the open source alternatives that are particularly interesting for including curated experimental MS/MS data. Some extensive metabolome databases such as Metlin [191], MassBank [192], MoNA, European MassBank, Global Natural Products Social Molecular Networking resource (GNPS) [28], and Human Metabolome Database (HMDB) [193] include experimental MS/MS data from a large number of natural products from different sources. Additionally, other databases oriented toward specific organisms such as ECMDB for *E. coli*, YMDB for yeast, and ResPect for plants, also provide access to extensive collections of experimental MS/MS data [194]. Most of these alternatives have their own inner structures, hence the functionalities for querying and extracting data vary. An interesting feature that is worth noting for large scale metabolomics is a database integration with other data processing and analysis pipelines. Metlin provides the largest individual collection of MS/MS experimental data acquired in multiple collision energies and despite being the only one of the larger databases mentioned to impose restrictions on downloading data, it is directly integrated with the XCMS online processing platform [191]. MZmine 2 incorporates functions to directly query Kyoto Encyclopedia of Genes and Genomes (KEGG), HMDB, Yeast Metabolome Database (YMDB), LipidMaps, the European MassBank, Chemspider, and Metacyc [184]. MS-DIAL allows for the uploading and automatic matching of spectra against user-defined libraries in the msp. format, the commonly used text-based format for metabolomics spectra [179]. It also provides multiple libraries for download in this format including MassBank, ReSpect, and GNPS, among others.

One of the major obstacles for mass spectrometry based annotation in metabolomics is the lack of characterized standards for the majority of the different compounds found in nature. Therefore, several alternatives have been developed to cope with this limitation. Most large metabolomics databases such as Metlin and HMDB have significantly expanded their coverage by including in silico generated MS/MS spectra of known compounds for which no experimental data are available, usually based on machine learning or quantum mechanics calculations [195]. Similar approaches have also been used to predict putative compounds based on the MS/MS spectra. These functionalities are provided in tools such as CSI:fingerID [196] and MS-FINDER [197]. Another interesting and popular approach provided via the GNPS database is to construct networks based on spectra similarity and extend annotations based on the assumption that related metabolites exhibit similar second order spectra [28].

Additionally to the mass spectra, LC-MS provides another measure with intrinsic structural information, which is the retention time. However, as previously mentioned, there are multiple factors affecting retention time in liquid chromatography that are hard to predict and control to a similar extent of what is done in gas chromatography. Therefore, few approaches for metabolite annotation and cataloguing include retention time information. It is worth mentioning a few recent efforts in integrating this extra information in spectral libraries such as the WEIZMASS [97] as well as strategies for in silico prediction of retention indexes as that provided by PredRet [198].

Finally, there are also approaches for compound annotation that are independent of the acquisition of tandem mass spectra fragmentation. Most untargeted metabolomics platforms nowadays rely on high resolution and high accuracy *m/z* measurements provided mainly by QTOF and Orbitrap-based systems. Directly matching putative molecular ion masses against databases is not as reliable as MS/MS spectral matching, it can however, still bring valuable insights in putative structures and the higher accuracy measurements provided by those systems can provide a significantly smaller search space. This is particularly useful when genomics and taxonomic data can be integrated into the search to even further restrict the space of reasonable matches. Unfortunately, this information is still often scarce and not well structured, with few databases providing extensive collections of curated data. One exception worth mentioning is the KNApSack database, which provides the distribution of over 50,000 compounds across over 20,000 different species [199].

5. Successful Applications of Metabolomics in Natural Products Discovery

5.1. Metabolomics for Secondary Metabolites Identification

In the science of natural product discovery using metabolomics, the separation of metabolites is usually performed using GC or HPLC, which is generally coupled with on-line MS detection. MS fragmentation spectra provide precise information on the structures, and it is thus common to record multiple-stage MS data with the spectra of two or more products (MS2, MSn, where n is the number of production stages) [200]. As above-mentioned, several MS-based databases and software tools are now applied for natural products identification; these databases include the Golm Metabolome Database (GMD) that uses GC retention indices and electron impact (EI) mass spectra, all acquired under defined conditions as mass spectral tags (MSTs) for both neat authentic standards as well as plant extracts in GC-MS data [78]. When GC-MS data of a plant extract are obtained, the MST of its components can be matched to the GMD to record either known metabolites or unique identifiers for unidentified peaks in complex plant extracts. There is also the decision tree search, which predicts substructures of unknown metabolites, even if the full structure cannot be predicted [78]. METLIN, a MS/MS database, includes about 62,000 spectra representing more than 12,000 metabolites. These spectra were acquired under standardized conditions using electrospray ionization in positive and negative ionization modes on a quadrupole time-of-flight (QTOF) mass spectrometer [201]. The limitations of this database are the lack of chromatographic data and closed design. Meanwhile, MassBank is a database of high resolution MS-based platforms (GC-MS and HPLC-MS). All MassBank has spectral information, and some have chromatographic information, but there are no retention times (or indices) data [192]. In contrast, ReSpect is an MS2 database specific to plant metabolites. The main advantage of this database is that spectral records are annotated with taxonomic information about the species from which a particular metabolite has been extracted and to which structural class the metabolite belongs [202]. Similarly, the Global Natural Products Social Molecular Networking resource (GNPS) is an MS2 database for natural products [28]. Several databases are considered as an important tool to facilitate the dereplication process [80,203].

NMR is the most powerful technique with respect to structure elucidation, as it is highly reproducible, quantitative, requires simple sample preparation, and is able to measure analytes over a wide range of solvent conditions [204]. It is, however, limited by being slower and less sensitive than MS, and as such, it cannot easily be used as a routine technology in metabolomics. Instead, its application is mainly limited to measuring the most abundant metabolites in the sample [205,206]. In natural product discovery, fractions with the metabolites of interest are separated through chromatographic runs and then subjected to off-line NMR analysis. It is more common in applications of metabolomics to make NMR analysis of plant extracts or biological fluids without chromatographic separation [80].

The applications and limitations of NMR databases that are particularly useful for natural products discovery are discussed in Table 1. To search in one of the mentioned NMR databases, the obtained spectra should be processed and interpreted using external software and entered into a web interface as a peak table. The rNMR software interfaces convert spectral peaks to a searchable file format, so can be used in the MMCD database [207]. On the other hand, SpinAssign is used for the identification of compounds using the 13C-HSQC spectra of complex cell extracts [208], while the MetaboHunter and COLMAR tools are designed to search the NMR spectra of mixtures against those of a subset of metabolites in HMDB and BMRB [209,210].

Table 1. Nuclear magnetic resonance (NMR) databases and computational methods.

Database	Applications	Notes	References
NAPROC-13	Is useful in cases of dereplication as it has more than 20.000 of natural product spectra and inclusion of metabolite classification	Its closed design, there is no user access to spectral data. ^{13}C-NMR spectra, molecular formula or predicted molecular weight can only be searched separately.	[211]
NMRShiftDB	Not limited to NP and has 51,000 collected spectra. It accepts submissions	It lists NMR chemical shifts, but not peak size.	[212,213]
SDBS	Not limited to NP and has 29,000 collected spectra. It uses multiple kinds of spectra in a single search.	Does not accept submissions.	[214]
BML-NMR	The spectral depth of 16 different one- and two-dimensional experiments for each compound provides high quality references	Covers only 208 compounds, but each compound is measured with 16 different NMR parameter sets which provides high quality references. These metabolites were selected based on their importance within metabolic pathways and their detection potential by NMR. They were analyzed at pH 6.6, 7.0, and 7.4.	[215]
BMRB database	Contains NMR data for various biomolecules with a focus on protein, peptide, and nucleic acid spectra. Spectra are available for downloading in raw and processed data formats	The database mainly covers the compounds involved in the lignin biosynthesis which is a plant cell wall constituent and the compounds obtained by plant cell wall deconstruction.	[216]

NAPROC-13: Carbon-13 Database of Natural Products and Related Substances; DB: Database; SDBS: Spectral Database for Organic Compounds; BML-NMR: Birmingham Metabolite Library; BMRB: Biological Magnetic Resonance Data Bank.

There is an increasing tendency in metabolomics to analyze the same sample by both NMR and MS. Taking advantage of both methods enables a better coverage of the metabolome [217,218]. NMR allows the identification of trends in metabolic alteration across core metabolic pathways, while MS provides the identification of the minor metabolites. The integration of NMR and MS can increase the size of the obtained dataset with the added complexity of the simultaneous processing, analysis, and interpretation of two dissimilar data types [218,219]. Indeed, as described below, LC-online-NMR has been proposed as a route toward improving coverage of the metabolome. However, despite the advocacy of this approach, relatively few examples of its utility have been published to date [106,114].

5.2. Metabolomics in Defining or Refining the Pathway Structure of Plant Natural Products

Given that this subject has been extensively reviewed [3], we will keep this section short, suffice to say that the chemistry of many medicinal plants has remained largely unexplored. Despite the advances of synthetic biology for the production of medicinal compounds in heterologous hosts (described in the section below), the native plant species often remain the most reliable and economic source for their production. It is thus of fundamental importance to investigate the metabolic composition of medicinal plants to characterize their natural metabolic diversity and to define their in planta biosynthetic routes in order to develop strategies to further increase their content. In this vein, the cases of benzoisoquinoline and monoterpenoid indole alkaloids, cannabinoids, caffeine, ginsenosides, withanolides, artemisinin, and taxol represent interesting case studies in which the contribution of recent integrative metabolomics and genomic approaches in augmenting earlier biochemical strategies have helped to elucidate their metabolic pathways and cellular compartmentation [3].

5.3. Metabolomics to Aid Metabolic Engineering of Secondary Metabolites Production

A recent approach in the field of metabolomics is its use as a generic debugging tool for engineered microbial production systems that arise as a result of advanced synthetic biology. The synthetic biology of secondary metabolite production is not restricted to the awakening of the cryptic metabolite production encoded in newly sequenced genomes, but it also serves in modifying and recombining the modular biosynthesis apparatus found in nature, which leads to the generation of novel chemistry, in addition to the more ambitious re-engineering strategies such as refactoring, which means replacing the native regulatory machinery by fine-tuned designer systems or the improvement of production levels through the redeployment of primary metabolism [220].

In a metabolomics study of the consequences of an overexpression of ncRNA-based regulatory element in *Streptomyces coelicolor* [221], where ncRNA affected not only the nearby metabolites, but also the metabolite levels that showed rapid changes throughout the metabolic network of the organism. The reproducibility of the metabolite dynamics was ensured by repeated metabolite profiling. Another application is the detection of the accumulated toxic side products and intermediates, or the depletion of required precursors, which is considered as metabolic bottlenecks. In these cases, metabolomics provides an unbiased overview of the metabolic status of a system and its changes due to the overproduction of compounds of interest, which in combination with systems biological modeling can drive cycles of refined engineering [222]. The *Streptomyces lydicus* metabolome was characterized in various cultures where the critical precursors of streptolydigin were identified through the depletion in overproducing cultures. Here, the added precursors (glutamic acid and proline) led to an increase in streptolydigin production, with no effect on the strain growth. Therefore, metabolomics approaches enabled the identification of the major bottleneck in the system with no need for understanding the underlying metabolic and regulatory network [223].

Given the massive diversity of secondary metabolism in plants, a recent approach to metabolic engineering has been combinatorial biosynthesis, an engineering strategy that allows for the generation of novel, structurally similar, but distinct compounds, which is achieved by combining biosynthetic genes of related metabolic pathways from different organisms in a single host [224–226]. This requires the use of enzymes that have relaxed specificity and thus are able to incorporate derivatives of their natural substrates. Other than pioneering studies on the synthesis of novel carotenoids in nonphotosynthetic *Escherichia coli* [227], a few other reports of successful targeted combinatorial biosynthesis of (novel) terpenoids in plants exist. For example, the biosynthesis of monoterpenoid indole alkaloids (MIAs) in the Madagascar periwinkle (*Catharanthus roseus*) was carried out following the expression of two different bacterial halogenases, yielding new-to-nature halogenated MIA derivatives [228]. Similarly, biosynthetic genes of five different plants were combined in yeast to produce a monoglycosylated triterpene saponin that does not occur in any of the parent plant species [229]. Moreover, artemisinin was recently transferred from the medicinal plant of origin into the crop plant tobacco [230]. In all of the examples above, metabolomics was instrumental in confirming the novel natural products. It is important to note that this subject area is currently in bloom and it seems reasonable to anticipate that many more examples will follow shortly. Furthermore, the labelling approach in metabolomics such as ^{13}C-based metabolomics studies, can also be applied for discovering novel secondary metabolism [231–233]. Although this technique needs a single source of carbon or nitrogen, this can be achieved by tracing the distribution of a class-specific precursor to side branches of the metabolic network [234].

5.4. Metabolomics and Dereplication

Another application of metabolomics is the dereplication of the natural product biosynthesis at different development stages through the use of various analytical methods, the bioassay guided isolation, thus rapid dereplication of known activities is efficiently delivered [235]. High resolution Fourier transform mass spectrometry (HRFTMS) is usually used for dereplication of secondary metabolites together with LTQ-Orbitrap and high resolution NMR. Identification of high resolution MS and NMR of the metabolomes is undertaken using the online and in-house databases. Multivariate analysis allows the Fourier transformation of the FID (free induction decay) of multiple samples data, in order to statistically validate the parameters to produce biologically novel secondary metabolites [236,237]. Combining PCA with dereplication and molecular networking was applied on microbial extracts for the discovery of novel compounds where the molecular networking provided a chance to compare the metabolic profiles of complex crude fermentation extracts, leading to efficient chemical dereplication [238].

The dereplication process was summarized by Tawfike et al. (2013, Figure 2) where the extracts, fractions, and purified secondary metabolites were first tested for possible biological activity. After that, the chemical profiling of the active extract or fraction was analyzed using high resolution mass spectrometry and NMR spectroscopy to identify the constituents of interest at an early stage to minimize the efforts by focusing on the active principles. Following the application of various metabolomic tools and multivariate statistics, the obtained huge datasets could be utilized for the identification of the biomarkers of the biologically-active extracts. The identified biomarkers can then be employed in the production of the novel secondary metabolites to provide sustainability of the interesting metabolites in genetic and metabolic engineering [113].

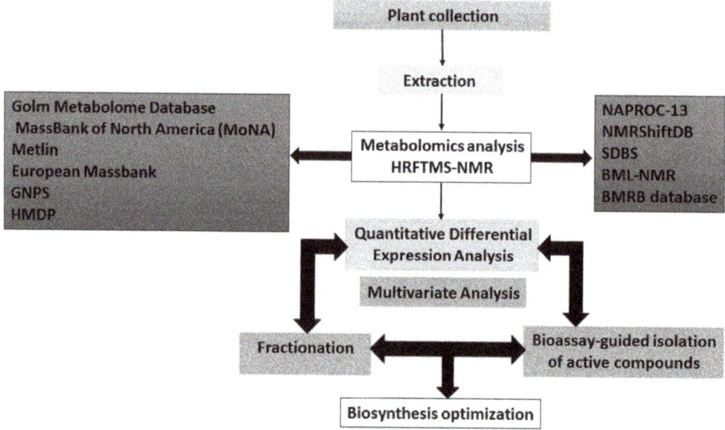

Figure 2. A flowchart showing the application of metabolomics in dereplication.

In this section, we will discuss some examples for the application of metabolomics in dereplication. Dereplication was applied for the identification and quantification of tocopherol using HPLC analysis of Brazil oil nut from different geographic locations in Brazil. This study proved the presence of the different amounts of α- and β-tocopherol in some samples with the absence of tocopherols in other samples [239]. Dereplication was also used to select superior banana genotypes for breeding based on the phenolic contents and carotenoid profiles, where the HPLC analysis showed an appreciable amount of pro-vitamin A carotenoids in the active germosplasm samples compared with the main cultivars that are currently marketed [240]. Coffee genotypes of three different regions in Brazil were distinguished using GC-MS coupled with statistical analysis where 44 metabolites were characterized as the chemomarkers for differentiation of the origin and genotype [241]. Design of experiments (DoE) and partial least squares (PLS) were used to prove that the physicochemical properties of the solvent have a significant effect on its extraction capacity [242]. Moreover, dereplication was used to detect bromotyrosine-derived metabolites in 14 species of sponge belonging to genus *Aplysina* through measuring their UV absorption using LC-photodiode array detector (PDA)-MS analysis [243]. Additionally, dereplication could differentiate between green and brown Brazilian propolis bead on their terpenoid content using GC-MS [244] as well as the discovery of biomarkers and novel active constituents that may be active against different neglected diseases [245]. UHPLC-(TOF)-MS, together with hierarchical clustering analyses (HCA), was applied to differentiate between six *Lippia* species based on their chemical constituents [246]. In addition, dereplication was successfully applied to determine 11 anti-inflammatory biomarkers in 57 extracts of Asteraceae leaves using HPLC-MS among 1241 peaks [247]. Similarly, 44 metabolites were identified as chemomarkers to determine the geographical origin and genotype of coffee in Brazil using GC-Q-MS [241].

5.5. Metabolomics for Quality Control of Natural Products

Metabolomics is also finding utility in the quality control of natural products, being used to monitor the variation of metabolic profiles among individuals, environmental alterations during growth and harvesting, post harvesting treatment, extraction, and method of isolation. The change in the chemical profiles have a significant effect on the efficacy of phytomedicines prepared from these herbs [18]. Unsupervised principal component analysis (PCA) and supervised partial least square analysis/partial least square analysis with discriminant analysis (PLS/PLS-DA), combined with ^{1}H-NMR, are the most common techniques used in quality control. These methods were used to study the effect of location on the percentage of various constituents of chamomile (*Matricaria recutita* L.) [248]. Determination of Δ^9-tetrahydrocannabinolic acid (THCA) and cannabidiolic acid (CBDA) was carried out to discriminate between different *Cannabis sativa* [249]. Similarly, discrimination of different samples of St John's Wort (*Hypericum perforatum*) from different batches of the same supplier, according to the variation of the content of the antidepressant flavonoids, has also been reported [250]. Metabolomics has also been applied to detect adulterations of herbal preparations with similar species but with low levels of constituents responsible for the activity. A company produced antimalarial capsules containing *Artemisia afra*, which is free from artemisinin instead of *Artemisia annua*, and this could be easily detected by ^{1}H-NMR and PCA [251]. Metabolomics can additionally be used to profile the biofluids of individuals who have been medicated with natural products and a vast literature describes such studies [30,252,253].

6. Conclusions and Future Perspectives

Metabolomics experiments offer an improved expedited route for plant natural product drug discovery and is increasingly applied in a bioactivity-guided approach from natural extracts. This review provides an overview for the most widely employed analytical tools in metabolomics experiments for natural products' drug ability viz. GC-MS, LC-MS, CE-MS, and NMR with recent advances in their applications. Furthermore, different sample preparation protocols, data processing tools, and some successful applications of metabolomics in natural products discovery are outlined, highlighting its advantages and limitations. Although metabolomics has clearly transcended the classical approach of lead compound discovery as previously mentioned, the main challenge in such experiments appears to simultaneously explore the extreme complexity and the huge chemical diversity of metabolites present in any natural organism. In addition, the annotation of metabolites in such complicated environments remains another obstacle that needs to be tackled. The establishment of organism databases and sharing of spectral data in public repositories will aid in that regard, and avoid dereplication of already existing metabolites in bioactivity guided studies for active agents. Bioinformatic tools that can link the metabolomics dataset with bioactivity results have yet to be implemented for all spectral datasets. Finally, we believe that there is urgent need to apply the integrative approaches of genomics and metabolomics to understanding the metabolism of natural products in plants and elucidate their metabolic pathways.

Funding: M.A.F. thanks the Alexander von Humboldt Foundation for financial support and jesour grant number 30 from the Academy of Scientific Research & Technology (ASRT), Egypt. S.A. and ARF acknowledge the PlantaSYST project by the European Union Horizon 2020 Research and Innovation Programme (SGA-CSA no. 664621 and no. 739582 under FPA no. 664620).

Conflicts of Interest: The authors declare no conflict of interest.

References

1. Cragg, G.M.; Newman, D.J. Natural Product Drug Discovery in the Next Millennium. *Pharm. Biol.* **2001**, *39*, 8–17. [CrossRef] [PubMed]

2. Newman, D.J.; Cragg, G.M. Natural Products as Sources of New Drugs from 1981 to 2014. *J. Nat. Prod.* **2016**, *79*, 629–661. [CrossRef]

3. Scossa, F.; Benina, M.; Alseekh, S.; Zhang, Y.; Fernie, A.R. The Integration of Metabolomics and Next-Generation Sequencing Data to Elucidate the Pathways of Natural Product Metabolism in Medicinal Plants. *Planta Med.* **2018**, *84*, 855–873. [CrossRef]

4. Skirycz, A.; Kierszniowska, S.; Méret, M.; Willmitzer, L.; Tzotzos, G. Medicinal Bioprospecting of the Amazon Rainforest: A Modern Eldorado? *Trends Biotechnol.* **2016**, *34*, 781–790. [CrossRef] [PubMed]

5. Yoshioka, T.; Nagatomi, Y.; Harayama, K.; Bamba, T. Development of an analytical method for polycyclic aromatic hydrocarbons in coffee beverages and dark beer using novel high-sensitivity technique of supercritical fluid chromatography/mass spectrometry. *J. Biosci. Bioeng.* **2018**, *126*, 126–130. [CrossRef] [PubMed]

6. Morales, D.; Piris, A.J.; Ruiz-Rodriguez, A.; Prodanov, M.; Soler-Rivas, C. Extraction of bioactive compounds against cardiovascular diseases from *Lentinula edodes* using a sequential extraction method. *Biotechnol. Prog.* **2018**, *34*, 746–755. [CrossRef]

7. Joana Gil-Chávez, G.; Villa, J.A.; Fernando Ayala-Zavala, J.; Basilio Heredia, J.; Sepulveda, D.; Yahia, E.M.; González-Aguilar, G.A. Technologies for Extraction and Production of Bioactive Compounds to be Used as Nutraceuticals and Food Ingredients: An Overview. *Compr. Rev. Food Sci. Food Saf.* **2013**, *12*, 5–23. [CrossRef]

8. Gan, Z.; Liang, Z.; Chen, X.; Wen, X.; Wang, Y.; Li, M.; Ni, Y. Separation and preparation of 6-gingerol from molecular distillation residue of Yunnan ginger rhizomes by high-speed counter-current chromatography and the antioxidant activity of ginger oils in vitro. *J. Chromatogr. B* **2016**, *1011*, 99–107. [CrossRef]

9. Williams, S.R.; Oatley, D.L.; Abdrahman, A.; Butt, T.; Nash, R. Membrane Technology for the Improved Separation of Bioactive Compounds. *Procedia Eng.* **2012**, *44*, 2112–2114. [CrossRef]

10. Salam, A.M.; Lyles, J.T.; Quave, C.L. Methods in the Extraction and Chemical Analysis of Medicinal Plants. In *Methods and Techniques in Ethnobiology and Ethnoecology*; Albuquerque, U.P., de Lucena, R.F.P., Cruz da Cunha, L.V.F., Alves, R.R.N., Eds.; Springer: New York, NY, USA, 2019; pp. 257–283. [CrossRef]

11. Thomford, N.E.; Senthebane, D.A.; Rowe, A.; Munro, D.; Seele, P.; Maroyi, A.; Dzobo, K. Natural Products for Drug Discovery in the 21st Century: Innovations for Novel Drug Discovery. *Int. J. Mol. Sci.* **2018**, *19*, 1578. [CrossRef] [PubMed]

12. Robinette, S.L.; Brüschweiler, R.; Schroeder, F.C.; Edison, A.S. NMR in metabolomics and natural products research: Two sides of the same coin. *Acc. Chem. Res.* **2012**, *45*, 288–297. [CrossRef] [PubMed]

13. Sharma, G.; Tyagi, A.K.; Singh, R.P.; Chan, D.C.F.; Agarwal, R. Synergistic Anti-Cancer Effects of Grape Seed Extract and Conventional Cytotoxic Agent Doxorubicin Against Human Breast Carcinoma Cells. *Breast Cancer Res. Treat.* **2004**, *85*, 1–12. [CrossRef] [PubMed]

14. Chusri, S.; Siriyong, T.; Na-Phatthalung, P.; Voravuthikunchai, S.P. Synergistic effects of ethnomedicinal plants of Apocynaceae family and antibiotics against clinical isolates of Acinetobacter baumannii. *Asian Pac. J. Trop. Med.* **2014**, *7*, 456–461. [CrossRef]

15. Skroza, D.; Generalić Mekinić, I.; Svilović, S.; Šimat, V.; Katalinić, V. Investigation of the potential synergistic effect of resveratrol with other phenolic compounds: A case of binary phenolic mixtures. *J. Food Compos. Anal.* **2015**, *38*, 13–18. [CrossRef]

16. Shyur, L.-F.; Yang, N.-S. Metabolomics for phytomedicine research and drug development. *Curr. Opin. Chem. Biol.* **2008**, *12*, 66–71. [CrossRef] [PubMed]

17. Goodacre, R.; Vaidyanathan, S.; Dunn, W.B.; Harrigan, G.G.; Kell, D.B. Metabolomics by numbers: Acquiring and understanding global metabolite data. *Trends Biotechnol.* **2004**, *22*, 245–252. [CrossRef] [PubMed]

18. Wang, M.; Lamers, R.J.; Korthout, H.A.; van Nesselrooij, J.H.; Witkamp, R.F.; van der Heijden, R.; Voshol, P.J.; Havekes, L.M.; Verpoorte, R.; van der Greef, J. Metabolomics in the context of systems biology: Bridging traditional Chinese medicine and molecular pharmacology. *Phytother. Res.* **2005**, *19*, 173–182. [CrossRef]

19. Ulrich-Merzenich, G.; Zeitler, H.; Jobst, D.; Panek, D.; Vetter, H.; Wagner, H. Application of the "-Omic-" technologies in phytomedicine. *Phytomedicine* **2007**, *14*, 70–82. [CrossRef]

20. Tebani, A.; Afonso, C.; Bekri, S. Advances in metabolome information retrieval: Turning chemistry into biology. Part I: Analytical chemistry of the metabolome. *J. Inherit. Metab. Dis.* **2018**, *41*, 379–391. [CrossRef]

21. Alseekh, S.; Fernie, A.R. Metabolomics 20 years on: What have we learned and what hurdles remain? *Plant J.* **2018**, *94*, 933–942. [CrossRef]

22. Pinto, R.C. Chemometrics Methods and Strategies in Metabolomics. In *Metabolomics: From Fundamentals to Clinical Applications*; Sussulini, A., Ed.; Springer International Publishing: Cham, Switzerland, 2017; pp. 163–190. [CrossRef]

23. Farrag, A.R.H.; Abdallah, H.M.I.; Khattab, A.R.; Elshamy, A.I.; Gendy, A.E.-N.G.E.; Mohamed, T.A.; Farag, M.A.; Efferth, T.; Hegazy, M.-E.F. Antiulcer activity of Cyperus alternifolius in relation to its UPLC-MS metabolite fingerprint: A mechanistic study. *Phytomedicine* **2019**, *62*, 152970. [CrossRef]

24. Yang, S.Y.; Kim, H.K.; Lefeber, A.W.M.; Erkelens, C.; Angelova, N.; Choi, Y.H.; Verpoorte, R. Application of Two-Dimensional Nuclear Magnetic Resonance Spectroscopy to Quality Control of Ginseng Commercial Products. *Planta Med.* **2006**, *72*, 364–369. [CrossRef]

25. Zeng, Z.-D.; Liang, Y.-Z.; Chau, F.-T.; Chen, S.; Daniel, M.K.-W.; Chan, C.-O. Mass spectral profiling: An effective tool for quality control of herbal medicines. *Anal. Chim. Acta* **2007**, *604*, 89–98. [CrossRef]

26. Dunn, W.B.; Erban, A.; Weber, R.J.M.; Creek, D.J.; Brown, M.; Breitling, R.; Hankemeier, T.; Goodacre, R.; Neumann, S.; Kopka, J.; et al. Mass appeal: Metabolite identification in mass spectrometry-focused untargeted metabolomics. *Metabolomics* **2013**, *9*, 44–66. [CrossRef]

27. Perez de Souza, L.; Naake, T.; Tohge, T.; Fernie, A.R. From chromatogram to analyte to metabolite. How to pick horses for courses from the massive web resources for mass spectral plant metabolomics. *GigaScience* **2017**, *6*, gix037. [CrossRef] [PubMed]

28. Wang, M.; Carver, J.J.; Phelan, V.V.; Sanchez, L.M.; Garg, N.; Peng, Y.; Nguyen, D.D.; Watrous, J.; Kapono, C.A.; Luzzatto-Knaan, T.; et al. Sharing and community curation of mass spectrometry data with Global Natural Products Social Molecular Networking. *Nat. Biotechnol.* **2016**, *34*, 828. [CrossRef]

29. Olivon, F.; Elie, N.; Grelier, G.; Roussi, F.; Litaudon, M.; Touboul, D. MetGem Software for the Generation of Molecular Networks Based on the t-SNE Algorithm. *Anal. Chem.* **2018**, *90*, 13900–13908. [CrossRef]

30. Riekeberg, E.; Powers, R. New frontiers in metabolomics: From measurement to insight. *F1000Research* **2017**, *6*, 1148. [CrossRef]

31. Sumner, L.W.; Amberg, A.; Barrett, D.; Beale, M.H.; Beger, R.; Daykin, C.A.; Fan, T.W.; Fiehn, O.; Goodacre, R.; Griffin, J.L.; et al. Proposed minimum reporting standards for chemical analysis Chemical Analysis Working Group (CAWG) Metabolomics Standards Initiative (MSI). *Metabolomics* **2007**, *3*, 211–221. [CrossRef] [PubMed]

32. Vuckovic, D. Current trends and challenges in sample preparation for global metabolomics using liquid chromatography-mass spectrometry. *Anal. Bioanal. Chem.* **2012**, *403*, 1523–1548. [CrossRef] [PubMed]

33. Cajka, T.; Fiehn, O. Toward Merging Untargeted and Targeted Methods in Mass Spectrometry-Based Metabolomics and Lipidomics. *Anal. Chem.* **2016**, *88*, 524–545. [CrossRef] [PubMed]

34. Blainey, P.; Krzywinski, M.; Altman, N. Replication. *Nat. Methods* **2014**, *11*, 879–880. [CrossRef] [PubMed]

35. Kim, H.K.; Verpoorte, R. Sample preparation for plant metabolomics. *Phytochem. Anal.* **2010**, *21*, 4–13. [CrossRef]

36. Ernst, M.; Silva, D.B.; Silva, R.R.; Vencio, R.Z.; Lopes, N.P. Mass spectrometry in plant metabolomics strategies: From analytical platforms to data acquisition and processing. *Nat. Prod. Rep.* **2014**, *31*, 784–806. [CrossRef]

37. Fernie, A.R.; Trethewey, R.N.; Krotzky, A.J.; Willmitzer, L. Innovation—Metabolite profiling: From diagnostics to systems biology. *Nat. Rev. Mol. Cell Biol.* **2004**, *5*, 763–769. [CrossRef]

38. Dunn, W.B.; Winder, C.L. Sample Preparation Related to the Intracellular Metabolome of Yeast: Methods for Quenching, Extraction, and Metabolite Quantitation. *Method Enzym.* **2011**, *500*, 277–297. [CrossRef]

39. Bylda, C.; Thiele, R.; Kobold, U.; Volmer, D.A. Recent advances in sample preparation techniques to overcome difficulties encountered during quantitative analysis of small molecules from biofluids using LC-MS/MS. *Analyst* **2014**, *139*, 2265–2276. [CrossRef]

40. Mushtaq, M.Y.; Choi, Y.H.; Verpoorte, R.; Wilson, E.G. Extraction for Metabolomics: Access to the Metabolome. *Phytochem. Anal.* **2014**, *25*, 291–306. [CrossRef]

41. Salem, M.A.; Giavalisco, P. Semi-targeted Lipidomics of Plant Acyl Lipids Using UPLC-HR-MS in Combination with a Data-Independent Acquisition Mode. In *Plant Metabolomics: Methods and Protocols, Methods in Molecular Biology*; António, C., Ed.; Springer: New York, NY, USA; Humana Press: Totowa, NJ, USA, 2018; Volume 1778, pp. 137–155.

42. Salem, M.; Bernach, M.; Bajdzienko, K.; Giavalisco, P. A Simple Fractionated Extraction Method for the Comprehensive Analysis of Metabolites, Lipids, and Proteins from a Single Sample. *J. Vis. Exp.* **2017**, *124*, e55802. [CrossRef] [PubMed]

43. Raterink, R.-J.; Lindenburg, P.W.; Vreeken, R.J.; Ramautar, R.; Hankemeier, T. Recent developments in sample-pretreatment techniques for mass spectrometry-based metabolomics. *TrAC Trends Anal. Chem.* **2014**, *61*, 157–167. [CrossRef]

44. Bligh, E.G.; Dyer, W.J. A rapid method of total lipid extraction and purification. *Can. J. Biochem. Physiol.* **1959**, *37*, 911–917. [CrossRef] [PubMed]

45. Folch, J.; Lees, M.; Stanley, G.H.S. A Simple Method for the Isolation and Purification of Total Lipides from Animal Tissues. *J. Biol. Chem.* **1957**, *226*, 497–509. [PubMed]

46. Weckwerth, W.; Wenzel, K.; Fiehn, O. Process for the integrated extraction, identification and quantification of metabolites, proteins and RNA to reveal their co-regulation in biochemical networks. *Proteomics* **2004**, *4*, 78–83. [CrossRef] [PubMed]

47. Fiehn, O.; Kopka, J.; Trethewey, R.N.; Willmitzer, L. Identification of uncommon plant metabolites based on calculation of elemental compositions using gas chromatography and quadrupole mass spectrometry. *Anal. Chem.* **2000**, *72*, 3573–3580. [CrossRef]

48. Fiehn, O.; Kopka, J.; Dormann, P.; Altmann, T.; Trethewey, R.N.; Willmitzer, L. Metabolite profiling for plant functional genomics. *Nat. Biotechnol.* **2000**, *18*, 1157–1161. [CrossRef]

49. Lisec, J.; Schauer, N.; Kopka, J.; Willmitzer, L.; Fernie, A.R. Gas chromatography mass spectrometry-based metabolite profiling in plants. *Nat. Protoc.* **2006**, *1*, 387–396. [CrossRef]

50. Giavalisco, P.; Li, Y.; Matthes, A.; Eckhardt, A.; Hubberten, H.M.; Hesse, H.; Segu, S.; Hummel, J.; Kohl, K.; Willmitzer, L. Elemental formula annotation of polar and lipophilic metabolites using ^{13}C, ^{15}N and ^{34}S isotope labelling, in combination with high-resolution mass spectrometry. *Plant J. Cell Mol. Biol.* **2011**, *68*, 364–376. [CrossRef]

51. Bozek, K.; Wei, Y.; Yan, Z.; Liu, X.; Xiong, J.; Sugimoto, M.; Tomita, M.; Paabo, S.; Sherwood, C.C.; Hof, P.R.; et al. Organization and evolution of brain lipidome revealed by large-scale analysis of human, chimpanzee, macaque, and mouse tissues. *Neuron* **2015**, *85*, 695–702. [CrossRef]

52. Khrameeva, E.E.; Bozek, K.; He, L.; Yan, Z.; Jiang, X.; Wei, Y.; Tang, K.; Gelfand, M.S.; Prufer, K.; Kelso, J.; et al. Neanderthal ancestry drives evolution of lipid catabolism in contemporary Europeans. *Nat. Commun.* **2014**, *5*, 3584. [CrossRef]

53. Bozek, K.; Wei, Y.; Yan, Z.; Liu, X.; Xiong, J.; Sugimoto, M.; Tomita, M.; Paabo, S.; Pieszek, R.; Sherwood, C.C.; et al. Exceptional evolutionary divergence of human muscle and brain metabolomes parallels human cognitive and physical uniqueness. *PLoS Biol.* **2014**, *12*, e1001871. [CrossRef]

54. Kuyukina, M.S.; Ivshina, I.B.; Philp, J.C.; Christofi, N.; Dunbar, S.A.; Ritchkova, M.I. Recovery of Rhodococcus biosurfactants using methyl tertiary-butyl ether extraction. *J. Microbiol. Methods* **2001**, *46*, 149–156. [CrossRef]

55. Matyash, V.; Liebisch, G.; Kurzchalia, T.V.; Shevchenko, A.; Schwudke, D. Lipid extraction by methyl-tert-butyl ether for high-throughput lipidomics. *J. Lipid Res.* **2008**, *49*, 1137–1146. [CrossRef] [PubMed]

56. Salem, M.A.; Li, Y.; Bajdzienko, K.; Fisahn, J.; Watanabe, M.; Hoefgen, R.; Schottler, M.A.; Giavalisco, P. RAPTOR Controls Developmental Growth Transitions by Altering the Hormonal and Metabolic Balance. *Plant Physiol.* **2018**, *177*, 565–593. [CrossRef] [PubMed]

57. Farag, M.A.; Meyer, A.; Ali, S.E.; Salem, M.A.; Giavalisco, P.; Westphal, H.; Wessjohann, L.A. Comparative Metabolomics Approach Detects Stress-Specific Responses during Coral Bleaching in Soft Corals. *J. Proteome Res.* **2018**, *17*, 2060–2071. [CrossRef] [PubMed]

58. Alhajturki, D.; Muralidharan, S.; Nurmi, M.; Rowan, B.A.; Lunn, J.E.; Boldt, H.; Salem, M.A.; Alseekh, S.; Jorzig, C.; Feil, R.; et al. Dose-dependent interactions between two loci trigger altered shoot growth in BG-5 x Krotzenburg-0 (Kro-0) hybrids of Arabidopsis thaliana. *New Phytol.* **2018**, *217*, 392–406. [CrossRef] [PubMed]

59. Salem, M.A.; Li, Y.; Wiszniewski, A.; Giavalisco, P. Regulatory-associated protein of TOR (RAPTOR) alters the hormonal and metabolic composition of Arabidopsis seeds, controlling seed morphology, viability and germination potential. *Plant J. Cell Mol. Biol.* **2017**, *92*, 525–545. [CrossRef]

60. Salem, M.A.; Juppner, J.; Bajdzienko, K.; Giavalisco, P. Protocol: A fast, comprehensive and reproducible one-step extraction method for the rapid preparation of polar and semi-polar metabolites, lipids, proteins, starch and cell wall polymers from a single sample. *Plant Methods* **2016**, *12*, 45. [CrossRef]

61. Doppler, M.; Kluger, B.; Bueschl, C.; Schneider, C.; Krska, R.; Delcambre, S.; Hiller, K.; Lemmens, M.; Schuhmacher, R. Stable Isotope-Assisted Evaluation of Different Extraction Solvents for Untargeted Metabolomics of Plants. *Int. J. Mol. Sci.* **2016**, *17*, 1017. [CrossRef]

62. Roessner, U.; Dias, D.A. Plant Tissue Extraction for Metabolomics. *Metab. Tools Nat. Prod. Discov. Methods Protoc.* **2013**, *1055*, 21–28. [CrossRef]

63. De Vos, R.C.; Moco, S.; Lommen, A.; Keurentjes, J.J.; Bino, R.J.; Hall, R.D. Untargeted large-scale plant metabolomics using liquid chromatography coupled to mass spectrometry. *Nat. Protoc.* **2007**, *2*, 778–791. [CrossRef]

64. Cajka, T.; Fiehn, O. Comprehensive analysis of lipids in biological systems by liquid chromatography-mass spectrometry. *Trends Anal. Chem. TrAC* **2014**, *61*, 192–206. [CrossRef] [PubMed]

65. Fernie, A.R.; Aharoni, A.; Willmitzer, L.; Stitt, M.; Tohge, T.; Kopka, J.; Carroll, A.J.; Saito, K.; Fraser, P.D.; DeLuca, V. Recommendations for reporting metabolite data. *Plant Cell* **2011**, *23*, 2477–2482. [CrossRef] [PubMed]

66. Dixon, R.A.; Strack, D. Phytochemistry meets genome analysis, and beyond. *Phytochemistry* **2003**, *62*, 815–816. [CrossRef]

67. Rai, A.; Saito, K.; Yamazaki, M. Integrated omics analysis of specialized metabolism in medicinal plants. *Plant J. Cell Mol. Biol.* **2017**, *90*, 764–787. [CrossRef] [PubMed]

68. Kruger, N.J.; Troncoso-Ponce, M.A.; Ratcliffe, R.G. 1H NMR metabolite fingerprinting and metabolomic analysis of perchloric acid extracts from plant tissues. *Nat. Protoc.* **2008**, *3*, 1001–1012. [CrossRef] [PubMed]

69. Tohge, T.; Fernie, A.R. Combining genetic diversity, informatics and metabolomics to facilitate annotation of plant gene function. *Nat. Protoc.* **2010**, *5*, 1210–1227. [CrossRef]

70. Sumner, L.W.; Mendes, P.; Dixon, R.A. Plant metabolomics: Large-scale phytochemistry in the functional genomics era. *Phytochemistry* **2003**, *62*, 817–836. [CrossRef]

71. Lu, W.; Su, X.; Klein, M.S.; Lewis, I.A.; Fiehn, O.; Rabinowitz, J.D. Metabolite Measurement: Pitfalls to Avoid and Practices to Follow. *Annu. Rev. Biochem.* **2017**, *86*, 277–304. [CrossRef] [PubMed]

72. Fiehn, O. Metabolomic—The link between genotypes and phenotypes. *Plant Mol. Biol.* **2002**, *48*, 155–171. [CrossRef] [PubMed]

73. Kopka, J.; Fernie, A.; Weckwerth, W.; Gibon, Y.; Stitt, M. Metabolite profiling in plant biology: Platforms and destinations. *Genome Biol.* **2004**, *5*, 109. [CrossRef]

74. Obata, T.; Fernie, A.R. The use of metabolomics to dissect plant responses to abiotic stresses. *Cell. Mol. Life Sci.* **2012**, *69*, 3225–3243. [CrossRef]

75. Roessner, U.; Luedemann, A.; Brust, D.; Fiehn, O.; Linke, T.; Willmitzer, L.; Fernie, A. Metabolic profiling allows comprehensive phenotyping of genetically or environmentally modified plant systems. *Plant Cell* **2001**, *13*, 11–29. [CrossRef]

76. Halket, J.M.; Przyborowska, A.; Stein, S.E.; Mallard, W.G.; Down, S.; Chalmers, R.A. Deconvolution gas chromatography/mass spectrometry of urinary organic acids-potential for pattern recognition and automated identification of metabolic disorders. *Rapid Commun. Mass Spectrom.* **1999**, *13*, 279–284. [CrossRef]

77. Saito, K.; Matsuda, F. Metabolomics for Functional Genomics, Systems Biology, and Biotechnology. *Annu. Rev. Plant Biol.* **2010**, *61*, 463–489. [CrossRef]

78. Kopka, J.; Schauer, N.; Krueger, S.; Birkemeyer, C.; Usadel, B.; Bergmuller, E.; Dormann, P.; Weckwerth, W.; Gibon, Y.; Stitt, M.; et al. GMD@CSB.DB: The Golm Metabolome Database. *Bioinformatics* **2005**, *21*, 1635–1638. [CrossRef] [PubMed]

79. Schauer, N.; Steinhauser, D.; Strelkov, S.; Schomburg, D.; Allison, G.; Moritz, T.; Lundgren, K.; Roessner-Tunali, U.; Forbes, M.G.; Willmitzer, L.; et al. GC-MS libraries for the rapid identification of metabolites in complex biological samples. *FEBS Lett.* **2005**, *579*, 1332–1337. [CrossRef] [PubMed]

80. Johnson, S.R.; Lange, B.M. Open-access metabolomics databases for natural product research: Present capabilities and future potential. *Front. Bioeng. Biotechnol.* **2015**, *3*, 22. [CrossRef] [PubMed]

81. Liu, Z.; Sirimanne, S.R.; Patterson, D.G., Jr.; Needham, L.L. Comprehensive two-dimensional gas chromatography for the fast separation and determination of pesticides extracted from human serum. *Anal. Chem.* **1994**, *66*, 3086–3092. [CrossRef]

82. Mondello, L.; Casillia, A.; Tranchida, P.Q.; Dugo, G.; Dugo, P. Comprehensive two-dimensional gas chromatography in combination with rapid scanning quadrupole mass spectrometry in perfume analysis. *J. Chromatogr. A* **2005**, *1067*, 235–243. [CrossRef]

83. Yu, Z.; Huang, H.; Reim, A.; Charles, P.D.; Northage, A.; Jackson, D.; Parry, I.; Kessler, B.M. Optimizing 2D gas chromatography mass spectrometry for robust tissue, serum and urine metabolite profiling. *Talanta* **2017**, *165*, 685–691. [CrossRef]

84. Egert, B.; Weinert, C.H.; Kulling, S.E. A peaklet-based generic strategy for the untargeted analysis of comprehensive two-dimensional gas chromatography mass spectrometry data sets. *J. Chromatogr. A* **2015**, *1405*, 168–177. [CrossRef] [PubMed]

85. Wong, Y.F.; Hartmann, C.; J Marriott, P. Multidimensional gas chromatography methods for bioanalytical research. *Bioanalysis* **2014**, *6*, 2461–2479. [CrossRef] [PubMed]

86. Mostafa, A.; Edwards, M.; Gorecki, T. Optimization aspects of comprehensive two-dimensional gas chromatography. *J. Chromatogr. A* **2012**, *1255*, 38–55. [CrossRef]

87. Hill, C.B.; Roessner, U. Metabolic Profiling of Plants by GC–MS. In *The Handbook of Plant Metabolomics*; Weckwerth, W., Kahl, G., Eds.; Wiley-VCH: Weinheim, Germany, 2013; pp. 1–23. [CrossRef]

88. Iijima, Y.; Nakamura, Y.; Ogata, Y.; Tanaka, K.; Sakurai, N.; Suda, K.; Suzuki, T.; Suzuki, H.; Okazaki, K.; Kitayama, M.; et al. Metabolite annotations based on the integration of mass spectral information. *Plant J. Cell Mol. Biol.* **2008**, *54*, 949–962. [CrossRef] [PubMed]

89. Matsuda, F.; Hirai, M.Y.; Sasaki, E.; Akiyama, K.; Yonekura-Sakakibara, K.; Provart, N.J.; Sakurai, T.; Shimada, Y.; Saito, K. AtMetExpress development: A phytochemical atlas of Arabidopsis development. *Plant Physiol.* **2010**, *152*, 566–578. [CrossRef]

90. Matsuda, F.; Yonekura-Sakakibara, K.; Niida, R.; Kuromori, T.; Shinozaki, K.; Saito, K. MS/MS spectral tag-based annotation of non-targeted profile of plant secondary metabolites. *Plant J. Cell Mol. Biol.* **2009**, *57*, 555–577. [CrossRef] [PubMed]

91. Okazaki, Y.; Shimojima, M.; Sawada, Y.; Toyooka, K.; Narisawa, T.; Mochida, K.; Tanaka, H.; Matsuda, F.; Hirai, A.; Hirai, M.Y.; et al. A chloroplastic UDP-glucose pyrophosphorylase from Arabidopsis is the committed enzyme for the first step of sulfolipid biosynthesis. *Plant Cell* **2009**, *21*, 892–909. [CrossRef] [PubMed]

92. Salem, M.A.; Giavalisco, P. Semi-targeted Lipidomics of Plant Acyl Lipids Using UPLC-HR-MS in Combination with a Data-Independent Acquisition Mode. *Methods Mol. Biol.* **2018**, *1778*, 137–155. [CrossRef] [PubMed]

93. Gu, L.; Jones, A.D.; Last, R.L. LC-MS/MS assay for protein amino acids and metabolically related compounds for large-scale screening of metabolic phenotypes. *Anal. Chem.* **2007**, *79*, 8067–8075. [CrossRef]

94. Bocker, S.; Rasche, F. Towards de novo identification of metabolites by analyzing tandem mass spectra. *Bioinformatics* **2008**, *24*, i49–i55. [CrossRef]

95. Keurentjes, J.J.; Fu, J.; de Vos, C.H.; Lommen, A.; Hall, R.D.; Bino, R.J.; van der Plas, L.H.; Jansen, R.C.; Vreugdenhil, D.; Koornneef, M. The genetics of plant metabolism. *Nat. Genet.* **2006**, *38*, 842–849. [CrossRef] [PubMed]

96. Giavalisco, P.; Kohl, K.; Hummel, J.; Seiwert, B.; Willmitzer, L. 13C isotope-labeled metabolomes allowing for improved compound annotation and relative quantification in liquid chromatography-mass spectrometry-based metabolomic research. *Anal. Chem.* **2009**, *81*, 6546–6551. [CrossRef] [PubMed]

97. Shahaf, N.; Rogachev, I.; Heinig, U.; Meir, S.; Malitsky, S.; Battat, M.; Wyner, H.; Zheng, S.; Wehrens, R.; Aharoni, A. The WEIZMASS spectral library for high-confidence metabolite identification. *Nat. Commun.* **2016**, *7*, 12423. [CrossRef] [PubMed]

98. Giavalisco, P.; Hummel, J.; Lisec, J.; Inostroza, A.C.; Catchpole, G.; Willmitzer, L. High-resolution direct infusion-based mass spectrometry in combination with whole 13C metabolome isotope labeling allows unambiguous assignment of chemical sum formulas. *Anal. Chem.* **2008**, *80*, 9417–9425. [CrossRef]

99. Nakabayashi, R.; Kusano, M.; Kobayashi, M.; Tohge, T.; Yonekura-Sakakibara, K.; Kogure, N.; Yamazaki, M.; Kitajima, M.; Saito, K.; Takayama, H. Metabolomics-oriented isolation and structure elucidation of 37 compounds including two anthocyanins from Arabidopsis thaliana. *Phytochemistry* **2009**, *70*, 1017–1029. [CrossRef]

100. Tohge, T.; Fernie, A.R. Web-based resources for mass-spectrometry-based metabolomics: A user's guide. *Phytochemistry* **2009**, *70*, 450–456. [CrossRef]

101. Ishii, N.; Nakahigashi, K.; Baba, T.; Robert, M.; Soga, T.; Kanai, A.; Hirasawa, T.; Naba, M.; Hirai, K.; Hoque, A.; et al. Multiple high-throughput analyses monitor the response of *E. coli* to perturbations. *Science* **2007**, *316*, 593–597. [CrossRef]

102. Fernie, A.R.; Tohge, T. The Genetics of Plant Metabolism. *Annu. Rev. Genet.* **2017**, *51*, 287–310. [CrossRef]

103. Williams, B.J.; Cameron, C.J.; Workman, R.; Broeckling, C.D.; Sumner, L.W.; Smith, J.T. Amino acid profiling in plant cell cultures: An inter-laboratory comparison of CE-MS and GC-MS. *Electrophoresis* **2007**, *28*, 1371–1379. [CrossRef]

104. Soga, T.; Imaizumi, M. Capillary electrophoresis method for the analysis of inorganic anions, organic acids, amino acids, nucleotides, carbohydrates and other anionic compounds. *Electrophoresis* **2001**, *22*, 3418–3425. [CrossRef]

105. Ren, J.-L.; Zhang, A.-H.; Kong, L.; Wang, X.-J. Advances in mass spectrometry-based metabolomics for investigation of metabolites. *RSC Adv.* **2018**, *8*, 22335–22350. [CrossRef]

106. Kim, H.K.; Choi, Y.H.; Verpoorte, R. NMR-based plant metabolomics: Where do we stand, where do we go? *Trends Biotechnol.* **2011**, *29*, 267–275. [CrossRef] [PubMed]

107. Kim, H.K.; Saifullah; Khan, S.; Wilson, E.G.; Kricun, S.D.P.; Meissner, A.; Goraler, S.; Deelder, A.M.; Choi, Y.H.; Verpoorte, R. Metabolic classification of South American Ilex species by NMR-based metabolomics. *Phytochemistry* **2010**, *71*, 773–784. [CrossRef] [PubMed]

108. Mahrous, E.A.; Farag, M.A. Two dimensional NMR spectroscopic approaches for exploring plant metabolome: A review. *J. Adv. Res.* **2015**, *6*, 3–15. [CrossRef] [PubMed]

109. Grkovic, T.; Pouwer, R.H.; Vial, M.-L.; Gambini, L.; Noël, A.; Hooper, J.N.A.; Wood, S.A.; Mellick, G.D.; Quinn, R.J. NMR Fingerprints of the Drug-like Natural-Product Space Identify Iotrochotazine A: A Chemical Probe to Study Parkinson's Disease. *Angew. Chem. Int. Ed.* **2014**, *53*, 6070–6074. [CrossRef] [PubMed]

110. Starks, C.M.; Williams, R.B.; Norman, V.L.; Lawrence, J.A.; O'Neil-Johnson, M.; Eldridge, G.R. Phenylpropanoids from Phragmipedium calurum and their antiproliferative activity. *Phytochemistry* **2012**, *82*, 172–175. [CrossRef]

111. Cai, S.-S.; Short, L.C.; Syage, J.A.; Potvin, M.; Curtis, J.M. Liquid chromatography–atmospheric pressure photoionization-mass spectrometry analysis of triacylglycerol lipids—Effects of mobile phases on sensitivity. *J. Chromatogr. A* **2007**, *1173*, 88–97. [CrossRef]

112. Kim, H.K.; Choi, Y.H.; Verpoorte, R. NMR-based metabolomic analysis of plants. *Nat. Protoc.* **2010**, *5*, 536. [CrossRef]

113. Tawfike, A.F.; Viegelmann, C.; Edrada-Ebel, R. Metabolomics and Dereplication Strategies in Natural Products. In *Metabolomics Tools for Natural Product Discovery: Methods and Protocols*; Roessner, U., Dias, D.A., Eds.; Humana Press: Totowa, NJ, USA, 2013; pp. 227–244. [CrossRef]

114. Zhang, L.; Hatzakis, E.; Patterson, A.D. NMR-Based Metabolomics and Its Application in Drug Metabolism and Cancer Research. *Curr. Pharmacol. Rep.* **2016**, *2*, 231–240. [CrossRef]

115. Simmler, C.; Napolitano, J.G.; McAlpine, J.B.; Chen, S.-N.; Pauli, G.F. Universal quantitative NMR analysis of complex natural samples. *Curr. Opin. Biotechnol.* **2014**, *25*, 51–59. [CrossRef]

116. Nagana Gowda, G.A.; Raftery, D. Can NMR solve some significant challenges in metabolomics? *J. Magn. Reson.* **2015**, *260*, 144–160. [CrossRef] [PubMed]

117. Halabalaki, M.; Vougogiannopoulou, K.; Mikros, E.; Skaltsounis, A.L. Recent advances and new strategies in the NMR-based identification of natural products. *Curr. Opin. Biotechnol.* **2014**, *25*, 1–7. [CrossRef] [PubMed]

118. Kovacs, H.; Moskau, D.; Spraul, M. Cryogenically cooled probes—A leap in NMR technology. *Prog. Nucl. Magn. Reson. Spectrosc.* **2005**, *46*, 131–155. [CrossRef]

119. Dona, A.C. CHAPTER 1 Instrumental Platforms for NMR-based Metabolomics. In *NMR-Based Metabolomics*; The Royal Society of Chemistry: London, UK, 2018; pp. 1–21.

120. Deborde, C.; Moing, A.; Roch, L.; Jacob, D.; Rolin, D.; Giraudeau, P. Plant metabolism as studied by NMR spectroscopy. *Prog. Nucl. Magn. Reson. Spectrosc.* **2017**, *102–103*, 61–97. [CrossRef] [PubMed]

121. Anklin, C. Chapter 3 Small-volume NMR: Microprobes and Cryoprobes. In *Modern NMR Approaches to the Structure Elucidation of Natural Products: Volume 1: Instrumentation and Software*; The Royal Society of Chemistry: London, UK, 2016; Volume 1, pp. 38–57.

122. Ardenkjær-Larsen, J.H.; Fridlund, B.; Gram, A.; Hansson, G.; Hansson, L.; Lerche, M.H.; Servin, R.; Thaning, M.; Golman, K. Increase in signal-to-noise ratio of >10,000 times in liquid-state NMR. *Proc. Natl. Acad. Sci. USA* **2003**, *100*, 10158–10163. [CrossRef] [PubMed]

123. Forseth, R.R.; Schroeder, F.C. NMR-spectroscopic analysis of mixtures: From structure to function. *Curr. Opin. Chem. Biol.* **2011**, *15*, 38–47. [CrossRef] [PubMed]

124. Novoa-Carballal, R.; Fernandez-Megia, E.; Jimenez, C.; Riguera, R. NMR methods for unravelling the spectra of complex mixtures. *Nat. Prod. Rep.* **2011**, *28*, 78–98. [CrossRef]

125. Rabenstein, D.L.; Fan, S. Proton nuclear magnetic resonance spectroscopy of aqueous solutions: Complete elimination of the water resonance by spin-spin relaxation. *Anal. Chem.* **1986**, *58*, 3178–3184. [CrossRef]

126. Johnson, C.S. Diffusion ordered nuclear magnetic resonance spectroscopy: Principles and applications. *Prog. Nucl. Magn. Reson. Spectrosc.* **1999**, *34*, 203–256. [CrossRef]

127. Morris, K.F.; Johnson, C.S. Diffusion-ordered two-dimensional nuclear magnetic resonance spectroscopy. *J. Am. Chem. Soc.* **1992**, *114*, 3139–3141. [CrossRef]

128. Balayssac, S.; Trefi, S.; Gilard, V.; Malet-Martino, M.; Martino, R.; Delsuc, M.-A. 2D and 3D DOSY 1H NMR, a useful tool for analysis of complex mixtures: Application to herbal drugs or dietary supplements for erectile dysfunction. *J. Pharm. Biomed. Anal.* **2009**, *50*, 602–612. [CrossRef] [PubMed]

129. Politi, M.; Zloh, M.; Pintado, M.E.; Castro, P.M.L.; Heinrich, M.; Prieto, J.M. Direct metabolic fingerprinting of commercial herbal tinctures by nuclear magnetic resonance spectroscopy and mass spectrometry. *Phytochem. Anal.* **2009**, *20*, 328–334. [CrossRef] [PubMed]

130. Tsuda, M.; Yasuda, T.; Fukushi, E.; Kawabata, J.; Sekiguchi, M.; Fromont, J.; Kobayashi, J. Agesamides A and B, Bromopyrrole Alkaloids from Sponge Agelas Species: Application of DOSY for Chemical Screening of New Metabolites. *Org. Lett.* **2006**, *8*, 4235–4238. [CrossRef] [PubMed]

131. Ludwig, C.; Viant, M.R. Two-dimensional J-resolved NMR spectroscopy: Review of a key methodology in the metabolomics toolbox. *Phytochem. Anal.* **2010**, *21*, 22–32. [CrossRef]

132. Georgiev, M.I.; Ali, K.; Alipieva, K.; Verpoorte, R.; Choi, Y.H. Metabolic differentiations and classification of Verbascum species by NMR-based metabolomics. *Phytochemistry* **2011**, *72*, 2045–2051. [CrossRef] [PubMed]

133. Aguilar, J.A.; Faulkner, S.; Nilsson, M.; Morris, G.A. Pure Shift 1H NMR: A Resolution of the Resolution Problem? *Angew. Chem.* **2010**, *122*, 3993–3995. [CrossRef]

134. Lopez, J.M.; Cabrera, R.; Maruenda, H. Ultra-Clean Pure Shift 1H-NMR applied to metabolomics profiling. *Sci. Rep.* **2019**, *9*, 6900. [CrossRef]

135. Breton, R.C.; Reynolds, W.F. Using NMR to identify and characterize natural products. *Nat. Prod. Rep.* **2013**, *30*, 501–524. [CrossRef]

136. Parkinson, J.A. Chapter 2—NMR Spectroscopy Methods in Metabolic Phenotyping. In *The Handbook of Metabolic Phenotyping*; Lindon, J.C., Nicholson, J.K., Holmes, E., Eds.; Elsevier: Amsterdam, The Netherlands, 2019; pp. 53–96. [CrossRef]

137. Koskela, H.; Heikkilä, O.; Kilpeläinen, I.; Heikkinen, S. Quantitative two-dimensional HSQC experiment for high magnetic field NMR spectrometers. *J. Magn. Reson.* **2010**, *202*, 24–33. [CrossRef]

138. Dossey, A.T.; Walse, S.S.; Conle, O.V.; Edison, A.S. Parectadial, a Monoterpenoid from the Defensive Spray of Parectatosoma mocquerysi. *J. Nat. Prod.* **2007**, *70*, 1335–1338. [CrossRef]

139. Farag, M.A.; Porzel, A.; Al-Hammady, M.A.; Hegazy, M.-E.F.; Meyer, A.; Mohamed, T.A.; Westphal, H.; Wessjohann, L.A. Soft Corals Biodiversity in the Egyptian Red Sea: A Comparative MS and NMR Metabolomics Approach of Wild and Aquarium Grown Species. *J. Proteome Res.* **2016**, *15*, 1274–1287. [CrossRef] [PubMed]

140. Farag, M.A.; Mahrous, E.A.; Lübken, T.; Porzel, A.; Wessjohann, L. Classification of commercial cultivars of Humulus lupulus L. (hop) by chemometric pixel analysis of two dimensional nuclear magnetic resonance spectra. *Metabolomics* **2014**, *10*, 21–32. [CrossRef]

141. Schroeder, F.C.; Gibson, D.M.; Churchill, A.C.L.; Sojikul, P.; Wursthorn, E.J.; Krasnoff, S.B.; Clardy, J. Differential Analysis of 2D NMR Spectra: New Natural Products from a Pilot-Scale Fungal Extract Library. *Angew. Chem. Int. Ed.* **2007**, *46*, 901–904. [CrossRef] [PubMed]

142. Pungaliya, C.; Srinivasan, J.; Fox, B.W.; Malik, R.U.; Ludewig, A.H.; Sternberg, P.W.; Schroeder, F.C. A shortcut to identifying small molecule signals that regulate behavior and development in *Caenorhabditis elegans*. *Proc. Natl. Acad. Sci. USA* **2009**, *106*, 7708–7713. [CrossRef]

143. Butcher, R.A.; Schroeder, F.C.; Fischbach, M.A.; Straight, P.D.; Kolter, R.; Walsh, C.T.; Clardy, J. The identification of bacillaene, the product of the PksX megacomplex in Bacillus subtilis. *Proc. Natl. Acad. Sci. USA* **2007**, *104*, 1506–1509. [CrossRef]

144. McCarney, E.R.; Dykstra, R.; Galvosas, P. Evaluation of benchtop NMR Diffusion Ordered Spectroscopy for small molecule mixture analysis. *Magn. Reson. Imaging* **2019**, *56*, 103–109. [CrossRef]

145. Blümich, B. Low-field and benchtop NMR. *J. Magn. Reson.* **2019**, *306*, 27–35. [CrossRef]

146. Gouilleux, B.; Marchand, J.; Charrier, B.; Remaud, G.S.; Giraudeau, P. High-throughput authentication of edible oils with benchtop Ultrafast 2D NMR. *Food Chem.* **2018**, *244*, 153–158. [CrossRef]

147. Kruk, J.; Doskocz, M.; Jodłowska, E.; Zacharzewska, A.; Łakomiec, J.; Czaja, K.; Kujawski, J. NMR Techniques in Metabolomic Studies: A Quick Overview on Examples of Utilization. *Appl. Magn. Reson.* **2017**, *48*, 1–21. [CrossRef] [PubMed]

148. Pérez, E.M.S.; Iglesias, M.J.; Ortiz, F.L.; Pérez, I.S.; Galera, M.M. Study of the suitability of HRMAS NMR for metabolic profiling of tomatoes: Application to tissue differentiation and fruit ripening. *Food Chem.* **2010**, *122*, 877–887. [CrossRef]

149. Daolio, C.; Beltrame, F.L.; Ferreira, A.G.; Cass, Q.B.; Cortez, D.A.G.; Ferreira, M.M.C. Classification of commercial Catuaba samples by NMR, HPLC and chemometrics. *Phytochem. Anal.* **2008**, *19*, 218–228. [CrossRef] [PubMed]

150. Broberg, A.; Kenne, L. Use of High-Resolution Magic Angle Spinning Nuclear Magnetic Resonance Spectroscopy for in Situ Studies of Low-Molecular-Mass Compounds in Red Algae. *Anal. Biochem.* **2000**, *284*, 367–374. [CrossRef] [PubMed]

151. Miglietta, M.L.; Lamanna, R. 1H HR-MAS NMR of carotenoids in aqueous samples and raw vegetables. *Magn. Reson. Chem.* **2006**, *44*, 675–685. [CrossRef] [PubMed]

152. Sekiyama, Y.; Chikayama, E.; Kikuchi, J. Profiling Polar and Semipolar Plant Metabolites throughout Extraction Processes Using a Combined Solution-State and High-Resolution Magic Angle Spinning NMR Approach. *Anal. Chem.* **2010**, *82*, 1643–1652. [CrossRef] [PubMed]

153. Van der Graaf, M.; Heerschap, A. Effect of Cation Binding on the Proton Chemical Shifts and the Spin–Spin Coupling Constant of Citrate. *J. Magn. Reson. Ser. B* **1996**, *112*, 58–62. [CrossRef]

154. Jaroszewski, J.W. Hyphenated NMR Methods in Natural Products Research, Part 1: Direct Hyphenation. *Planta Med.* **2005**, *71*, 691–700. [CrossRef]

155. Seger, C.; Sturm, S.; Stuppner, H. Mass spectrometry and NMR spectroscopy: Modern high-end detectors for high resolution separation techniques—State of the art in natural product HPLC-MS, HPLC-NMR, and CE-MS hyphenations. *Nat. Prod. Rep.* **2013**, *30*, 970–987. [CrossRef]

156. Gebretsadik, T.; Linert, W.; Thomas, M.; Berhanu, T.; Frew, R. LC-NMR for Natural Products Analysis: A Journey from an Academic Curiosity to a Robust Analytical Tool. *Sci* **2019**, *1*, 31. [CrossRef]

157. Keifer, P.A. Chemical-shift referencing and resolution stability in methanol:water gradient LC–NMR. *J. Magn. Reson.* **2010**, *205*, 130–140. [CrossRef]

158. Bringmann, G.; Wohlfarth, M.; Rischer, H.; Heubes, M.; Saeb, W.; Diem, S.; Herderich, M.; Schlauer, J. A Photometric Screening Method for Dimeric Naphthylisoquinoline Alkaloids and Complete On-Line Structural Elucidation of a Dimer in Crude Plant Extracts, by the LC–MS/LC–NMR/LC–CD Triad. *Anal. Chem.* **2001**, *73*, 2571–2577. [CrossRef]

159. Waridel, P.; Wolfender, J.-L.; Lachavanne, J.-B.; Hostettmann, K. ent-Labdane glycosides from the aquatic plant Potamogeton lucens and analytical evaluation of the lipophilic extract constituents of various Potamogeton species. *Phytochemistry* **2004**, *65*, 945–954. [CrossRef] [PubMed]

160. Zehl, M.; Braunberger, C.; Conrad, J.; Crnogorac, M.; Krasteva, S.; Vogler, B.; Beifuss, U.; Krenn, L. Identification and quantification of flavonoids and ellagic acid derivatives in therapeutically important Drosera species by LC–DAD, LC–NMR, NMR, and LC–MS. *Anal. Bioanal. Chem.* **2011**, *400*, 2565–2576. [CrossRef]

161. Clarkson, C.; Sibum, M.; Mensen, R.; Jaroszewski, J.W. Evaluation of on-line solid-phase extraction parameters for hyphenated, high-performance liquid chromatography–solid-phase extraction–nuclear magnetic resonance applications. *J. Chromatogr. A* **2007**, *1165*, 1–9. [CrossRef] [PubMed]

162. Griffiths, L.; Horton, R. Optimization of LC–NMR. III—Increased signal-to-noise ratio through column trapping. *Magn. Reson. Chem.* **1998**, *36*, 104–109. [CrossRef]

163. Xu, Y.-J.; Capistrano, R.; Dhooghe, L.; Foubert, K.; Lemière, F.; Maregesi, S.; Baldé, A.; Apers, S.; Pieters, L. Herbal Medicines and Infectious Diseases: Characterization by LC-SPE-NMR of Some Medicinal Plant Extracts Used against Malaria. *Planta Med.* **2011**, *77*, 1139–1148. [CrossRef] [PubMed]

164. Castro, A.; Moco, S.; Coll, J.; Vervoort, J. LC-MS-SPE-NMR for the Isolation and Characterization of neo-Clerodane Diterpenoids from Teucrium luteum subsp. flavovirens. *J. Nat. Prod.* **2010**, *73*, 962–965. [CrossRef]

165. Gao, H.; Zehl, M.; Kaehlig, H.; Schneider, P.; Stuppner, H.; Moreno, Y.; Banuls, L.; Kiss, R.; Kopp, B. Rapid Structural Identification of Cytotoxic Bufadienolide Sulfates in Toad Venom from Bufo melanosticus by LC-DAD-MSn and LC-SPE-NMR. *J. Nat. Prod.* **2010**, *73*, 603–608. [CrossRef]

166. Chen, C.-K.; Lin, F.-H.; Tseng, L.-H.; Jiang, C.-L.; Lee, S.-S. Comprehensive Study of Alkaloids from Crinum asiaticum var. sinicum Assisted by HPLC-DAD-SPE-NMR. *J. Nat. Prod.* **2011**, *74*, 411–419. [CrossRef]

167. Lai, Y.-C.; Chen, C.-K.; Lin, W.-W.; Lee, S.-S. A comprehensive investigation of anti-inflammatory diarylheptanoids from the leaves of Alnus formosana. *Phytochemistry* **2012**, *73*, 84–94. [CrossRef]

168. Clarkson, C.; Hansen, S.H.; Smith, P.J.; Jaroszewski, J.W. Identification of Major and Minor Constituents of Harpagophytum procumbens (Devil's Claw) Using HPLC-SPE-NMR and HPLC-ESIMS/APCIMS. *J. Nat. Prod.* **2006**, *69*, 1280–1288. [CrossRef]

169. Nyberg, N.T.; Baumann, H.; Kenne, L. Solid-Phase Extraction NMR Studies of Chromatographic Fractions of Saponins from Quillaja saponaria. *Anal. Chem.* **2003**, *75*, 268–274. [CrossRef] [PubMed]

170. Kuhlisch, C.; Pohnert, G. Metabolomics in chemical ecology. *Nat. Prod. Rep.* **2015**, *32*. [CrossRef] [PubMed]

171. Kongstad, K.T.; Özdemir, C.; Barzak, A.; Wubshet, S.G.; Staerk, D. Combined Use of High-Resolution α-Glucosidase Inhibition Profiling and High-Performance Liquid Chromatography–High-Resolution Mass Spectrometry–Solid-Phase Extraction–Nuclear Magnetic Resonance Spectroscopy for Investigation of Antidiabetic Principles in Crude Plant Extracts. *J. Agric. Food Chem.* **2015**, *63*, 2257–2263. [CrossRef] [PubMed]

172. Agnolet, S.; Jaroszewski, J.W.; Verpoorte, R.; Staerk, D. 1H NMR-based metabolomics combined with HPLC-PDA-MS-SPE-NMR for investigation of standardized Ginkgo biloba preparations. *Metabolomics* **2010**, *6*, 292–302. [CrossRef]

173. Tang, H.; Xiao, C.; Wang, Y. Important roles of the hyphenated HPLC-DAD-MS-SPE-NMR technique in metabonomics. *Magn. Reson. Chem.* **2009**, *47*, S157–S162. [CrossRef] [PubMed]

174. Staerk, D.; Kesting, J.R.; Sairafianpour, M.; Witt, M.; Asili, J.; Emami, S.A.; Jaroszewski, J.W. Accelerated dereplication of crude extracts using HPLC–PDA–MS–SPE–NMR: Quinolinone alkaloids of Haplophyllum acutifolium. *Phytochemistry* **2009**, *70*, 1055–1061. [CrossRef] [PubMed]

175. Lima, R.d.C.L.; Gramsbergen, S.M.; Van Staden, J.; Jäger, A.K.; Kongstad, K.T.; Staerk, D. Advancing HPLC-PDA-HRMS-SPE-NMR Analysis of Coumarins in Coleonema album by Use of Orthogonal Reversed-Phase C18 and Pentafluorophenyl Separations. *J. Nat. Prod.* **2017**, *80*, 1020–1027. [CrossRef]

176. Liu, B.; Kongstad, K.T.; Qinglei, S.; Nyberg, N.T.; Jäger, A.K.; Staerk, D. Dual High-Resolution α-Glucosidase and Radical Scavenging Profiling Combined with HPLC-HRMS-SPE-NMR for Identification of Minor and Major Constituents Directly from the Crude Extract of Pueraria lobata. *J. Nat. Prod.* **2015**, *78*, 294–300. [CrossRef]

177. Luedemann, A.; Strassburg, K.; Erban, A.; Kopka, J. TagFinder for the quantitative analysis of gas chromatography—Mass spectrometry (GC-MS)-based metabolite profiling experiments. *Bioinformatics* **2008**, *24*, 732–737. [CrossRef]

178. Cuadros-Inostroza, Á.; Caldana, C.; Redestig, H.; Kusano, M.; Lisec, J.; Peña-Cortés, H.; Willmitzer, L.; Hannah, M.A. TargetSearch—A Bioconductor package for the efficient preprocessing of GC-MS metabolite profiling data. *BMC Bioinform.* **2009**, *10*, 428. [CrossRef]

179. Tsugawa, H.; Cajka, T.; Kind, T.; Ma, Y.; Higgins, B.; Ikeda, K.; Kanazawa, M.; VanderGheynst, J.; Fiehn, O.; Arita, M. MS-DIAL: Data-independent MS/MS deconvolution for comprehensive metabolome analysis. *Nat. Methods* **2015**, *12*, 523. [CrossRef] [PubMed]

180. Matsuo, T.; Tsugawa, H.; Miyagawa, H.; Fukusaki, E. Integrated Strategy for Unknown EI–MS Identification Using Quality Control Calibration Curve, Multivariate Analysis, EI–MS Spectral Database, and Retention Index Prediction. *Anal. Chem.* **2017**, *89*, 6766–6773. [CrossRef] [PubMed]

181. Beale, D.J.; Pinu, F.R.; Kouremenos, K.A.; Poojary, M.M.; Narayana, V.K.; Boughton, B.A.; Kanojia, K.; Dayalan, S.; Jones, O.A.H.; Dias, D.A. Review of recent developments in GC–MS approaches to metabolomics-based research. *Metabolomics* **2018**, *14*, 152. [CrossRef] [PubMed]

182. Fiehn, O. Metabolomics by Gas Chromatography—Mass Spectrometry: Combined Targeted and Untargeted Profiling. *Curr. Protoc. Mol. Biol.* **2016**, *114*, 30.4.1–30.4.32. [CrossRef] [PubMed]

183. Smith, C.A.; Want, E.J.; O'Maille, G.; Abagyan, R.; Siuzdak, G. XCMS: Processing Mass Spectrometry Data for Metabolite Profiling Using Nonlinear Peak Alignment, Matching, and Identification. *Anal. Chem.* **2006**, *78*, 779–787. [CrossRef]

184. Pluskal, T.; Castillo, S.; Villar-Briones, A.; Orešič, M. MZmine 2: Modular framework for processing, visualizing, and analyzing mass spectrometry-based molecular profile data. *BMC Bioinform.* **2010**, *11*, 395. [CrossRef] [PubMed]

185. Röst, H.L.; Sachsenberg, T.; Aiche, S.; Bielow, C.; Weisser, H.; Aicheler, F.; Andreotti, S.; Ehrlich, H.-C.; Gutenbrunner, P.; Kenar, E.; et al. OpenMS: A flexible open-source software platform for mass spectrometry data analysis. *Nat. Methods* **2016**, *13*, 741–748. [CrossRef]

186. Schiffman, C.; Petrick, L.; Perttula, K.; Yano, Y.; Carlsson, H.; Whitehead, T.; Metayer, C.; Hayes, J.; Rappaport, S.; Dudoit, S. Filtering procedures for untargeted LC-MS metabolomics data. *BMC Bioinform.* **2019**, *20*, 334. [CrossRef] [PubMed]

187. Nothias, L.-F.; Nothias-Esposito, M.; da Silva, R.; Wang, M.; Protsyuk, I.; Zhang, Z.; Sarvepalli, A.; Leyssen, P.; Touboul, D.; Costa, J.; et al. Bioactivity-Based Molecular Networking for the Discovery of Drug Leads in Natural Product Bioassay-Guided Fractionation. *J. Nat. Prod.* **2018**, *81*, 758–767. [CrossRef]

188. Kuhl, C.; Tautenhahn, R.; Böttcher, C.; Larson, T.R.; Neumann, S. CAMERA: An Integrated Strategy for Compound Spectra Extraction and Annotation of Liquid Chromatography/Mass Spectrometry Data Sets. *Anal. Chem.* **2012**, *84*, 283–289. [CrossRef]

189. Li, H.; Cai, Y.; Guo, Y.; Chen, F.; Zhu, Z.-J. MetDIA: Targeted Metabolite Extraction of Multiplexed MS/MS Spectra Generated by Data-Independent Acquisition. *Anal. Chem.* **2016**, *88*, 8757–8764. [CrossRef] [PubMed]

190. Wolfender, J.-L.; Nuzillard, J.-M.; van der Hooft, J.J.J.; Renault, J.-H.; Bertrand, S. Accelerating Metabolite Identification in Natural Product Research: Toward an Ideal Combination of Liquid Chromatography–High-Resolution Tandem Mass Spectrometry and NMR Profiling, in Silico Databases, and Chemometrics. *Anal. Chem.* **2019**, *91*, 704–742. [CrossRef] [PubMed]

191. Guijas, C.; Montenegro-Burke, J.R.; Domingo-Almenara, X.; Palermo, A.; Warth, B.; Hermann, G.; Koellensperger, G.; Huan, T.; Uritboonthai, W.; Aisporna, A.E.; et al. METLIN: A Technology Platform for Identifying Knowns and Unknowns. *Anal. Chem.* **2018**, *90*, 3156–3164. [CrossRef]

192. Horai, H.; Arita, M.; Kanaya, S.; Nihei, Y.; Ikeda, T.; Suwa, K.; Ojima, Y.; Tanaka, K.; Tanaka, S.; Aoshima, K.; et al. MassBank: A public repository for sharing mass spectral data for life sciences. *J. Mass Spectrom.* **2010**, *45*, 703–714. [CrossRef]

193. Wishart, D.S.; Feunang, Y.D.; Marcu, A.; Guo, A.C.; Liang, K.; Vázquez-Fresno, R.; Sajed, T.; Johnson, D.; Li, C.; Karu, N.; et al. HMDB 4.0: The human metabolome database for 2018. *Nucleic Acids Res.* **2017**, *46*, D608–D617. [CrossRef] [PubMed]

194. Scalbert, A.; Brennan, L.; Manach, C.; Andres-Lacueva, C.; Dragsted, L.O.; Draper, J.; Rappaport, S.M.; van der Hooft, J.J.; Wishart, D.S. The food metabolome: A window over dietary exposure. *Am. J. Clin. Nutr.* **2014**, *99*, 1286–1308. [CrossRef] [PubMed]

195. Blaženović, I.; Kind, T.; Torbašinović, H.; Obrenović, S.; Mehta, S.S.; Tsugawa, H.; Wermuth, T.; Schauer, N.; Jahn, M.; Biedendieck, R.; et al. Comprehensive comparison of in silico MS/MS fragmentation tools of the CASMI contest: Database boosting is needed to achieve 93% accuracy. *J. Cheminformatics* **2017**, *9*, 32. [CrossRef]

196. Dührkop, K.; Shen, H.; Meusel, M.; Rousu, J.; Böcker, S. Searching molecular structure databases with tandem mass spectra using CSI:FingerID. *Proc. Natl. Acad. Sci. USA* **2015**, *112*, 12580–12585. [CrossRef]

197. Tsugawa, H.; Kind, T.; Nakabayashi, R.; Yukihira, D.; Tanaka, W.; Cajka, T.; Saito, K.; Fiehn, O.; Arita, M. Hydrogen Rearrangement Rules: Computational MS/MS Fragmentation and Structure Elucidation Using MS-FINDER Software. *Anal. Chem.* **2016**, *88*, 7946–7958. [CrossRef]

198. Stanstrup, J.; Neumann, S.; Vrhovšek, U. PredRet: Prediction of Retention Time by Direct Mapping between Multiple Chromatographic Systems. *Anal. Chem.* **2015**, *87*, 9421–9428. [CrossRef]

199. Afendi, F.M.; Okada, T.; Yamazaki, M.; Hirai-Morita, A.; Nakamura, Y.; Nakamura, K.; Ikeda, S.; Takahashi, H.; Altaf-Ul-Amin, M.; Darusman, L.K.; et al. KNApSAcK Family Databases: Integrated Metabolite–Plant Species Databases for Multifaceted Plant Research. *Plant Cell Physiol.* **2011**, *53*, e1. [CrossRef]

200. Murray, K.K.; Boyd, R.K.; Eberlin, M.N.; Langley, G.J.; Li, L.; Naito, Y. Definitions of terms relating to mass spectrometry (IUPAC Recommendations 2013). *Pure Appl. Chem.* **2013**, *85*, 1515–1609. [CrossRef]

201. Smith, C.; O'Maille, G.; Want, E.J.; Qin, C.; Trauger, S.A.; Brandon, T.R.; Custodio, D.E.; Abagyan, R.; Siuzdak, G. METLIN: A metabolite mass spectral database. *Ther. Drug Monit.* **2005**, *27*, 747–751. [CrossRef]

202. Sawada, Y.; Nakabayashi, R.; Yamada, Y.; Suzuki, M.; Sato, M.; Sakata, A.; Akiyama, K.; Sakurai, T.; Matsuda, F.; Aoki, T. RIKEN tandem mass spectral database (ReSpect) for phytochemicals: A plant-specific MS/MS-based data resource and database. *Phytochemistry* **2012**, *82*, 38–45. [CrossRef] [PubMed]

203. Mohimani, H.; Gurevich, A.; Mikheenko, A.; Garg, N.; Nothias, L.F.; Ninomiya, A.; Takada, K.; Dorrestein, P.C.; Pevzner, P.A. Dereplication of peptidic natural products through database search of mass spectra. *Nat. Chem. Biol.* **2017**, *13*, 30–37. [CrossRef] [PubMed]

204. Pan, Z.; Raftery, D. Comparing and combining NMR spectroscopy and mass spectrometry in metabolomics. *Anal. Bioanal. Chem.* **2007**, *387*, 525–527. [CrossRef]

205. Gowda, G.N.; Zhang, S.; Gu, H.; Asiago, V.; Shanaiah, N.; Raftery, D. Metabolomics-based methods for early disease diagnostics. *Expert Rev. Mol. Diagn.* **2008**, *8*, 617–633. [CrossRef]

206. Marshall, D.D.; Powers, R. Beyond the paradigm: Combining mass spectrometry and nuclear magnetic resonance for metabolomics. *Prog. Nucl. Magn. Reson. Spectrosc.* **2017**, *100*, 1–16. [CrossRef]

207. Lewis, I.A.; Schommer, S.C.; Markley, J.L. rNMR: Open source software for identifying and quantifying metabolites in NMR spectra. *Magn. Reson. Chem.* **2009**, *47*, S123–S126. [CrossRef]

208. Chikayama, E.; Sekiyama, Y.; Okamoto, M.; Nakanishi, Y.; Tsuboi, Y.; Akiyama, K.; Saito, K.; Shinozaki, K.; Kikuchi, J. Statistical indices for simultaneous large-scale metabolite detections for a single NMR spectrum. *Anal. Chem.* **2010**, *82*, 1653–1658. [CrossRef]

209. Bingol, K.; Bruschweiler-Li, L.; Li, D.-W.; Bruschweiler, R. Customized metabolomics database for the analysis of NMR 1H–1H TOCSY and 13C–1H HSQC-TOCSY spectra of complex mixtures. *Anal. Chem.* **2014**, *86*, 5494–5501. [CrossRef] [PubMed]

210. Bingol, K.; Zhang, F.; Bruschweiler-Li, L.; Bruschweiler, R. TOCCATA: A customized carbon total correlation spectroscopy NMR metabolomics database. *Anal. Chem.* **2012**, *84*, 9395–9401. [CrossRef] [PubMed]

211. López-Pérez, J.L.; Therón, R.; del Olmo, E.; Díaz, D. NAPROC-13: A database for the dereplication of natural product mixtures in bioassay-guided protocols. *Bioinformatics* **2007**, *23*, 3256–3257. [CrossRef] [PubMed]

212. Steinbeck, C.; Krause, S.; Kuhn, S. NMRShiftDB constructing a free chemical information system with open-source components. *J. Chem. Inf. Comput. Sci.* **2003**, *43*, 1733–1739. [CrossRef] [PubMed]

213. Steinbeck, C.; Kuhn, S. NMRShiftDB–compound identification and structure elucidation support through a free community-built web database. *Phytochemistry* **2004**, *65*, 2711–2717. [CrossRef]

214. Yamamoto, O.; Someno, K.; Wasada, N.; Hiraishi, J.; Hayamizu, K.; Tanabe, K.; Tamura, T.; Yanagisawa, M. An integrated spectral data base system including IR, MS, 1H-NMR, 13C-NMR, ESR and Raman spectra. *Anal. Sci.* **1988**, *4*, 233–239. [CrossRef]

215. Ludwig, C.; Easton, J.M.; Lodi, A.; Tiziani, S.; Manzoor, S.E.; Southam, A.D.; Byrne, J.J.; Bishop, L.M.; He, S.; Arvanitis, T.N. Birmingham metabolite library: A publicly accessible database of 1-D ^{1}H and 2-D ^{1}H J-resolved NMR spectra of authentic metabolite standards (BML-NMR). *Metabolomics* **2012**, *8*, 8–18. [CrossRef]

216. Ulrich, E.; Akutsu, H.; Doreleijers, J.; Harano, Y.; Ioannidis, Y.; Lin, J.; Livny, M.; Mading, S.; Maziuk, D.; Miller, Z. BioMagResBank. *Nucleic Acids Res.* **2008**, *36*, D402–D408. [CrossRef] [PubMed]

217. Yanshole, V.V.; Snytnikova, O.A.; Kiryutin, A.S.; Yanshole, L.V.; Sagdeev, R.Z.; Tsentalovich, Y.P. Metabolomics of the rat lens: A combined LC-MS and NMR study. *Exp. Eye Res.* **2014**, *125*, 71–78. [CrossRef]

218. Bingol, K.; Bruschweiler-Li, L.; Yu, C.; Somogyi, A.; Zhang, F.; Brüschweiler, R. Metabolomics beyond spectroscopic databases: A combined MS/NMR strategy for the rapid identification of new metabolites in complex mixtures. *Anal. Chem.* **2015**, *87*, 3864–3870. [CrossRef]

219. Bingol, K.; Brüschweiler, R. Two elephants in the room: New hybrid nuclear magnetic resonance and mass spectrometry approaches for metabolomics. *Curr. Opin. Clin. Nutr. Metab. Care* **2015**, *18*, 471. [CrossRef]

220. Wohlleben, W.; Mast, Y.; Muth, G.; Röttgen, M.; Stegmann, E.; Weber, T. Synthetic biology of secondary metabolite biosynthesis in actinomycetes: Engineering precursor supply as a way to optimize antibiotic production. *FEBS Lett.* **2012**, *586*, 2171–2176. [CrossRef] [PubMed]

221. Jankevics, A.; Merlo, M.E.; de Vries, M.; Vonk, R.J.; Takano, E.; Breitling, R. Metabolomic analysis of a synthetic metabolic switch in Streptomyces coelicolor A3 (2). *Proteomics* **2011**, *11*, 4622–4631. [CrossRef] [PubMed]

222. Breitling, R.; Achcar, F.; Takano, E. Modeling challenges in the synthetic biology of secondary metabolism. *ACS Synth. Biol.* **2013**, *2*, 373–378. [CrossRef] [PubMed]

223. Cheng, J.-S.; Liang, Y.-Q.; Ding, M.-Z.; Cui, S.-F.; Lv, X.-M.; Yuan, Y.-J. Metabolic analysis reveals the amino acid responses of Streptomyces lydicus to pitching ratios during improving streptolydigin production. *Appl. Microbiol. Biotechnol.* **2013**, *97*, 5943–5954. [CrossRef]

224. Arendt, P.; Pollier, J.; Callewaert, N.; Goossens, A. Synthetic biology for production of natural and new-to-nature terpenoids in photosynthetic organisms. *Plant J. Cell Mol. Biol.* **2016**, *87*, 16–37. [CrossRef]

225. Julsing, M.K.; Koulman, A.; Woerdenbag, H.J.; Quax, W.J.; Kayser, O. Combinatorial biosynthesis of medicinal plant secondary metabolites. *Biomol. Eng.* **2006**, *23*, 265–279. [CrossRef]

226. Pollier, J.; Moses, T.; Goossens, A. Combinatorial biosynthesis in plants: A (p)review on its potential and future exploitation. *Nat. Prod. Rep.* **2011**, *28*, 1897–1916. [CrossRef]

227. Umeno, D.; Arnold, F.H. A C35 carotenoid biosynthetic pathway. *Appl. Environ. Microbiol.* **2003**, *69*, 3573–3579. [CrossRef] [PubMed]

228. Runguphan, W.; Qu, X.; O'Connor, S.E. Integrating carbon-halogen bond formation into medicinal plant metabolism. *Nature* **2010**, *468*, 461–464. [CrossRef]

229. Moses, T.; Papadopoulou, K.K.; Osbourn, A. Metabolic and functional diversity of saponins, biosynthetic intermediates and semi-synthetic derivatives. *Crit. Rev. Biochem. Mol. Biol.* **2014**, *49*, 439–462. [CrossRef]

230. Fuentes, P.; Zhou, F.; Erban, A.; Karcher, D.; Kopka, J.; Bock, R. A new synthetic biology approach allows transfer of an entire metabolic pathway from a medicinal plant to a biomass crop. *eLife* **2016**, *5*, e13664. [CrossRef]

231. Hiller, K.; Metallo, C.M.; Kelleher, J.K.; Stephanopoulos, G. Nontargeted elucidation of metabolic pathways using stable-isotope tracers and mass spectrometry. *Anal. Chem.* **2010**, *82*, 6621–6628. [CrossRef] [PubMed]

232. Creek, D.J.; Chokkathukalam, A.; Jankevics, A.; Burgess, K.E.; Breitling, R.; Barrett, M.P. Stable isotope-assisted metabolomics for network-wide metabolic pathway elucidation. *Anal. Chem.* **2012**, *84*, 8442–8447. [CrossRef] [PubMed]

233. Bueschl, C.; Krska, R.; Kluger, B.; Schuhmacher, R. Isotopic labeling-assisted metabolomics using LC–MS. *Anal. Bioanal. Chem.* **2013**, *405*, 27–33. [CrossRef] [PubMed]

234. Ellis, D.I.; Goodacre, R. Metabolomics-assisted synthetic biology. *Curr. Opin. Biotechnol.* **2012**, *23*, 22–28. [CrossRef] [PubMed]

235. Moldenhauer, J.; Chen, X.H.; Borriss, R.; Piel, J. Biosynthesis of the antibiotic bacillaene, the product of a giant polyketide synthase complex of the trans-AT family. *Angew. Chem. Int. Ed.* **2007**, *46*, 8195–8197. [CrossRef] [PubMed]

236. Blunt, J.W.; Munro, M.H. Data, 1 H-NMR databases, data manipulation. *Phytochem. Rev.* **2013**, *12*, 435–447. [CrossRef]

237. Buckingham, J. *Dictionary of Natural Products on DVD*; Version 21: 1; Chapman & Hall/CRC Press: Boca Raton, FL, USA, 2012.

238. Purves, K.; Macintyre, L.; Brennan, D.; Hreggviðsson, G.; Kuttner, E.; Ásgeirsdóttir, M.; Young, L.; Green, D.; Edrada-Ebel, R.; Duncan, K. Using molecular networking for microbial secondary metabolite bioprospecting. *Metabolites* **2016**, *6*, 2. [CrossRef]

239. Funasaki, M.; Menezes, I.S.; dos Santos BARROSO, H.; Zanotto, S.P.; Carioca, C.R.F. Tocopherol profile of Brazil nut oil from different geographic areas of the Amazon region. *Acta Amaz.* **2012**, *43*. [CrossRef]

240. Borges, C.V.; de Oliveira Amorim, V.B.; Ramlov, F.; da Silva Ledo, C.A.; Donato, M.; Maraschin, M.; Amorim, E.P. Characterisation of metabolic profile of banana genotypes, aiming at biofortified *Musa* spp. cultivars. *Food Chem.* **2014**, *145*, 496–504. [CrossRef] [PubMed]

241. da Silva Taveira, J.H.; Borém, F.M.; Figueiredo, L.P.; Reis, N.; Franca, A.S.; Harding, S.A.; Tsai, C.-J. Potential markers of coffee genotypes grown in different Brazilian regions: A metabolomics approach. *Food Res. Int.* **2014**, *61*, 75–82. [CrossRef]

242. Pilon, A.C.; Carnevale Neto, F.; Freire, R.T.; Cardoso, P.; Carneiro, R.L.; Da Silva Bolzani, V.; Castro-Gamboa, I. Partial least squares model and design of experiments toward the analysis of the metabolome of Jatropha gossypifolia leaves: Extraction and chromatographic fingerprint optimization. *J. Sep. Sci.* **2016**, *39*, 1023–1030. [CrossRef] [PubMed]

243. Silva, M.M.; Bergamasco, J.; Lira, S.P.; Lopes, N.P.; Hajdu, E.; Peixinho, S.; Berlinck, R.G. Dereplication of bromotyrosine-derived metabolites by LC-PDA-MS and analysis of the chemical profile of 14 aplysina sponge specimens from the Brazilian coastline. *Aust. J. Chem.* **2010**, *63*, 886–894. [CrossRef]

244. Bittencourt, M.L.; Ribeiro, P.R.; Franco, R.L.; Hilhorst, H.W.; de Castro, R.D.; Fernandez, L.G. Metabolite profiling, antioxidant and antibacterial activities of Brazilian propolis: Use of correlation and multivariate analyses to identify potential bioactive compounds. *Food Res. Int.* **2015**, *76*, 449–457. [CrossRef] [PubMed]

245. Castro, C.B.; Pires, D.O. Brazilian coral reefs: What we already know and what is still missing. *Bull. Mar. Sci.* **2001**, *69*, 357–371.

246. Funari, C.S.; Eugster, P.J.; Martel, S.; Carrupt, P.-A.; Wolfender, J.-L.; Silva, D.H.S. High resolution ultra high pressure liquid chromatography–time-of-flight mass spectrometry dereplication strategy for the metabolite profiling of Brazilian Lippia species. *J. Chromatogr. A* **2012**, *1259*, 167–178. [CrossRef]

247. Chagas-Paula, D.A.; Oliveira, T.B.; Zhang, T.; Edrada-Ebel, R.; Da Costa, F.B. Prediction of anti-inflammatory plants and discovery of their biomarkers by machine learning algorithms and metabolomic studies. *Planta Med.* **2015**, *81*, 450–458. [CrossRef]

248. Wang, Y.; Tang, H.; Nicholson, J.K.; Hylands, P.J.; Sampson, J.; Whitcombe, I.; Stewart, C.G.; Caiger, S.; Oru, I.; Holmes, E. Metabolomic strategy for the classification and quality control of phytomedicine: A case study of chamomile flower (*Matricaria recutita* L.). *Planta Med.* **2004**, *70*, 250–255.

249. Choi, Y.H.; Kim, H.K.; Hazekamp, A.; Erkelens, C.; Lefeber, A.W.; Verpoorte, R. Metabolomic Differentiation of Cannabis s ativa Cultivars Using 1H NMR Spectroscopy and Principal Component Analysis. *J. Nat. Prod.* **2004**, *67*, 953–957. [CrossRef] [PubMed]

250. Rasmussen, B.; Cloarec, O.; Tang, H.; Stærk, D.; Jaroszewski, J.W. Multivariate analysis of integrated and full-resolution 1H-NMR spectral data from complex pharmaceutical preparations: St. John's wort. *Planta Med.* **2006**, *72*, 556–563. [CrossRef] [PubMed]

251. Van der Kooy, F.; Verpoorte, R.; Meyer, J.M. Metabolomic quality control of claimed anti-malarial *Artemisia afra* herbal remedy and *A. afra* and *A. annua* plant extracts. *S. Afr. J. Bot.* **2008**, *74*, 186–189. [CrossRef]

252. Johnson, C.H.; Ivanisevic, J.; Siuzdak, G. Metabolomics: Beyond biomarkers and towards mechanisms. *Nat. Rev. Mol. Cell Biol.* **2016**, *17*, 451–459. [CrossRef] [PubMed]

253. Zhao, Q.; Zhang, J.-L.; Li, F. Application of Metabolomics in the Study of Natural Products. *Nat. Prod. Bioprospecting* **2018**, *8*, 1–14. [CrossRef] [PubMed]

 *metabolites* 

Review

Experimental Design and Sample Preparation in Forest Tree Metabolomics

Ana M. Rodrigues [1], Ana I. Ribeiro-Barros [1,2] and Carla António [1,*

[1] Plant Metabolomics Laboratory, Instituto de Tecnologia Química e Biológica António Xavier, Universidade Nova de Lisboa (ITQB NOVA), 2780-157 Oeiras, Portugal; amrodrigues@itqb.unl.pt (A.M.R.); aribeiro@isa.ulisboa.pt (A.I.R.-B.)

[2] Plant Stress and Biodiversity Laboratory, Linking Landscape, Environment, Agriculture and Food (LEAF), Instituto Superior de Agronomia, Universidade de Lisboa (ISA/ULisboa), 1349-017 Lisboa, Portugal

* Correspondence: antonio@itqb.unl.pt

Received: 4 October 2019; Accepted: 20 November 2019; Published: 22 November 2019

Abstract: Appropriate experimental design and sample preparation are key steps in metabolomics experiments, highly influencing the biological interpretation of the results. The sample preparation workflow for plant metabolomics studies includes several steps before metabolite extraction and analysis. These include the optimization of laboratory procedures, which should be optimized for different plants and tissues. This is particularly the case for trees, whose tissues are complex matrices to work with due to the presence of several interferents, such as oleoresins, cellulose. A good experimental design, tree tissue harvest conditions, and sample preparation are crucial to ensure consistency and reproducibility of the metadata among datasets. In this review, we discuss the main challenges when setting up a forest tree metabolomics experiment for mass spectrometry (MS)-based analysis covering all technical aspects from the biological question formulation and experimental design to sample processing and metabolite extraction and data acquisition. We also highlight the importance of forest tree metadata standardization in metabolomics studies.

Keywords: plant metabolomics; forestry; trees; mass spectrometry; metabolite extraction; GC-MS; LC-MS; metadata standardization; databases

---

1. Introduction

Metabolomics is an "omics" technology used to obtain comprehensive information on the metabolome: a diverse pool of low molecular weight molecules (metabolites), present in a cell or organism, and at a particular physiological or developmental stage [1]. For the past 20 years, the number of mass spectrometry (MS)-based metabolomics studies in plants has grown exponentially and plant metabolomics has established itself as a powerful tool to address biological questions related to plant growth and development and plant responses to environmental perturbations [2,3]. Despite the continuous advances in MS technology, the coverage of the plant metabolome is a major challenge in plant metabolomics research mainly due to the high chemical diversity, broad dynamic range of concentration, and specific cellular compartmentalization of metabolites. In addition, no single analytical technology can cover the entire plant metabolome, and different extraction techniques and combinations of complementary analytical technologies are often employed [4]. In general, preparing a plant sample for a metabolomics study involves the establishment of a good experimental design, followed by several standard steps for sample preparation, namely: harvest immediately followed by quenching, aliquot weighing, metabolite extraction, pre-analytical procedures (if required, e.g., chemical derivatization), and finally, metabolite analysis [2,5–7]. The standardization of these metabolomics workflows ensures data consistency and allows the reproducibility of the generated data and metadata (information about data origins). Although most steps are common to any metabolomics experiment, the optimization of laboratory procedures is often adopted, according to the requirement of the sample (species or tissue) under study. This is particularly the case of metabolomics studies on tree species. Forest tree metabolomics represents additional challenges when compared to other plant metabolomics studies. These include an experimental design that takes into account the long life cycle and the genetic variability of forest tree species as well the presence of interferents that can require additional steps during sample preparation (e.g., additional concentration steps) [8]. In this review, we highlight the major challenges when setting up an MS-based forest tree metabolomics experiment. Although this review is focused mainly on forest tree species, the methodology here reviewed can be applied to other woody species.

2. Experimental Design for Forest Tree Metabolomics

In a plant metabolomics study, after the formulation of the biological question, experimental design planning is the first crucial step of the metabolomics workflow. The experimental design includes the complete planning of the experiment, including plant growth conditions and the treatments to be applied to the plants. In this section, all the critical steps and important decisions for a good experimental design are discussed.

2.1. Biological Question Formulation

A plant metabolomics experiment starts with the formulation of a good hypothesis (i.e., biological question) to plan an appropriate experimental design, sample preparation, and statistical strategies for data analysis. Without a clear biological question, the observed changes can be misinterpreted or have multiple possible interpretations that would not reveal important information related to the biological system. Thus, it is absolutely crucial to understand the biological system under study to not only select the suitable tissue(s) for analysis but also the appropriate controls. Understanding the biological system will allow the elaboration of an accurate experimental design, and ultimately, to answer the biological question. It is important to highlight that frequently (and wrongly) experiments are designed for other "omics" technologies (i.e., transcriptomics, proteomics), and the leftover samples are later used for metabolomics analysis. This can extensively compromise the entire metabolomics analysis because the objective of the study might be different, the number of replicates may not be sufficient, or the sample storage conditions were not ideal, thereby affecting the stability of metabolites within the sample [9].

Forests represent a crucial driver to achieve the sustainable development goals (SDG) from Agenda 2030 of the United Nations through the provision of a wide range of ecosystem goods and services with a direct impact on socio-economic development and environmental balance [10,11]. In addition to the direct economic benefits provided by tree species, i.e., timber and non-timber products, gaming and tourism, forests have an immensurable ecological value, being the major determinants for water, oxygen, carbon, and energy balance and can be seen as a major opportunity to mitigate climate change effects [12], i.e., continued drought, increased soil and water salinization and acidification, and intensification of extreme temperatures [13]. In forest tree metabolomics research, most biological questions are indeed related to the responses towards the acclimation and adaptation to a permanently changing environment [14–26] as well as to the identification of potentially active components in tree species of pharmacological, agricultural, environmental, or industrial importance [27–33].

2.2. Experimental Design

The experimental design should ensure that the analytical data derived from the collected biological material would allow answering the initially proposed biological question through a reliable statistical analysis. Therefore, the experimental design (Figure 1) typically includes all variables of the experiment, from the plant growth and treatments (e.g., plant growth conditions, randomization, replicates, controls), sample preparation conditions (e.g., harvested tissue, quenching method, pool material or not, metabolite extraction protocol), and analytical platform (e.g., GC-MS, LC-MS, mass spectrometry imaging, targeted or untargeted approach) to statistical treatments [7,9,34]. Added to these factors, all sources of additional variation (e.g., genotype, sample size, tissue selection, developmental stage, environmental conditions, batch/block effect) should be investigated and minimized to avoid misleading conclusions [7,9,35]. The experimental design should also take into account the time frame of the metabolomics experiment. Because metabolites are highly dynamic (in time and space), a metabolomics study can reflect the steady state (or instant snap-shot) of the metabolism or its dynamic time-course evaluation [9,36–38]. In plant metabolomics, due to the destructive nature of the sampling procedure, most conducted studies are transversal (i.e., cross-sectional), where different samples are used for each time point, whereas in human metabolomics, longitudinal studies are fairly common [39,40]. Even if the harvesting procedure is not completely destructive, the wounding effect in plants should be taken into account as it can affect metabolite profiles. Longitudinal studies in plant metabolomics include the analysis of volatile organic compounds (VOCs) through non-destructive headspace techniques (further details in Section 3).

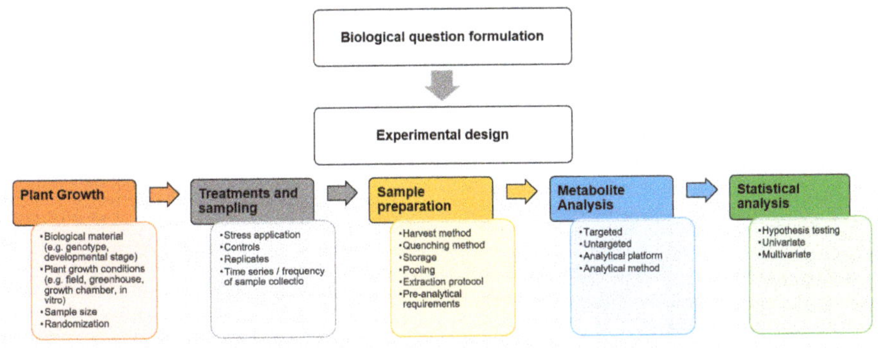

Figure 1. Experimental design and workflow in a plant metabolomics experiment.

Forest tree experiments are particularly difficult to execute, mainly because of the tree's long life cycle and lack of genomic tools [41], which in turn leads to highly costly, and time-consuming long-term studies. Thus, a rigorously elaborated experimental design can help to control time and costs and assure that the experiment and respective derived data are reliable and reproducible [42].

2.2.1. Experimental Conditions

The experimental design should clearly define the experimental conditions of the study (i.e., plant growth conditions and treatment(s) to be applied). Plants can be grown under controlled environmental conditions (e.g., growth chambers, nurseries, greenhouses) or in field conditions. From growth chamber to field conditions, there is a gradual decrease in the level of environmental control and a gradual increase in its complexity. Therefore, most metabolomics studies are essentially comparative, i.e., controls (healthy and/or mock treatments in the case of plant–pathogen interactions) vs. treated samples, and always provided that plants are grown under the same conditions [35]. However, in the field, plants are subjected to uncontrolled variations in the environment (e.g., variations in light intensity, temperature, water availability). Despite the plant's metabolism degree of plasticity acting as a buffer against sudden fluctuations in the environment, this complex set of variables deeply impact the plant's physiology and metabolism [9], which is often the case of forest tree long-term research. Hence, care must be taken when making comparisons amongst field-grown individuals or even when extrapolating results or establishing correlations between trees grown under controlled and uncontrolled (i.e., field) environmental conditions [43].

In vitro assays are an alternative biotechnological approach to in vivo field studies as it drastically reduces the time needed for the experiment to be conducted and eliminates environmental related fluctuations, allowing the manipulation of single variables in a controlled environment, which is impossible to achieve in field or greenhouse conditions. In vitro cultures have been applied to forest tree research, namely in the establishment of co-cultures to study plant–pathogen interactions, namely co-cultures of *Pinus pinaster* and *Bursaphelenchus xylophilus* as an alternative biotechnological approach to study the pine wilt disease [44] or for rapid clonal propagation of *Populus* spp. [45]. The analysis of plant–pathogen interactions poses a particular challenge in metabolomics studies due to the difficulty in discriminating between plant and pathogen. In this case, in vitro cell co-cultures can be regarded as an alternative dual metabolomics approach to study such metabolite responses in plant–pathogen interactions. This technique allows discrimination between plant cells and pathogens and can further be applied to compare the metabolite response to different pathogenic strains [46]. However, this biotechnological approach should be regarded as a preliminary tool for forest tree research; it is crucial to assess if these systems reflect the real physiological conditions of the plant to only later extrapolate the findings to the whole organism [9]. Another approach to study plant–pathogen interactions, without the need for cell cultures but also allowing the assessment of the spatial distribution of metabolites in plant and pathogen, is with mass spectrometry imaging [47]. However, this technique has not yet been applied in forest tree metabolomics research.

2.2.2. Replicates and Randomization

To compensate for quantitative and qualitative variations in metabolomics analyses, biological replicates are essential for powerful statistical analysis and reliable biological interpretation of the results. Technical replicates can compensate for protocol or instrumental variations but do not improve the statistical analysis of the results [9,48]. In plant metabolomics, the minimum acceptable number of biological replicates should be six [2,49,50]. The biological replicates should be representative of the population under study. For a stronger and significant statistical analysis, the number of replicates needed can be established by power analysis determined from the degree of analytical variance within the populations under study [50]. Statistical power analysis relates sample size, effect size (i.e., the difference of two group means divided by the pooled standard deviation) and significance level to the chance of detecting an effect in a dataset, and thus, should be performed before conducting the experiment as a key step in the experimental design [51,52]. Information for power analysis can be obtained through pilot studies or extrapolated from the literature [51]. Sample size determination modules can be found in bioinformatic tools for metabolite data analysis, such as MetaboAnalyst 3.0, based on the Bioconductor R package Sample Size and Power Analysis (SSPA) and using data from a pilot metabolomic study [52]. However, power analysis is often avoided, and sample size determination becomes driven by sample availability [51]. In the case of limited amounts of sample and high biological variation, pooling samples is a common procedure [7,9,38]. However, this information should be taken into account when performing the data analysis as it may compromise the quality of the data (e.g., a pool containing an odd sample or individual not grown under the same exact conditions).

Randomization is critical for reducing experimental error and biological variability. If working under controlled environmental conditions (e.g., growth chamber), plants should be rotated during the course of the experiment to compensate for variations in light intensity or ventilation that can ultimately affect metabolism and the reproducibility of the data [7,9,53,54]. If plants are grown in a greenhouse or field conditions, variation in environmental conditions is likely to be observed. In all cases, it is crucial to keep a record of all observed changes in the course of the experiment and include them in the metadata and storage databases to ensure data reusability [49,54]. A common strategy to compensate for the impossibility of performing randomization (especially when working with a high number of individuals) is to arrange plants in a block design [9]. In a block design, the individuals are divided into homogeneous groups (i.e., blocks), and treatments are assigned randomly within the block. Treatment comparisons are then performed within blocks because the variability within each block is lower than the variability between blocks. Additionally, harvest should be performed randomly in each block to minimize block effect. The use of an appropriate design is particularly important in forest tree studies because of the long-term nature of the experiments. The variation between blocks (i.e., block factors), such as time, operator, or location, can be later included in the analysis. The randomized complete block design is the standard design pattern because of its simplicity. In this block design, the same number of individuals from each treatment and/or genotype is randomly allocated per block, and act as biological replicates. This block design is most effective when the site is relatively uniform; however, this is rarely the case in forestry studies. To overcome this limitation, other designs, such as the spatially-balanced complete block design [55] or the incomplete block design [56], that allow for better control of heterogeneity are becoming widely popular.

3. Sample Preparation for Forest Tree Metabolomics

In metabolomics, as in any analytical science, the sample preparation protocol has a crucial impact on the obtained analytical data. The workflow includes the harvest of the biological material and immediate quenching of metabolism and storing prior sample homogenization and metabolite extraction [2,6,7,34,53,57]. Sample preparation must be meticulously planned to identify potential sources of experimental variation and errors that might compromise data analysis, re-usability of the data, or biological interpretation of the results [6]. To obtain a standard protocol, the sample preparation method should be validated for the plant tissue under study using technical replicate extractions to determine the method precision and quantitative reproducibility [57]. In this section, the importance and challenges of performing harvest and quenching of tree material, especially in field conditions, are discussed, followed by the most common metabolomic workflows in forest tree metabolomics.

3.1. Harvest and Quenching

The precise time and process of sampling, or harvest, is a decisive step in a metabolomics experiment because it determines the "metabolic snapshot" of the organism to be analyzed, which directly influences the biological interpretation of the results [6,7,38]. In addition, the harvest should be performed as quickly as possible to avoid diurnal variations and the loss of metabolites with high turnover rates [2,6,7,37,50]. When working with forest tree species, samples are, in most cases, collected in the field and should be properly stored until lab processing. Ideally, biological samples should be immediately frozen in liquid nitrogen to avoid loss or degradation of biomolecules. However, this method is practically impossible to apply to samples harvested in natural ecosystems. In these cases, the best approach is to use silica gel to dehydrate the samples, thus stopping biochemical reactions [58]. Nevertheless, volatile compounds are often difficult to recover. In addition to the sample storage under field conditions, data relative to the exact geographical location and edaphic–climatic conditions should be described in as much detail as possible, to provide a more complete characterization of the provenance [58].

After the harvest, the second step in sample preparation is to instantly quench metabolism, usually by flash-freeze, using liquid nitrogen (shock freezing). Quenching is a crucial step in metabolomics workflows to immediately stop the metabolism and avoid further changes occurring in the sample, such as metabolite degradation or variations in their concentration, chemical, or physical properties [2,6,34]. Other methods include freeze-drying or the use of ice-cold methanol. Despite the risk of lower extraction reproducibility when working with frozen fresh samples, freeze-drying is a slower process that can lead to the production of artifacts, and potentially lead to the irreversible adsorption of metabolites on cell walls and membranes [36,59]. Freeze-drying can be a convenient method when weighing large sample sets, but it has been reported to reduce extraction yields by 25% [60]. In a comparative phytohormone quantification study between fresh-frozen and freeze-dried plant material, namely needles of *P. pinaster*, leaves of *Eucalyptus globulus*, and cotyledons of *P. pinea*, higher recoveries were obtained when using fresh material [61]. Freeze drying methods enhance lipid extraction by eliminating all the water, and consequently, generating a more complex matrix making phytohormone quantification in the presence of these interferences difficult. The appropriate sample treatment should always be evaluated according to the plant tissue under study.

In plant metabolomics, quenching is usually followed by homogenization of the sample, typically using a pestle and mortar or ball mill for plant–cell–wall breakage and sample weighing. These steps are always performed under liquid nitrogen to prevent tissue from thawing [2,6,7].

3.2. Metabolite Extraction

The choice of a metabolite extraction protocol is extremely important in metabolomics studies as it can directly affect the metabolite coverage and metabolite concentration. Ideally, a metabolite extraction protocol aims to (i) efficiently isolate metabolites from the sample in a high-throughput manner; (ii) be as non-selective as possible to ensure adequate metabolite coverage; (iii) prevent metabolite loss or degradation; (iv) be reproducible; (v) remove interferents that can affect the analysis; (vi) be compatible with the chosen analytical technique; and, when necessary, (vii) concentrate low abundance metabolites before analysis [2,34,59,62]. Typically, a metabolomics experiment follows two alternative approaches, targeted or untargeted. In targeted metabolomics, a well-defined group of known annotated metabolites is identified, whereas an untargeted approach aims to provide an overview of all the measurable analytes in the biological sample, including unknown compounds. However, it is important to highlight that, due to the vast variety of metabolites present in the plant metabolome, at different concentration levels and with distinct physical–chemical properties, it is impossible to extract the whole range of metabolites using a single extraction protocol [2,6,38].

In a metabolomics extraction protocol, several aspects have to be considered, namely the choice of an appropriate solvent system, solvent solubility, solvent to sample ratio, duration, and temperature of the extraction [2,5,6,36,63]. The choice of the appropriate solvent depends not only on the metabolite properties to be extracted but also has to meet the specific requirements of the analytical platform to be used (e.g., GC-MS, LC-MS). The exception is the use of headspace extraction (e.g., solid-phase microextraction, SPME) for the extraction of volatile components without the need for solvents [5,38]. For LC-MS, the only main limitation is that the solvent in which the sample is injected must be miscible and should be similar to the LC mobile phases used. For the typical reverse-phase separations, solvents used are generally aqueous eluents with 5–50% of an organic solvent (e.g., methanol, acetonitrile) [6]. Moreover, the addition of stable isotopically labeled internal standards to the extraction buffer (e.g., ^{15}N and ^{13}C labeling strategies) in targeted plant metabolomics approach is an excellent tool (i) to monitor extraction reproducibility; (ii) to compensate for ionization suppression/enhancement effects, accuracy, precision, and matrix effects of an analytical method or during method validation; and (iii) for normalization in data analysis [9,62,64].

3.2.1. GC-MS Metabolite Profiling

In forest tree research, GC coupled to either a time-of-flight MS (GC-TOF-MS) or a fast scanning quadrupole MS (GC-qMS) have often been employed for high-throughput plant primary metabolite profiling allowing the measurement of complex mixtures of primary metabolites (e.g., organic acids, sugars, sugar alcohols, amino acids) in a single extract [8]. GC-TOF-MS shows numerous advantages over GC-qMS, namely, higher mass accuracy, higher duty cycles, and faster acquisition rates that ultimately contribute to a better deconvolution of overlapping peaks and higher sample throughput [2,3,65].

Despite the low number of publications in forest tree metabolomics, when compared to other omics studies [66], GC-TOF-MS has been the method of choice for the primary metabolite profiling of forest tree responses to abiotic and biotic stresses [24,25,58,67–69] as well as other plant growth-related processes [17,26,70–77]. In these forest tree metabolomics studies, as for plant metabolomics in general, primary metabolites for GC-TOF-MS analysis are commonly extracted using the well-established chloroform:methanol:water extraction protocol, with minor optimization variations across studies (e.g., time of extraction, temperature, solvent ratio, or addition order), and further derivatized with

N-methyl-*N*-(trimethylsilyl)trifloracetamide (MSTFA), containing a mixture of fatty acid methyl esters (FAMEs) with different chain length as time standards (i.e., standard for retention time calibration) [2,34,50,63]. This two-phase solvent system has the advantage of fractionating the metabolites from a single sample into a polar aqueous phase (methanol:water) and a lipophilic organic phase (chloroform), which can be further analyzed separately [63].

Additional GC-qMS studies in forest tree species include the profile of the volatile fraction, namely volatile organic compounds (VOCs) and essential oil (EO). This volatile fraction is mainly dominated by terpenoids, phenylpropanoids/benzenoids, fatty acid derivatives, and amino acid derivatives [78]. Despite the similarity in their qualitative chemical composition, the relative amounts of metabolites found in these two volatile fractions can differ greatly due to the distinct extraction processes involved [79]. VOCs are commonly collected with headspace techniques (e.g., SPME), while EOs are obtained exclusively with hydro-, steam- or dry-distillation, or in the case of citrus fruits, mechanically without heating [80]. EO screening studies are popular among forest tree species, namely *Eucalyptus* and *Pinus* spp., mainly due to the occurrence of EO chemotypes [81–84]. The non-destructive nature of headspace techniques, such as SPME, allow for time-course evaluation of VOCs emission and have been widely applied in forest tree research not only for plant chemotype classification but also for plant–pathogen interactions or plant–insect communication [85,86]. This technique simply requires the optimization of the type of fiber, exposure time and temperature, and desorption time and temperature [87].

3.2.2. LC-MS Metabolite Profiling

In forest tree research, LC-MS instruments have also been used for untargeted secondary metabolite profiling and phytohormone quantification studies. The focus of these studies was related with abiotic stress responses [19–24,67,88,89]; and to a smaller extent to biotic stress responses [90,91] and plant growth and developmental processes [77,92,93].

Metabolite extraction for LC-MS untargeted analysis are usually performed using a simple protocol based on methanol [19] or methanol:water 80:20, *v/v* [20,91] or 50:50, *v/v* [90] as extraction solvents. However, most untargeted secondary metabolite profiling studies in forest tree metabolomics research are performed in combination with GC-qMS or GC-TOF-MS primary metabolite profiling, ultimately allowing for more comprehensive coverage of the tree metabolome. In these cases, the chloroform:methanol:water two-phase solvent system is used, and the polar phase is evaporated to dryness and used for both GC-MS (after derivatization) and LC-MS metabolite profiling (after reconstitution in methanol:water) [21,77,88,89]. To take full advantage of the chloroform:methanol:water two-phase solvent system, the non-polar metabolites (lipophilic fraction) are analyzed by GC-MS after derivatization, for example, with tertmethyl–butyl–ether (MTBE) and trimethylsulfoniumhydroxide (TMSH) [23,92].

LC coupled to triple quadrupole-MS (LC-QqQ-MS) has been often employed in method development to quantify phytohormones in plant tissues [94]. Delatorre and co-workers [61] developed and validated an LC-QqQ-MS analytical method to quantify 20 phytohormones in forest tree species tissues, using 2-propanal:water:hydrochloric acid (2:1:0.002 *v/v/v*) and dichloromethane for a two-phase solvent system, a protocol originally described by Pan and co-workers [95]. Other metabolite extraction protocols for phytohormone quantification in forest tree species include a modified version of the well-established solvent extraction protocol described by Bieleski [96], namely methanol:water:formic acid (15:4:1, *v/v/v*) [97], methanol:water (80:20, *v/v*) [98,99], methanol:water:acetic acid (90:9:1, *v/v/v*) [16], or water:diethyl ether:acetic acid [22]. The Bieleski solvent extraction protocol [96] has been used for the extraction of phytohormones, particularly cytokines [100].

3.3. Pre-Analytical Requirements

In targeted and untargeted metabolomic approaches, the presence of matrix interferents can hinder the MS-based metabolite analysis by adding further complexity in regards to metabolite ionization and interfering molecules that influence the signal response of the metabolites under study. Although in targeted approaches, the use of stable isotopically labeled internal standards can compensate for matrix-induced ionization effects, the availability of standards in untargeted approaches can be very limited. In both cases, metabolite concentration and further sample clean-up steps to remove interferents might be necessary. Solid-phase extraction (SPE) is the method of choice in these cases [5]. SPE is a simple sample preparation technique based on the removal of analytes from a liquid sample by retention on a solid sorbent (e.g., silica, alkylated silica), based on the functional group interactions of the analytes, flowing solvent, and the solid sorbent [5,6,101]. The retained analytes are subsequently eluted from the sorbent using a solvent or solvent mixture with sufficient elution strength. Because it involves the use of large amounts of solvents and requires several steps of concentration, this technique is time-consuming and not high-throughput, resulting in an added risk of losing components during the process. To improve sample high-throughput and reproducibility, 96-well solid-phase extraction plates, and robotic SPE-MS systems have been developed and are commercially available [101]. An SPE step is often included in sample preparation for MS-based phytohormone analysis in forest tree studies to remove interfering components from the matrix, such as pigments, resinic acids, terpenes, carotenoids, flavonoids, cellulose, and lipids, and increase the recovery rates of the phytohormones under study [61,64,99].

4. The Importance of Forest Tree Metadata Standardization

Advances in high-throughput MS-based platforms have been responsible for the generation of extremely large metabolomics datasets. For a comprehensive understanding of biochemical pathways and regulatory networks involved in different plant processes and responses, metabolomics datasets can be further integrated with other "omics" studies (e.g., transcriptomics, proteomics), providing the data is available in a standardized and reproducible way. Thus, the description of metabolomics studies should include all the information needed to allow the repetition of the experiment and the re-usability of the data.

To promote standardization of all stages of a metabolomics analysis (i.e., experimental design, biological context, chemical analysis, and data processing) and ensure metadata consistency, in 2007, members of the metabolomics community established the Metabolomics Standard Initiative (MSI) [102,103]. The MSI aimed at reporting standards and provide a clear description of the biological system under study and of the metabolomics analysis workflows to allow data to be efficiently applied, shared, and reused. A decade later, a set of guideline principles known as the FAIR principles (i.e., Findable, Accessible, Interoperable, and Re-usable) were also designed to assure good (meta)data management by data holders and data publishers [104]. ELIXIR, the European infrastructure for biological data (https://elixir-europe.org), has also brought together several communities (e.g., plant sciences, metabolomics) with the common interest of dealing with the increasing complexity of data, ultimately making data easier to find, analyze and share. Challenges currently faced by the metabolomics community, namely (i) minimum information standards and early data capture; (ii) global spectral databases; (iii) tools and standards registries; (iv) compound identifier mapping; (v) omics data integration, and (vi) metabolite identification, have also been reported by ELIXIR in a dedicated workshop [105]. However, despite these initiatives, the compliance with these reporting standards still varies greatly across public repositories [106], and data and metadata sharing remain a critical issue in metabolomics publications [107,108].

Within the field of plant metabolomics, forest tree metabolomics studies present additional challenges concerning the standardization of metadata. Forest trees are species with long life cycles, and details of the experimental metadata (e.g., parental original or field growth conditions) are often not described. To re-use data derived from these studies, the description of the metadata should include detailed information of the harvested material (e.g., geographical location, growth conditions, biological growth stages, and phenological parameters) [8]. These parameters might reflect adaptive traits mediated by epigenetic changes that affect the material under study [109,110]. Thus, as epigenetic changes affect the transcriptome, proteome, and ultimately the metabolome, the integration of these "omics" data strongly depends on the availability of detailed information of the harvested material. Plant phenotyping has been developed significantly over the past years due to the progress in novel sensors, automation tools, and quantitative data analysis methods. Yet, the consequent increase of data generation is still a struggle for the standardization of data acquisition and its re-usability [111].

The urge for metadata standardization for plant phenotyping experiments has been addressed by community-driven projects, for example, the MIAPPE project (Minimal Information About a Phenotypic Experiment), the ISA framework (investigation, study, assay) [112,113] or the GnpIS data repository (genetic and genomic information system) [114]. MIAPPE is available as a checklist of metadata to adequately describe a plant phenotyping experiment and as software to validate, store, and disseminate MIAPPE-compliant data (https://www.miappe.org/). In early 2019, MIAPPE version 1.1 was released as an extended version to include woody plants and compatibility with other phenotyping frameworks (e.g., ISA framework), which represents an important step towards the standardization of forest tree metadata. By providing curated databases, and in accordance with the FAIR principles, these platforms allow data and metadata to be easier to find, integrate, and analyze. Due to the amount of data generated in forest tree studies, dedicated databases or extensions to existing databases have been developed. Dedicated tree databases (e.g., PlantGenIE, TreeGenes, and Hardwood Genomics Project) covering mostly genetic data have now the goal of associating phenotypic and environmental data [115].

Integrated into crop ontology, the woody plant ontology has been established as an additional platform for the annotation of forest tree metadata [116]. The woody plant ontology adds a set of definitions to the existing crop ontology to describe specific tree traits (e.g., secondary growth in wood and cork formation). Studies with forest tree species, particularly in field experiments, can take several years to develop, and it is crucial to adequately annotate all metadata across the tree's long life cycles. Such case studies that use machine-accessible metadata can be found in the literature. One example is the forest growth measurements from individual *Picea abies* trees over the course of 109 years [117]. The metadata file describing the reported data is openly available in an ISA-tab format and can be further used to analyze and validate forest growth.

5. Conclusion

Metabolomics studies are often regarded as the ultimate response of biological systems to genetic or environmental alterations. Although most MS-based plant metabolomics research is performed on crop and non-tree model species, in recent years, studies on forest tree species have generated particular interest, especially after major genomics breakthroughs in forest tree research (e.g., availability of the *Populus trichocarpa* reference genome in 2006). In this area, MS-based metabolomics represents a unique opportunity to explore the forest tree's adaptation to environmental fluctuations as well as other economic and ecological relevant developmental processes. However, and as previously discussed, to successfully obtain significant data from metabolomics analyses, it is crucial to have a well-planned experimental design and an appropriate sample preparation. Any metabolomics study should include, in great detail, a clear description of the design of the experiment as well as of other technical parameters. Despite the struggles, continuous efforts from the metabolomics scientific community have been made to ensure data and metadata reproducibility between laboratories and to promote the availability of curated databases and repositories containing high-quality data (including dedicated woody species platforms).

Author Contributions: C.A. conceived the idea and contributed to the organization of the initial manuscript; A.M.R. contributed to Sections 1–5; A.I.R.-B. contributed to Sections 2 and 3. All authors contributed to the critical editing and organization of the final manuscript.

Funding: This research was funded by Fundação para a Ciência e Tecnologia (FCT) through the FCT Investigator Programme (contract IF/00376/2012/CP0165/CT0003, CA), the R&D unit GREEN-IT 'Bioresources for sustainability' (UID/Multi/04551/2013) and the LEAF R&D unit (UID/AGR/04129/2013). AMR acknowledges FCT for the

Ph.D. fellowship (PD/BD/114417/2016), and the ITQB NOVA international Ph.D. programme 'Plants for Life' (PD/00035/2013).

Conflicts of Interest: The authors declare no conflict of interest.

References

1. Oliver, S.G.; Winson, M.K.; Kell, D.B.; Baganz, F. Systematic functional analysis of the yeast genome. *Trends Biotechnol.* **1998**, *16*, 373–378. [CrossRef]
2. Jorge, T.F.; Rodrigues, J.A.; Caldana, C.; Schmidt, R.; van Dongen, J.T.; Thomas-Oates, J.; António, C. Mass spectrometry-based plant metabolomics: Metabolite responses to abiotic stress. *Mass Spectrom. Rev.* **2016**, *35*, 620–649. [CrossRef] [PubMed]
3. Alseekh, S.; Wu, S.; Brotman, Y.; Fernie, A.R. Guidelines for sample normalization to minimize batch variation for large-scale metabolic profiling of plant natural genetic variance. In *Plant Metabolomics*; António, C., Ed.; Humana Press: New York, NY, USA, 2018. [CrossRef]
4. Fernie, A.R.; Stitt, M. On the discordance of metabolomics with proteomics and transcriptomics: Coping with increasing complexity in logic, chemistry, and network interactions scientific correspondence. *Plant Physiol.* **2012**, *158*, 1139–1145. [CrossRef] [PubMed]
5. Dettmer, K.; Aronov, P.A.; Hammock, B.D. Mass spectrometry based-metabolomics. *Mass Spectrom. Rev.* **2007**, *26*, 51–78. [CrossRef] [PubMed]
6. Kim, H.K.; Verpoorte, R. Sample preparation for plant metabolomics. *Phytochem. Anal.* **2010**, *21*, 4–13. [CrossRef] [PubMed]
7. Allwood, J.W.; De Vos, R.C.H.; Moing, A.; Deborde, C.; Erban, A.; Kopka, J.; Goodacre, R.; Hall, R.D. Plant metabolomics and its potential for systems biology research: Background concepts, technology, and methodology. *Methods Enzymol.* **2011**, *500*, 299–336. [CrossRef]
8. Rodrigues, A.M.; Miguel, C.; Chaves, I.; António, C. Mass spectrometry-based forest tree metabolomics. *Mass Spectrom. Rev.* **2019**. [CrossRef]
9. Martins, M.C.M.; Caldana, C.; Wolf, L.D.; de Abreu, L.G.F. The Importance of Experimental Design, Quality Assurance, and Control in Plant Metabolomics Experiments. In *Plant Metabolomics*; António, C., Ed.; Humana Press: New York, NY, USA, 2018. [CrossRef]
10. Baumgartner, R.J. Sustainable development goals and the forest sector—A complex relationship. *Forests* **2019**, *10*, 152. [CrossRef]
11. Bončina, A.; Simončič, T.; Rosset, C. Assessment of the concept of forest functions in Central European forestry. *Environ. Sci. Policy* **2019**, *99*, 123–135. [CrossRef]
12. Loomis, J.J.; Knaus, M.; Dziedzic, M. Integrated quantification of forest total economic value. *Land Use Policy* **2019**, *84*, 335–346. [CrossRef]
13. IPCC Core Writing Team. Climate Change 2014: Synthesis Report. In *Contribution of Working Groups I, II and III to the Fifth Assessment Report of the Intergovernmental Panel on Climate Change*; IPCC: Geneva, Switzerland, 2014.
14. Janz, D.; Behnke, K.; Schnitzler, J.P.; Kanawati, B.; Schmitt-Kopplin, P.; Polle, A. Pathway analysis of the transcriptome and metabolome of salt sensitive and tolerant poplar species reveals evolutionary adaption of stress tolerance mechanisms. *BMC Plant Biol.* **2010**, *10*, 150. [CrossRef] [PubMed]
15. Warren, C.R.; Aranda, I.; Cano, F.J. Metabolomics demonstrates divergent responses of two *Eucalyptus* species to water stress. *Metabolomics* **2012**, *8*, 186–200. [CrossRef]
16. Correia, B.; Pintó-Marijuan, M.; Castro, B.B.; Brossa, R.; López-Carbonell, M.; Pinto, G. Hormonal dynamics during recovery from drought in two *Eucalyptus globulus* genotypes: From root to leaf. *Plant Physiol. Biochem.* **2014**, *82*, 151–160. [CrossRef] [PubMed]
17. Budzinski, I.G.; Moon, D.H.; Morosini, J.S.; Lindén, P.; Bragatto, J.; Moritz, T.; Labate, C.A. Integrated analysis of gene expression from carbon metabolism, proteome and metabolome, reveals altered primary metabolism in *Eucalyptus grandis* bark, in response to seasonal variation. *BMC Plant Biol.* **2016**, *16*, 149. [CrossRef]
18. Correia, B.; Valledor, L.; Hancock, R.D.; Renaut, J.; Pascual, J.; Soares, A.M.V.M.; Pinto, G. Integrated proteomics and metabolomics to unlock global and clonal responses of *Eucalyptus globulus* recovery from water deficit. *Metabolomics* **2016**, *12*, 141. [CrossRef]

19. De Miguel, M.; Guevara, M.A.; Sánchez-Gómez, D.; María, N.; Díaz, L.M.; Mancha, J.A.; de Símon, B.F.; Cadahía, E.; Desai, N.; Aranda, I.; et al. Organ-specific metabolic responses to drought in *Pinus pinaster* Ait. *Plant Physiol. Biochem.* **2016**, *102*, 17–26. [CrossRef]

20. Rivas-Ubach, A.; Barbeta, A.; Sardans, J.; Guenther, A.; Ogaya, R.; Oravec, M.; Urban, O.; Peñuelas, J. Topsoil depth substantially influences the responses to drought of the foliar metabolomes of Mediterranean forests. *Perspect. Plant Ecol. Syst.* **2016**, *21*, 41–54. [CrossRef]

21. De Símon, B.F.; Sanz, M.; Cervera, M.T.; Pinto, E.; Aranda, I.; Cadahía, E. Leaf metabolic response to water deficit in *Pinus pinaster* Ait. relies upon ontogeny and genotype. *Environ. Exp. Bot.* **2017**, *14*, 41–55. [CrossRef]

22. Correia, B.; Hancock, R.D.; Amaral, J.; Gomez-Cadenas, A.; Valledor, L.; Pinto, G. Combined drought and heat activates protective responses in *Eucalyptus globulus* that are not activated when subjected to drought or heat stress alone. *Front Plant Sci.* **2018**, *9*, 819. [CrossRef]

23. Escandón, M.; Meijón, M.; Valledor, L.; Pascual, J.; Pinto, G.; Cañal, M.J. Metabolome integrated analysis of high-temperature response in *Pinus radiata*. *Front. Plant Sci.* **2018**, *9*, 485. [CrossRef]

24. Mokochinski, J.B.; Mazzafera, P.; Sawaya, A.C.H.F.; Mumm, R.; de Vos, R.C.H.; Hall, R.D. Metabolic responses of *Eucalyptus* species to different temperature regimes. *J. Integr. Plant Biol.* **2018**, *60*, 397–411. [CrossRef] [PubMed]

25. Rodríguez-Calcerrada, J.; Rodrigues, A.M.; Perdiguero, P.; António, C.; Altkin, O.K.; Li, M.; Colada, C.; Gil, L. A molecular approach to drought-induced reduction in leaf CO_2 exchange in drought-resistant *Quercus ilex*. *Physiol. Plant.* **2018**, *162*, 394–408. [CrossRef] [PubMed]

26. Watanabe, M.; Netzer, F.; Tohge, T.; Orf, I.; Brotman, Y.; Dubbert, D.; Fernie, A.R.; Rennenberg, H.; Hoefgen, R.; Herschbach, C. Metabolome and lipidome profiles of *Populus canescens* twig tissues during annual growth show phospholipid-linked storage and mobilization of C, N, and S. *Front Plant Sci.* **2018**, *9*, 1292. [CrossRef] [PubMed]

27. Cadahía, E.; de Simón, B.F.; Aranda, I.; Sanz, M.; Sánchez-Gómez, D.; Pinto, E. Non-targeted metabolomic profile of *Fagus sylvatica* L. leaves using liquid chromatography with mass spectrometry and gas chromatography with mass spectrometry. *Phytochem. Anal.* **2015**, *26*, 171–182. [CrossRef] [PubMed]

28. Dhandapani, S.; Jin, J.; Sridhar, V.; Sarojam, R.; Chua, N.H.; Jang, I.C. Integrated metabolome and transcriptome analysis of *Magnolia champaca* identifies biosynthetic pathways for floral volatile organic compounds. *BMC Genom.* **2017**, *18*, 463. [CrossRef]

29. Dean, L.L. Targeted and non-targeted analyses of secondary metabolites in nut and seed processing. *Eur. J. Lipid Sci. Technol.* **2018**, *120*, 1700479. [CrossRef]

30. Li, S.S.; Wu, Q.; Yin, D.D.; Feng, C.Y.; Liu, Z.A.; Wang, L.S. Phytochemical variation among the traditional Chinese medicine Mu Dan Pi from *Paeonia suffruticosa* (tree peony). *Phytochemistry* **2018**, *146*, 16–24. [CrossRef]

31. Moura, I.; Duvane, J.A.; Silva, M.J.; Ribeiro, N.; Ribeiro-Barros, A.I. Woody species from the Mozambican Miombo woodlands: A review on their ethnomedicinal uses and pharmacological potential. *J. Med. Plant Res.* **2018**, *12*, 15–31. [CrossRef]

32. Farag, M.A.; El-Kersh, D.M.; Ehrlich, A.; Choucry, M.A.; El-Seedi, H.; Frolov, A.; Wessjohann, L.A. Variation in *Ceratonia siliqua* pod metabolome in context of its different geographical origin, ripening stage and roasting process. *Food Chem.* **2019**, *283*, 675–687. [CrossRef]

33. Wang, Z.; Zhu, C.; Liu, S.; He, C.; Chen, F.; Xiao, P. Comprehensive metabolic profile analysis of the root bark of different species of tree peonies (*Paeonia Sect. Moutan*). *Phytochemistry* **2019**, *163*, 118–125. [CrossRef]

34. Álvarez-Sánchez, B.; Priego-Capote, F.; de Castro, M.D.L. Metabolomics analysis II. Preparation of biological samples prior to detection. *Trends Anal. Chem.* **2010**, *29*, 120–127. [CrossRef]

35. Broadhurst, D.I.; Kell, D.B. Statistical strategies for avoiding false discoveries in metabolomics and related experiments. *Metabolomics* **2006**, *2*, 171–196. [CrossRef]

36. Fiehn, O. Metabolomics—The link between genotypes and phenotypes. *Plant Mol. Biol.* **2002**, *48*, 155–171. [CrossRef] [PubMed]

37. Stitt, M.; Fernie, A.R. From measurements of metabolites to metabolomics: An 'on the fly' perspective illustrated by recent studies of carbon–nitrogen interactions. *Curr. Opin. Biotechnol.* **2003**, *14*, 136–144. [CrossRef]

38. Hall, R.D. Plant metabolomics: From holistic hope, to hype, to hot topic. *New Phytol.* **2006**, *169*, 453–468. [CrossRef]

39. Ghimenti, S.; Tabucchi, S.; Lomonaco, T.; Di Francesco, F.; Fuoco, R.; Onor, M.; Lenzi, M.; Trivella, M.G. Monitoring breath during oral glucose tolerance tests. *J. Breath Res.* **2013**, *7*, 017115. [CrossRef]

40. Lomonaco, T.; Ghimenti, S.; Piga, I.; Biagini, D.; Onor, M.; Fuoco, R.; Paolicchi, A.; Ruocco, L.; Pellegrini, G.; Trivella, M.G.; et al. Monitoring of warfarin therapy: Preliminary results from a longitudinal pilot study. *Microchem. J.* **2018**, *136*, 170–176. [CrossRef]

41. Neale, D.B.; Kremer, A. Forest tree genomics: Growing resources and applications. *Nat. Rev. Genet.* **2011**, *12*, 111–122. [CrossRef]

42. Asbjornsen, H.; Campbell, J.; Jennings, K.; Vadeboncoeur, M.; McIntire, C.; Templer, P.; Phillips, R.; Bauerle, T.; Dietze, M.; Frey, S.; et al. Guidelines and considerations for designing field experiments for simulating precipitation extremes in forest ecosystems. *Methods Ecol. Evol.* **2018**, *9*, 2310–2325. [CrossRef]

43. Morgenstern, E.K. Geographic Variation in Forest Trees. In *Genetic Basis and Application of Knowledge in Silviculture*; UBC Press: Vancouver, BC, Canada, 1996; ISBN 0-7748-0579-X.

44. Faria, J.M.; Sena, I.; Vieira da Silva, I.; Ribeiro, B.; Barbosa, P.; Ascensão, L.; Bennett, R.N.; Mota, M.; Figueiredo, A.C. In vitro co-cultures of *Pinus pinaster* with *Bursaphelenchus xylophilus*: A biotechnological approach to study pine wilt disease. *Planta* **2015**, *241*, 1325–1336. [CrossRef]

45. Confalonieri, M.; Balestrazzi, A.; Bisoffi, S.; Carbonera, D. In vitro culture and genetic engineering of *Populus* spp.: Synergy for forest tree improvement. *Plant Cell Tissue Org. Cult.* **2003**, *72*, 109. [CrossRef]

46. Allwood, J.W.; Heald, J.; Lloyd, A.J.; Goodacre, R.; Mur, L.A.J. Separating the inseparable: The metabolomic analysis of plant-pathogen interactions. In *Plant Metabolomics*; Nigel, W.H., Robert, D.H., Eds.; Humana Press: New York, NY, USA, 2012; Volume 860, pp. 31–49. [CrossRef]

47. Boughton, B.A.; Thinagaran, D.; Sarabia, D.; Bacic, A.; Roessner, U. Mass spectrometry imaging for plant biology: A review. *Phytochem. Rev.* **2016**, *15*, 445–488. [CrossRef] [PubMed]

48. Blainey, P.; Krzywinski, M.; Altman, N. Points of significance: Replication. *Nat. Methods* **2014**, *11*, 879–880. [CrossRef] [PubMed]

49. Fiehn, O.; Wohlgemuth, G.; Scholz, M. Setup and Annotation of Metabolomic Experiments by Integrating Biological and Mass Spectrometric Metadata. In *Data Integration in the Life Sciences*; Ludäscher, B., Raschid, L., Eds.; Springer: Berlin/Heidelberg, Germany, 2005; Volume 3615. [CrossRef]

50. Lisec, J.; Schauer, N.; Kopka, J.; Willmitzer, L.; Fernie, A.R. Gas chromatography mass spectrometry–Based metabolite profiling in plants. *Nat. Protoc.* **2006**, *1*, 387–396. [CrossRef]

51. Blaise, B.J.; Correia, G.; Tin, A.; Young, J.H.; Vergnaud, A.-C.; Lewis, M.; Pearce, J.T.M.; Elliott, P.; Nicholson, J.K.; Holmes, E.; et al. Power analysis and sample size determination in metabolic phenotyping. *Anal. Chem.* **2016**, *88*, 5179–5188. [CrossRef]

52. Xia, J.; Wishart, D.S. Using MetaboAnalyst 3.0 for comprehensive metabolomics data analysis. *Curr. Protoc. Bioinform.* **2016**, *55*, 14.10.1–14.10.91. [CrossRef]

53. Fukusaki, E.; Kobayashi, A. Plant Metabolomics: Potential for Practical Operation. *J. Biosci. Bioeng.* **2005**, *100*, 347–354. [CrossRef]

54. Hannemann, J.; Poorter, H.; Usadel, B.; Bläsing, O.E.; Finck, A.; Tardieu, F.; Atkin, O.K.; Pons, T.; Stitt, M.; Gibon, Y. Xeml Lab: A tool that supports the design of experiments at a graphical interface and generates computer-readable metadata files, which capture information about genotypes, growth conditions, environmental perturbations and sampling strategy. *Plant Cell Environ.* **2009**, *32*, 1185–2000. [CrossRef]

55. Van Es, H.M.; Gomes, C.P.; Sellmann, M.; van Es, C.L. Spatially-Balanced Complete Block designs for field experiments. *Geoderma* **2007**, *140*, 346–352. [CrossRef]

56. Gezan, S.A.; White, T.L.; Huber, D.A. Comparison of Experimental Designs for Clonal Forestry Using Simulated Data. *For. Sci.* **2006**, *52*, 108–116. [CrossRef]

57. Fiehn, O.; Wohlgemuth, G.; Scholz, M.; Kind, T.; Lee, D.Y.; Lu, Y.; Moon, S.; Nikolau, B. Quality control for plant metabolomics: Reporting MSI-compliant studies. *Plant J.* **2008**, *53*, 691–704. [CrossRef] [PubMed]

58. Duvane, J.A.; Jorge, T.F.; Maquia, I.; Ribeiro, N.; Ribeiro-Barros, A.I.F.; António, C. Characterization of the primary metabolome of *Brachystegia boehmii* and *Colophospermum mopane* under different fire regimes in Miombo and Mopane African Woodlands. *Front. Plant Sci.* **2017**, *8*, 2130. [CrossRef] [PubMed]

59. T'Kindt, R.; Morreel, K.; Deforce, D.; Boerjan, W.; Van Bocxlaer, J. Joint GC–MS and LC–MS platforms for comprehensive plant metabolomics: Repeatability and sample pre-treatment. *J. Chromatogr. B* **2009**, *877*, 3572–3580. [CrossRef] [PubMed]

60. Forcat, S.; Bennett, M.H.; Mansfield, J.W.; Grant, M.R. A rapid and robust method for simultaneously measuring changes in the phytohormones ABA, JA and SA in plants following biotic and abiotic stress. *Plant Methods* **2008**, *4*, 16. [CrossRef]

61. Delatorre, C.; Rodríguez, A.; Rodríguez, L.; Majada, J.P.; Ordás, R.J.; Feito, I. Hormonal profiling: Development of a simple method to extract and quantify phytohormones in complex matrices by UHPLC–MS/MS. *J. Chromatogr. B* **2017**, *1040*, 239–249. [CrossRef]

62. Vuckovic, D. Current trends and challenges in sample preparation for global metabolomics using liquid chromatography–mass spectrometry. *Anal. Bioanal. Chem.* **2012**, *403*, 1523–1548. [CrossRef]

63. Gullberg, J.; Jonsson, P.; Nordström, A.; Sjöström, M.; Moritz, T. Design of experiments: An efficient strategy to identify factors influencing extraction and derivatization of Arabidopsis thaliana samples in metabolomic studies with gas chromatography/mass spectrometry. *Anal. Biochem.* **2004**, *331*, 283–295. [CrossRef]

64. Rodrigues, A.M.; António, C. Standard Key Steps in Mass Spectrometry-Based Plant Metabolomics Experiments: Instrument Performance and Analytical Method Validation. In *Plant Metabolomics*; António, C., Ed.; Humana Press: New York, NY, USA, 2018; Volume 1778. [CrossRef]

65. Jorge, T.F.; Mata, A.T.; António, C. Mass spectrometry as a quantitative tool in plant metabolomics. *Philos. Trans. R. Soc. A* **2016**, *374*, 20150370. [CrossRef]

66. Harfouche, A.; Meilan, R.; Altman, A. Molecular and physiological responses to abiotic stress in forest trees and their relevance to tree improvement. *Tree Physiol.* **2014**, *34*, 1181–1198. [CrossRef]

67. Srivastava, V.; Obudulu, O.; Bygdell, J.; Löfstedt, T.; Rydén, P.; Nilsson, R.; Ahnlund, M.; Johansson, A.; Jonsson, P.; Freyhult, E.; et al. OnPLS integration of transcriptomic, proteomic and metabolomic data shows multi-level oxidative stress responses in the cambium of transgenic hipI- superoxide dismutase Populus plants. *BMC Genom.* **2013**, *14*, 893. [CrossRef]

68. Angelcheva, L.; Mishra, Y.; Antti, H.; Kjellsen, T.; Funk, C.; Strimbeck, R.G.; Schröder, W.P. Metabolomic analysis of extreme freezing tolerance in Siberian spruce (*Picea obovata*). *New Phytol.* **2014**, *204*, 545–555. [CrossRef] [PubMed]

69. Wang, L.; Qu, L.; Hu, J.; Zhang, L.; Tang, F.; Lu, M. Metabolomics reveals constitutive metabolites that contribute resistance to fall webworm (*Hyphantria cunea*) in *Populus deltoides*. *Environm. Exp. Bot.* **2017**, *136*, 31–40. [CrossRef]

70. Andersson-Gunnerås, S.; Mellerowicz, E.J.; Love, J.; Segerman, B.; Ohmiya, Y.; Coutinho, P.M.; Nilsson, P.; Henrissat, B.; Moritz, T.; Sundberg, B. Biosynthesis of cellulose-enriched tension wood in Populus: Global analysis of transcripts and metabolites identifies biochemical and developmental regulators in secondary wall biosynthesis. *Plant J.* **2006**, *45*, 144–165. [CrossRef] [PubMed]

71. Druart, N.; Johansson, A.; Baba, K.; Schrader, J.; Sjödin, A.; Bhalerao, R.R.; Resman, L.; Trygg, J.; Moritz, T.; Bhalerao, R.P. Environmental and hormonal regulation of the activity-dormancy cycle in the cambial meristem involves stage-specific modulation of transcriptional and metabolic networks. *Plant J.* **2007**, *50*, 557–573. [CrossRef]

72. Hoffman, D.E.; Jonsson, P.; Bylesjö, M.; Trygg, J.; Antti, H.; Eriksson, M.E.; Moritz, T. Changes in diurnal patterns within the Populus transcriptome and metabolome in response to photoperiod variation. *Plant Cell Environ.* **2010**, *33*, 1298–1313. [CrossRef]

73. Kusano, M.; Jonsson, P.; Fukushima, A.; Gullberg, J.; Sjöström, M.; Trygg, J.; Moritz, T. Metabolite signature during short-day induced growth cessation in Populus. *Front. Plant Sci.* **2011**, *2*, 29. [CrossRef]

74. Businge, E.; Brackmann, K.; Moritz, T.; Egertsdotter, U. Metabolite profiling reveals clear metabolic changes during somatic embryo development of Norway spruce (*Picea abies*). *Tree Physiol.* **2012**, *32*, 232–244. [CrossRef]

75. Li, Q.F.; Wang, J.H.; Pulkkinen, P.; Kong, L.S. Changes in the metabolome of *Picea balfouriana* embryogenic tissues that were linked to different levels of 6-BAP by gas chromatography-mass spectrometry approach. *PLoS ONE* **2015**, *10*, e0141841. [CrossRef]

76. Guerra, F.P.; Richards, J.H.; Fiehn, O.; Famula, R.; Stanton, B.J.; Shuren, R.; Skykes, R.; Davis, M.F.; Neale, D.B. Analysis of the genetic variation in growth, ecophysiology, and chemical and metabolomic composition of wood of *Populus trichocarpa* provenances. *Tree Genet. Genomes* **2016**, *12*, 6. [CrossRef]

77. Dobrowolska, I.; Businge, E.; Abreu, I.N.; Moritz, T.; Egertsdotter, U. Metabolome and transcriptome profiling reveal new insights into somatic embryo germination in Norway spruce (*Picea abies*). *Tree Physiol.* **2017**, *37*, 1752–1766. [CrossRef]

78. Dudareva, N.; Negre, F.; Nagegowda, D.; Orlova, I. Plant volatiles: Recent advances and future perspectives. *Crit. Rev. Plant Sci.* **2006**, *25*, 417–440. [CrossRef]

79. Figueiredo, A.C. Biological properties of essential oils and volatiles: Sources of variability. *Nat. Volatiles Essent. Oils* **2017**, *4*, 1–13.

80. Council of Europe. *European Pharmacopoeia*, 7th ed.; European Directorate for the Quality of Medicines and Healthcare: Strasbourg, France, 2010.

81. Keszei, A.; Brubaker, C.L.; Foley, W.J. A molecular perspective on terpene variation in Australian Myrtaceae. *Aust. J. Bot.* **2008**, *56*, 197–213. [CrossRef]

82. Keszei, A.; Brubaker, C.L.; Carter, R.; Köllner, T.; Degenhardt, J.; Foley, W.J. Functional and evolutionary relationships between terpene synthases from Australian Myrtaceae. *Phytochemistry* **2010**, *71*, 844–852. [CrossRef] [PubMed]

83. Arrabal, C.; García-Vallejo, M.C.; Cadahia, E.; Cortijo, M.; de Símon, B.F. Characterization of two chemotypes of *Pinus pinaster* by their terpene and acid patterns in needles. *Plant Syst. Evol.* **2012**, *298*, 511–522. [CrossRef]

84. Rodrigues, A.M.; Mendes, M.D.; Lima, A.S.; Barbosa, P.M.; Ascensão, L.; Barroso, J.G.; Pedro, L.G.; Mota, M.M.; Figueiredo, A.C. *Pinus halepensis, P. pinaster, P. pinea* and *P. sylvestris* essential oils chemotypes and monoterpene hydrocarbon enantiomers, before and after inoculation with the pinewood nematode *Bursaphelenchus xylophilus. Chem. Biodivers.* **2017**, *14*, e1600153. [CrossRef] [PubMed]

85. Szmigielski, R.; Cieslak, M.; Rudziński, K.J.; Maciejewska, B. Identification of volatiles from *Pinus silvestris* attractive for *Monochamus galloprovincialis* using a SPME-GC/MS platform. *Environ. Sci. Pollut. Res. Int.* **2011**, *9*, 2860–2869. [CrossRef] [PubMed]

86. Gonçalves, E.; Figueiredo, A.C.; Barroso, J.G.; Henriques, J.; Sousa, E.; Bonifácio, L. Effect of *Monochamus galloprovincialis* feeding on *Pinus pinaster* and *Pinus pinea*, oleoresin and insect volatiles. *Phytochemistry* **2020**, *169*, 112159. [CrossRef]

87. Schäfer, B.; Hennig, P.; Engewald, W. Analysis of monoterpenes from conifer needles using solid phase microextraction. *J. High Res. Chromatogr.* **1995**, *18*, 587–592. [CrossRef]

88. Riikonen, J.; Kontunen-Soppela, S.; Ossipov, V.; Tervahauta, A.; Tuomainen, M.; Oksanen, E.; Vapaavouri, E.; Heinonen, J.; Kivimäenpää, M. Needle metabolome, freezing tolerance and gas exchange in Norway spruce seedlings exposed to elevated temperature and ozone concentration. *Tree Physiol.* **2012**, *32*, 1102–1112. [CrossRef]

89. De Símon, B.F.; Cadahía, E.; Aranda, I. Metabolic response to elevated CO_2 levels in *Pinus pinaster* Aiton needles in an ontogenetic and genotypic-dependent way. *Plant Physiol. Biochem.* **2018**, *132*, 202–212. [CrossRef] [PubMed]

90. Hantao, L.W.; Ribeiro, F.A.L.; Passador, M.M.; Furtado, E.L.; Poppi, R.J.; Gozzo, F.C.; Augusto, F. Metabolic profiling by ultra-performance liquid chromatography mass spectrometry and parallel factor analysis for the determination of disease biomarkers in Eucalyptus. *Metabolomics* **2014**, *13*, 1318–1325. [CrossRef]

91. Rivas-Ubach, A.; Sardans, J.; Hódar, J.A.; Garcia-Porta, J.; Guenther, A.; Paša-Tolić, L.; Oravec, M.; Urban, O.; Peñuelas, J. Close and distant: Contrasting the metabolism of two closely related subspecies of Scots pine under the effects of folivory and summer drought. *Ecol. Evol.* **2017**, *7*, 8976–8988. [CrossRef] [PubMed]

92. Meijón, M.; Feito, I.; Oravec, M.; Delatorre, C.; Weckwerth, W.; Majada, J.; Valledor, L. Exploring natural variation of *Pinus pinaster* Aiton using metabolomics: Is it possible to identify the region of origin of a pine from its metabolites? *Mol. Ecol.* **2016**, *25*, 959–976. [CrossRef]

93. Obudulu, O.; Mähler, N.; Skotare, T.; Bygdell, J.; Abreu, I.N.; Ahnlund, M.; Latha Gandla, M.; Petterle, A.; Moritz, T.; Hvidsten, T.R.; et al. A multi-omics approach reveals function of Secretory Carrier-Associated Membrane Proteins in wood formation of Populus trees. *BMC Genom.* **2018**, *19*, 11. [CrossRef]

94. Pan, X.; Wang, X. Profiling of plant hormones by mass spectrometry. *J. Chromatogr. B* **2009**, *877*, 280–2813. [CrossRef]

95. Pan, X.; Welti, R.; Wang, X. Quantitative analysis of major plant hormones in crude plant extracts by high-performance liquid chromatography-mass spectrometry. *Nat Protoc.* **2010**, *5*, 986–992. [CrossRef]

96. Bieleski, R.L. The problem of halting enzyme action when extracting plant tissues. *Anal. Biochem.* **1964**, *9*, 431–442. [CrossRef]

97. Kang, J.W.; Lee, H.; Lim, H.; Lee, W.Y. Identification of potential metabolic markers for the selection of a high-yield clone of *Quercus acutissima* in clonal seed orchard. *Forests* **2018**, *9*, 116. [CrossRef]

98. De Diego, N.; Pérez-Alfocea, F.; Cantero, E.; Lacuesta, M.; Moncaleán, P. Physiological response to drought in radiata pine: Phytohormone implication at leaf level. *Tree Physiol.* **2012**, *32*, 435–449. [CrossRef]

99. Ryu, M.; Mishra, R.C.; Jeon, J.; Lee, S.K.; Bae, H. Drought-induced susceptibility for *Cenangium ferruginosum* leads to progression of Cenangium-dieback disease in *Pinus koraiensis*. *Sci. Rep.* **2018**, *8*, 16368. [CrossRef] [PubMed]

100. Novák, O.; Hauserová, E.; Amakorová, P.; Dolezal, K.; Strnad, M. Cytokinin profiling in plant tissues using ultra-performance liquid chromatography-electrospray tandem mass spectrometry. *Phytochemistry* **2008**, *69*, 2214–2224. [CrossRef] [PubMed]

101. Raterink, R.J.; Lindenburg, P.W.; Vreeken, R.J.; Ramautar, R.; Hankemeier, T. Recent developments in sample-pretreatment techniques for mass spectrometry-based metabolomics. *Trends Anal. Chem.* **2014**, *61*, 157–167. [CrossRef]

102. Fiehn, O.; Robertson, D.; Griffin, J.; van der Werf, M.; Nikolau, B.; Morrison, N.; Sumner, L.W.; Goodacre, R.; Hardy, N.W.; Taylor, C.; et al. The metabolomics standards initiative (MSI). *Metabolomics* **2007**, *3*, 175–178. [CrossRef]

103. Fiehn, O.; Sumner, L.W.; Rhee, S.Y.; Ward, J.; Dickerson, J.; Lange, B.M.; Lane, G.; Roessner, U.; Last, R.; Nikolau, B. Minimum reporting standards for plant biology context information in metabolomic studies. *Metabolomics* **2007**, *3*, 195–201. [CrossRef]

104. Wilkinson, M.D.; Dumontier, M.; Aalbersberg, I.J.; Appleton, G.; Axton, M.; Baak, A.; Blomberg, N.; Boiten, J.W.; da Silva Santos, L.B.; Bourne, P.E.; et al. The FAIR Guiding Principles for scientific data management and stewardship. *Sci. Data* **2016**, *3*, 160018. [CrossRef]

105. Van Rijswijk, M.; Beirnaert, C.; Caron, C.; Cascante, M.; Dominguez, V.; Dunn, W.B.; Ebbels, T.M.D.; Giacomoni, F.; Gonzalez-Beltran, A.; Hankemeier, T.; et al. The future of metabolomics in ELIXIR. *F1000Res* **2017**, *6*, 1649. [CrossRef]

106. Spicer, R.A.; Salek, R.; Steinbeck, C. Compliance with minimum information guidelines in public metabolomics repositories. *Sci. Data* **2017**, *4*, 170137. [CrossRef]

107. Rocca-Serra, P.; Salek, R.M.; Arita, M.; Correa, E.; Dayalan, S.; Gonzalez-Beltran, A.; Ebbels, T.; Goodacre, R.; Hastings, J.; Haug, K.; et al. Data standards can boost metabolomics research, and if there is a will, there is a way. *Metabolomics* **2016**, *12*, 14. [CrossRef]

108. Spicer, R.A.; Steinbeck, C. A lost opportunity for science: Journals promote data sharing in metabolomics but do not enforce it. *Metabolomics* **2018**, *14*, 16. [CrossRef]

109. Yakovlev, I.A.; Fossdal, C.G.; Johnsen, Ø. MicroRNAs, the epigenetic memory and climatic adaptation in Norway spruce. *New Phytol.* **2010**, *187*, 1154–1169. [CrossRef] [PubMed]

110. Bräutigam, K.; Vining, K.J.; Lafon-Placette, C.; Fossdal, C.G.; Mirouze, M.; Marcos, J.G.; Fluch, S.; Fraga, M.F.; Guevara, M.A.; Abarca, D.; et al. Epigenetic regulation of adaptive responses of forest tree species to the environment. *Ecol. Evol.* **2013**, *3*, 399–415. [CrossRef] [PubMed]

111. Pieruschka, R.; Schurr, U. Plant Phenotyping: Past, Present, and Future. *Plant Phenomics* **2019**, 7507131. [CrossRef]

112. Krajewski, P.; Chen, D.; Ćwiek, H.; van Dijk, A.D.; Fiorani, F.; Kersey, P.; Klukas, C.; Lange, M.; Markiewicz, A.; Nap, J.P.; et al. Towards recommendations for metadata and data handling in plant phenotyping. *J. Exp. Bot.* **2015**, *66*, 5417–5427. [CrossRef]

113. Cwiek-Kupczynska, H.; Altmann, T.; Arend, D.; Arnaud, E.; Chen, D.; Cornut, G.; Fiorani, F.; Frohmberg, W.; Junker, A.; Klukas, C.; et al. Measures for interoperability of phenotypic data: Minimum information requirements and formatting. *Plant Methods* **2016**, *12*, 44. [CrossRef]

114. Pommier, C.; Michotey, C.; Cornut, G.; Roumet, P.; Duchêne, E.; Flores, R.; Lebreton, A.; Alaux, M.; Durand, S.; Kimmel, E.; et al. Applying FAIR Principles to Plant Phenotypic Data Management in GnpIS. *Plant Phenomics* **2019**, *2019*, 1671403. [CrossRef]

115. Wegrzyn, J.L.; Staton, M.A.; Street, N.R.; Main, D.; Grau, E.; Herndon, N.; Buehler, S.; Falk, T.; Zaman, S.; Ramnath, R.; et al. Cyberinfrastructure to improve forest health and productivity: The role of tree databases in connecting genomes, phenomes, and the environment. *Front. Plant Sci.* **2019**, *10*, 813. [CrossRef]

116. Michotey, C.; Chaves, I.; Anger, C.; Bastien, C.; Jorge, V.; Ehrenmann, F.; Adam-Blondon, A.-F.; Miguel, C. *Woody Plant Ontology—A New Ontology for Describing Woody Plant Traits COSTFA1306 Meeting 2018—Plant Phenotyping for Future Climate Challenges*; University of Leuven: Leuven, Belgium, 2018.
117. Kindermann, G.E.; Kristöfel, F.; Neumann, M.; Rössler, G.; Ledermann, T.; Schueler, S. 109 years of forest growth measurements from individual Norway spruce trees. *Sci. Data* **2018**, *5*, 180077. [CrossRef]

MDPI
St. Alban-Anlage 66
4052 Basel
Switzerland
Tel. +41 61 683 77 34
Fax +41 61 302 89 18
www.mdpi.com

Metabolites Editorial Office
E-mail: metabolites@mdpi.com
www.mdpi.com/journal/metabolites